Feminist Epistemology and Philosophy of Science

Heidi E. Grasswick
Editor

Feminist Epistemology and Philosophy of Science

Power in Knowledge

Springer

Editor
Prof. Heidi E. Grasswick
Department of Philosophy
Middlebury College
Middlebury, VT 05753, USA
grasswick@middlebury.edu

ISBN 978-1-4020-6834-8 e-ISBN 978-1-4020-6835-5
DOI 10.1007/978-1-4020-6835-5
Springer Dordrecht Heidelberg London New York

Library of Congress Control Number: 2011926688

Cover design: eStudio Calamar S.L.

Printed on acid-free paper

Springer is part of Springer Science+Business Media (www.springer.com)

Acknowledgments

There are many people who have helped in this project. Right from the volume's inception, series editor Libby Potter was a constant source of insightful advice and moral support. Ingrid van Laarhoven at Springer was always ready to answer yet another question as soon as it was asked. A large number of anonymous reviewers were also indispensible in making this volume what it is. The introduction benefitted immensely from Phyllis Rooney's careful reading. Undergraduate research assistant Stephanie Joyce assisted with formatting and compiling at crucial moments. I especially would like to thank Lorraine Code, Carla Fehr, Phyllis Rooney, and Ilya Storm for their friendship, encouragement and help with this project.

Contents

About the Contributors

Lorraine Code is Distinguished Research Professor Emerita of Philosophy at York University in Toronto Canada and a Fellow of the Royal Society of Canada. She was named the Distinguished Woman Philosopher for 2009 by the US Society for Women in Philosophy. Her principal area of research is feminist epistemology and the politics of knowledge and she has numerous publications in the field, including *What Can She Know? Feminist Theory and the Construction of Knowledge* (Cornell 1991) and *Rhetorical Spaces: Essays on Gendered Locations* (Routledge 1995). Her most recent book, largely written during her tenure of the Canada Council Killam Research Fellowship, is *Ecological Thinking: The Politics of Epistemic Location*, published by Oxford University Press in 2006. She is currently developing a moral epistemology sensitive to vulnerability, and working on questions generated by the new epistemologies of ignorance, on knowing across differences, and on the contestability of 'natural kinds.'

Nancy Daukas is Associate Professor of Philosophy at Guilford College, in Greensboro, NC. Her work focuses on questions in feminist epistemology, virtue epistemology, social epistemology, and the epistemology of testimony. A previous publication relevant to the work in this volume is 'Epistemic Trustworthiness and Social Location' (*Episteme: Journal of Social Epistemology,* Vol. 3 1–2 (2006), pp. 109–124). She has also published articles on teaching philosophy, epistemic contextualism, and skepticism.

Carla Fehr is Associate Professor of Philosophy at Iowa State University. She works in the philosophy of biology, feminist philosophy and feminist science studies. She is a co–Principal Investigator for ISU ADVANCE <http://www.advance.iastate.edu/>, a $3.3 million National Science Foundation grant, which is designed to test strategies for promoting the advancement and retention of women and minorities in science, technology, engineering and mathematics. Her research investigates the impact of culture on biological explanations of topics related to sex and gender and on the social structures of scientific communities that promote excellent research.

Samantha Frost is Associate Professor in the Department of Political Science, the Gender and Women's Studies Program, and the Unit for Criticism and Interpretive Theory at the University of Illinois, Urbana-Champaign. She is the author of *Lessons* *from a Materialist Thinker: Hobbesian Reflections on Ethics and Politics* (Stanford University Press, 2008) and co-editor, with Diana Coole, of *New Materialisms: Ontology, Ethics, and Politics* (Duke University Press 2010).

Heidi E. Grasswick is Professor of Philosophy at Middlebury College in Vermont. She is also a regular contributor to the Women's and Gender Studies Program at Middlebury. Her research focuses on feminist epistemology and social epistemology, particularly questions concerning the relationship between individual and communities as knowers and the role of trust in knowledge. She has published in such journals as *Hypatia, Synthese, and Social Epistemology.*

Sandra Harding is Professor in the Graduate School of Education and Information Studies at UCLA. Her teaching and research interests include feminist and postcolonial theory, epistemology, research methodology and philosophy of science. Harding is the editor or author of 16 books, including *The Science Question in Feminism* (1986), *Whose Science? Whose Knowledge?* (1991), *Is Science Multicultural?* (1998), *The Feminist Standpoint Theory Reader* (2004), *Science and Social Inequality* (2006), *Sciences From Below* (2008), and *The Postcolonial Science and Technology Studies Reader* (2011). She was the coeditor (with Kathyrn Norberg) of *Signs: Journal of Women in Culture and Society* from 2000 to 2005. She has also served as a consultant on epistemology and philosophy of science issues for several UN organizations, including the Pan American Health Organization; the United Nations Educational, Scientific, and Cultural Organization (UNESCO); the UN Development Fund for Women (UNIFEM); and the UN Commission on Science and Technology for Development.

Kristen Intemann is Associate Professor of Philosophy at Montana State University. Her research focuses on the roles of ethical, social, and political values in scientific research and the implications this has for scientific practices and conceptions of objectivity. She has published in journals including *Philosophy of Science*, *Biology & Philosophy*, *Hypatia*, *Social Epistemology*, *FASEB Journal*, and *The European Journal of Epidemiology.*

Nancy Arden McHugh is Professor of Philosophy and department chair at Wittenberg University. She is the author of Feminist Philosophies A-Z and articles in feminist epistemology and philosophy of science. She is completing a manuscript, Actions Which Change the Face of the World, from which the article in this volume is drawn. Nancy is one of the founding members and co-chair of FEMMSS (Feminist Epistemology, Metaphysics, Methodology and Sciences Studies). She has also worked with Wittenberg University philosophy majors to develop STEP (Stand-Up for Ethics Program), which holds workshops in ethics for high school students in the Springfield, OH schools.

Gaile Pohlhaus, Jr. is Assistant Professor of Philosophy and Affiliate of Women's, Gender, and Sexuality Studies at Miami University (Ohio). She has published in Social Epistemology, Political Theory, and the Journal for Peace and Justice Studies. She is currently working on a manuscript entitled Knowing (with) Others.

Kristina Rolin is Professor and Research Fellow at Aalto University, School of Economics, in Finland. Her main areas of research are philosophy of science and epistemology, with emphasis on social epistemology and feminist epistemology. She has published articles in *Cognitive Systems Research, Episteme, Hypatia, Perspectives on Science, Philosophy of Science, Philosophy of the Social Sciences, Science & Education, Science Studies, and Social Epistemology.*

Phyllis Rooney is Associate Professor of Philosophy at Oakland University in Michigan. She has published papers on various topics in feminist epistemology, including on reason and rationality, gender and cognition, pragmatist and naturalized epistemology, gender metaphors in philosophy, and on feminism and adversarial argumentation. Her paper included here is part of a recent project on feminist metaepistemology.

Alison Wylie is Professor of Philosophy and Anthropology at the University of Washington. She is a philosopher of science who works on philosophical issues raised by archaeological practice and by feminist research in the social sciences: ideals of objectivity, the role of contextual values in research practice, and models evidential reasoning. Her publications include *Thinking from Things: Essays in the Philosophy of Archaeology* (2002); edited volumes such as *Value-free Science?* (2007, with Kincaid and Dupré), *Epistemic Diversity and Dissent* (Special Issue of Episteme, 2006), and *Feminist Science Studies* (*Hypatia* 2004, with Hankinson Nelson); as well as essays that appear in *The Ethics of Cultural Appropriation* (2009), *Agnatology* (2008), *Evaluating Multiple Narratives* (Springer 2007), the *Sage Handbook of Feminist Research* (2007), *Science and other Cultures* (2003). She is currently Co-editor of *Hypatia*, and is working on a monograph entitled *Standpoint Matters, in Feminist Philosophy of Science*.

Introduction: Feminist Epistemology and Philosophy of Science in the Twenty-First Century

Heidi E. Grasswick

When I began graduate work in 1989, a mentor advised me that although it might make sense to specialize in feminist philosophy at the graduate level and beyond, the field of feminist epistemology, though already on the map of philosophy, was far too small and narrow to make a research career out of it. At a recent FEMMSS conference,[1] many senior feminist epistemologists commented that when they first began their careers, they were able to keep up on the literature in all of feminist philosophy – including social theory, ethics, epistemology, metaphysics and philosophy of science. But no longer. They noted that it was now a challenge just to keep up on the literature in feminist epistemology itself. My mentor could not have imagined the dramatic changes and the growth of the field that were to ensue.

Having enjoyed more than 25 years of development, feminist epistemology and philosophy of science are now thriving fields of inquiry, offering current scholars a rich tradition from which to draw. In 1983, when Harding and Hintikka published *Discovering Reality*, the first collection of essays in feminist epistemology and philosophy of science, one could find only a smattering of journal articles with any reference to the connections between gender and knowledge. Those that existed, such as Lorraine Code's 1981 'Is the Sex of the Knower Epistemically Significant?' and Sandra Harding's 1982 'Is Gender a Variable in Conceptions of Rationality: A Survey of Issues' were exploratory in nature, just beginning to consider how gender and knowledge might be linked. Today, the situation is remarkably different on many fronts. To be sure, feminist epistemologists continue to explore new ways of understanding the links between gender and knowledge. But the field is far from being appropriately characterized as 'exploratory.' Competing theories have been set out, filled out, critiqued, further developed and critiqued again. Work outside of

[1] FEMMSS (Association for Feminist Epistemologies, Methodologies, Metaphysics, and Science Studies) is an organization holding interdisciplinary conferences devoted to issues of feminist epistemology every 2 years. Its first conference was held in Fall 2004. The birth of FEMMSS is just one indication of the interest in and health of the field.

feminist epistemology, especially work outside of philosophy, now makes reference to and employs many of these theories. Key technical terminology and concepts have evolved that must be understood clearly by students of the field. A core body of works exists, much of it from the early 1990s when there was a burst of single-authored monographs in the field. Work in feminist epistemology can also now be found in a wide variety of venues. Monographs on topics related to feminist epistemology and volumes of collected feminist works are no longer restricted to just a few interested presses. In addition to the well-established feminist philosophy journal *Hypatia* and numerous interdisciplinary Women's Studies journals such as *Signs*, works in feminist epistemology are making appearances increasingly in both regular and special issues of mainstream philosophical journals.[2] The field is vibrant, varied, and continues to evolve.

The essays in this collection offer a sampling of recent work in feminist epistemology and philosophy of science. Many of the essays build on or apply canonical feminist theories of knowledge such as standpoint theory (Rooney, Daukas, Harding, Wylie) or the contextual empiricism of Helen Longino (Rolin, Daukas, Intemann, Fehr). Others seek to bridge the insights and programs of nonfeminist epistemologists with those of feminist epistemologists (Rolin, Daukas, McHugh), or seek to bridge the insights of feminist epistemology and other areas of feminist inquiry such as feminist metaphysics (Frost), feminist postcolonial studies (Harding), or feminist ethics (Code, Pohlhaus, Grasswick). Throughout the volume, many take up new topics of interest to feminists, pushing the field in new directions. Though far from a comprehensive collection of the different directions being pursued by feminist epistemologists today, these contributions represent some of the major problems feminist epistemologists are currently pursuing. To set the context for these contributions, and the contemporary state of feminist epistemology and philosophy of science, it is worth noting some of the key developments and directions of the field over the last 25 years.

In most general terms, feminist epistemology is a form of social epistemology (Anderson 1995; Grasswick 2008) in that it examines the relations between gender and knowledge, where gender is understood not as an attribute of individuals but rather as an axis of social relations. It is because society is structured significantly along the axis of gender that feminists take gender to be relevant to epistemology. Early on, feminists made a distinction between 'feminine epistemology' and 'feminist epistemology,' with the former capturing views that there exist specifically women's ways of knowing, and the latter representing views that examine the connection between the *power relations of gender* and knowledge.[3] Feminist epistemologists

[2] Some of the earlier examples of special issues of mainstream journals include *Synthese* (1995), *The Monist* (1994) and *Philosophical Topics* (1995). It is important to recognize the inroads feminists have made in mainstream venues, even though as Phyllis Rooney points out (this volume) feminist epistemology is still marginalized within mainstream epistemology.

[3] Though there were a few scholars developing feminine epistemology (for example, Belenky et al. 1986), feminist epistemology was the dominant force. In spite of this, many critics interpreted all feminist epistemology as a kind of feminine epistemology.

have been interested in how gendered power structures of society affect the shape of and possibilities for knowledge production and the exercise of epistemic agency. As epistemologists (and not just sociologists of knowledge) they have also been interested in offering normative correctives: how can we *know well* given the intersection of power structures and knowledge production? Several key developments have set the framework for further research as feminist epistemologists have sought to understand the relations between power and knowledge.

Broadening the Scope of Analysis

Focusing on power relations and their impact on the production of knowledge, feminist epistemologists early on broadened their scope beyond gender, incorporating such divisions of race, class and sexuality into their analyses and leading some to view their projects more broadly as 'liberatory epistemologies' encompassing any and all axes of oppression (Scheman 2001; Tuana 2001; Grasswick this volume). Feminist epistemologists recognized that wherever there is significant social stratification, there are likely to be epistemic effects, such as how cognitive authority gets distributed and whose background assumptions are taken for granted as starting points for inquiry. Perhaps more importantly, alongside other feminist theorists, they came to realize that gender itself cannot be understood in isolation, but must be understood 'as a component of complex interrelationships with other systems of identification and hierarchy' (Alcoff and Potter 1993, 3). Attempts to understand how gender is interwoven with these other systems of hierarchy have complicated feminist analyses of knowledge considerably. To give but one example, feminist standpoint theory, one of the landmarks of feminist epistemology, originally developed out of Marxian class analysis which posited two major social groups, the bourgeoisie and the proletariat. The socially underprivileged group was theorized as having an epistemic advantage originating in their relationship to the dominant class. According to this analysis, the socially underprivileged need to understand the world from the perspective of the dominant class who has the power to set the rules of engagement in order to survive. But the socially underprivileged also understand the world from the perspective of their experience of oppression. According to Marxian standpoint theory, this double perspective can lead to deeper epistemic insights on social relations than are available from the position of the dominant class. Early feminist standpoint theorists adapted these core ideas of Marxian standpoint theory to the sexual division of labor, arguing that the sexual division of labor was a primary social divider (Hartsock 1983). But in so doing, they relied on the original premise of society being bifurcated, with one group representing the epistemic inverse of the other. When feminists began to recognize the need to multiply the axes of power in their analyses, it became clear that standpoint theory (as well as other feminist theories) needed significant revision. Many feminist epistemologists went on to do that. For example, Patricia Hill Collins articulated a 'black women's standpoint' using 'outsider-within' language rather than strictly bifurcated class language (1990), and Sandra Harding's standpoint work evolved to attend more fully to postcolonial standpoints (especially1998). As seen in several of the contributions to this volume, whether employing standpoint theory or other frameworks, the complications arising from the multiplicity of marginalities continue to press feminist epistemologists as they seek to develop richer analyses of the intricate webs of power-infused social relations and their epistemic effects.

Challenges of Situated Knowing

Situated knowing is the single most influential concept to come out of feminist epistemology. According to those who stress the idea of 'situated knowing,' one's social location (gender being one such dimension) both shapes and limits one's knowing (Haraway 1988, 1991).[4] Although there are wide differences amongst them, feminist epistemologies based on situated knowing all stand in sharp contrast to the predominant approaches of traditional epistemology that characterize knowing as taking up the view-from-nowhere.[5] With its claim that knowing always involves a limited perspective, situated knowing helped feminists explain how masculine bias could have influenced our knowledge production practices so widely. Some (especially standpoint theorists) have interpreted situated knowing as explaining how people in marginalized positions might have better insights, based on their social location, that could be fostered to attain knowledge. While offering fruitful explanations such as these, the situated knowing thesis has also presented certain challenges that feminist epistemologists have recognized and taken up. For example, if knowledge is situated, then in what sense can knowledge be objective? Either feminists must abandon the idea of objective knowledge, or they must rework objectivity to prove consistent with their insights concerning situated knowledge. Furthermore, if social location limits one's knowledge, how can we know *across* social locations? Such questions have been central to discussions within feminist epistemology over the last 25 years.

In answering these questions today, many build on the theories and insights of feminist epistemologists developed in the 1990s. For example, one of the first systematic attempts to reconceptualize objectivity from within a feminist framework

[4] The term 'situated knowing', or 'situated knowledge' was first articulated by Donna Haraway in her article 'Situated Knowledges: The Science Question in Feminism and the Privilege of Partial Perspective' which was written as a response to Sandra Harding's work. The article originally appeared in *Feminist Studies* in 1988 and later appeared in her 1991 book *Simians, Cyborgs, and Women*.

[5] If predominant epistemological approaches have acknowledged any epistemic relevance of situation, they have only gone so far as recognizing that knowledge may be dependent on very generic human capacities such as perception and reasoning (so knowledge may represent a 'human' point of view). But this is a far cry from the feminist situated knowledge thesis claiming that the particularities of a person's social situation can shape and limit one's knowing.

was Helen Longino's contextual empiricism (Longino 1990).[6] According to Longino, objectivity is a function of communal practices and how well they foster transformative criticism. She set out four conditions of communal practices (recognized avenues for criticism, shared standards, community response and equality of intellectual authority[7]) arguing that when satisfied, these conditions ensure that the results of inquiry have survived a process of public scrutiny capable of weeding out whatever idiosyncratic background assumptions individuals may bring to the table. For Longino, objectivity is possible, in spite of the situatedness of knowing. Her work has been highly influential with both feminist and nonfeminist philosophers of science alike. As seen in several of the pieces in this volume, many feminist epistemologists and philosophers of science continue to work with Longino's framework: they set out to articulate its implications for individual agents (Daukas); they examine what cultural shifts need to occur in scientific communities to make good on her conditions (Fehr); they question whether all forms of dissent need to be taken seriously (Intemann); and they examine whether her conception of objectivity can be rescued from criticisms of a problematic intersubjectivism (Rolin).

Challenges of Communal Practices

One of the most significant features of Longino's contextual empiricism was its shift to thinking about knowledge production in terms of communal practices rather than as an activity of individuals. At the time, Longino was not alone in drawing attention to the usefulness of communal analyses for feminists. Most prominently, Lynn Hankinson Nelson built on a Quinean naturalized epistemology and argued that communities are the primary epistemological agents (Nelson 1990, 1993). Particularly for feminist philosophers of science, communal analyses of some form or another seemed necessary in order to explain how androcentrism and sexism in science only became visible after feminists entered the fray. Nelson, for example, argued that standards of evidence are communal and dynamic, and are influenced by our understanding of social relations. If that understanding of social relations changes, as happened with the growth of the feminist movement, then the standards of evidence will shift. Androcentric science that may have at one time satisfied the scientific community's existing standards of evidence can no longer be said to, and can now be criticized accordingly (Nelson 1993).

At the same time as a communal analysis offered many useful explanations to feminists, the turn to understanding knowledge production as communal practice

[6] In her 2002 book *The Fate of Knowledge* Longino finds 'critical contextual empiricism' a more accurate name for her position than her previously employed 'contextual empiricism'.

[7] In her 2002 book *The Fate of Knowledge* Longino modifies 'equality of intellectual authority' to 'tempered equality of intellectual authority' in order to account for differences in training or record (2002, 133).

opened up many new lines of inquiry. Relations both internal to communities and between communities, including wider social communities, needed to be examined for their epistemic import. This has led some theorists to take up issues of the economies of credibility, analyzing how those in marginalized positions often carry less credibility than warranted (Alcoff 2001). Others have focused on how to understand the epistemic responsibilities of individuals within those communities (Grasswick 2004; Daukas this volume). Still others have investigated how individuals must negotiate lines of accountability between multiple communities (McHugh, Pohlhaus this volume). Feminist epistemology and philosophy of science continue to work through the specific implications of a communal analysis of knowledge production.

Integrating Ethics and Epistemology

The implication of power relations in knowledge has also meant that, for many feminist epistemologists, it has been vitally important to investigate epistemological issues in conjunction with ethical issues, recognizing how deeply intertwined the two are. Lorraine Code's work has been paramount in this regard, attending to our rich and complex interactions with others and taking up the challenges of how to know well across social divides (Code 1995, 2006, this volume). In attempting to understand how to know well in an ethically charged way, many feminists have focused on an agent-centered approach to epistemology (Daukas this volume), moving away from the dominant trend in analytic epistemology towards externalism and reliabilism which understands justification (and knowledge) as the result of reliable processes of generating true beliefs, without the agent necessarily being aware of this reliability.[8] Additionally, feminists have played a significant role in developing arguments suggesting that ethical and political values cannot be eliminated from good epistemic practices, but rather play a legitimate epistemic role (Anderson 2004; Longino 1990; Nelson 1990). Such arguments then focus attention on how we can select appropriate values (Daukas, Intemann, Rolin this volume). Feminist epistemologists have commonly taken up questions concerning the kind of knowledge we produce, understanding these issues as inherently epistemological as well as ethical, and rejecting the traditional segregation of epistemological questions concerning the status of knowledge claims – their rationality and epistemic merit – from ethical questions concerning the direction of research and knowledge production. For example, feminist work on the epistemology of ignorance has stressed that ignorance is often the result not of a benign gap in our knowledge, but in deliberate choices to pursue certain kinds of knowledge while ignoring others

[8] This is not to say all feminist epistemologists have moved away from externalism and reliabilism. For example, Louise Antony's work on the bias paradox adopts a naturalized approach to knowledge and uses reliabilist reasoning to distinguish between good and bad biases (or values) in our knowledge-seeking (Antony 1993).

(Sullivan and Tuana 2007; Tuana and Sullivan 2006). For many feminists, responsible knowing is a matter of producing ethically sound knowledge, and we must therefore concern ourselves with our choices of knowledge production and who we take ourselves to be accountable to through those choices (Grasswick, McHugh, Pohlhaus this volume).

The developments I have sketched out give an indication of the range of research directions in which feminist epistemology and philosophy of science continue to head, in response to some of its key developments throughout its history. Many of these directions can be identified in the contributions to this volume.

Part I: Intersections: Feminism, Epistemology and Science Studies

With respect to the philosophical field of 'mainstream' epistemology as a whole, feminist epistemology has had and continues to have a vexed relationship. Much of the initial work formed as critiques of the existing prominent traditions of epistemology. Born out of the need for conceptual tools to explain the persistent gender bias that feminists were observing in knowledge production in various fields of inquiry, feminist epistemology quickly discovered that traditional epistemological resources were ill-equipped to correct such biases. Positivist philosophy of science with its commitment empirical testability and the need for value-neutrality in order to obtain high quality knowledge might have been able to explain gender bias as cases of researchers failing to adhere to the standards of good knowing, allowing their personal biases to influence their work when they should have been neutral and objective. But it could not explain why it was feminists who seemed to be the ones noticing the biases (Harding 1993; Nelson 1993). Furthermore, feminists began to recognize that the standard epistemologies themselves might be providing underlying frameworks that contributed to the continuation of gender bias. Whether it be the positivist ideals of value-neutrality dominant in the scientific realm, or an ideal of rationality carrying masculinist undertones yet understood to be necessary for knowing, or the standard abstract epistemological model of '*S* knows that *p*' where 'S' could be any knower, belying a commitment to the interchangeability of knowers, such frameworks make it easier to mask masculine bias as a neutral position. If knowledge is equated with a neutral point of view, then those who have the power to claim knowledge can mask (albeit unwittingly) their particular perspective as the neutral point of view.

But as Phyllis Rooney points out in her opening contribution, at the same time that feminist epistemology was developing, multiple approaches to epistemology were proliferating and gaining in stature; pragmatism, naturalism, contextualism, social epistemology and virtue epistemology all were taking hold as exciting directions to pursue within the field of epistemology. It became apparent to many that not all contemporary approaches to epistemology were equally inimical to feminist concerns and insights. While some feminist epistemologists aligned themselves with one or another approach, others adopted more of what Lorraine Code (this volume) has referred to as an epistemological 'scavenger approach,' gathering various resources for feminist projects wherever they may be found. The result of both strategies has been increasing intersections between feminist epistemology and these emergent epistemological approaches, particularly over the last 10 years. Not only are feminist epistemologists mining the resources of these approaches for their own projects, but their insights are also contributing significantly to the development of these approaches themselves (Schmitt 1994).[9]

Many of the essays in the first section of this volume are dedicated to an exploration of some of the more recent intersections between feminist epistemologies and other theories. But the section begins with a warning. As Phyllis Rooney's opening paper makes clear, despite the fact that feminist epistemology shares much in common with various contemporary approaches to epistemology, it still has not achieved acceptance within epistemology and has been marginalized more significantly than have the remaining variety of recent approaches. Rooney takes issue with the characterization of 'epistemology proper' with which feminist epistemology is often contrasted, noting that there is no agreement amongst even nonfeminist epistemologists concerning the central approaches and problems of epistemology. The idea of a unified 'epistemology proper' is a myth, but a powerful one when used to marginalize feminist epistemology. Rooney considers possible explanations for the continued marginalization of feminist epistemology such as residual sexism and the conceptual alignment of reason and masculinity, but she also looks at the situation more positively, arguing that feminist epistemologists are in a unique position to perceive some of the flaws of epistemology 'proper.' According to Rooney, the marginalization of feminist epistemology offers it a metaepistemological advantage, granting insights into the shortcomings of more mainstream epistemology. While Rooney's paper serves as a sobering reminder that feminists have far from achieved a settled place at the table of professional epistemologists, arguing instead that feminist epistemologists have failed to obtain the appropriate uptake for their ideas, her essay also upsets the idea of a unified epistemological approach that can be contrasted with a feminist approach, setting out instead a mosaic of epistemological approaches with which feminist epistemologies can interact.

Kristina Rolin's essay offers an excellent example of feminist epistemologists' engagement with nonfeminist epistemology when it has resources to offer. Rolin takes up the contextual empiricism of prominent feminist philosopher of science Helen Longino and examines its criticisms. Rolin articulates two core theses in Longino's work: contextual evidence and social objectivity. According to contextual evidence, epistemic justification is relative to a context of background assumptions. According to social objectivity, objectivity is satisfied to the degree that a community satisfies four norms of public criticism, uptake of criticism, shared standards, and tempered equality of authority. Armed with these two theses, Longino is able to

[9] For example, in his introduction to the 1994 collection *Socializing Epistemology: The Social Dimensions of Knowledge*, Frederick Schmitt acknowledges feminist philosophers of science as a major source of inspiration for the developments of social epistemology since 1980 (1994, 3).

explain how feminist values could have an appropriate place in scientific inquiry, and how sexist values managed to persist in apparently 'good' science. Yet critics have found Longino's norms inadequate and Rolin discusses several criticisms, among them, the argument that Longino falls into a kind of 'intersubjective relativism' at the communal level, without the resources to distinguish between good and bad biases. Rolin turns to the resources of Michael Williams's contextualism as a way of responding to Longino's critics. She argues that the contextualism of Williams, originally set out as a response to skepticism, can be understood as an extension of Longino's contextual empiricism. According to Williams's contextualism, one can be justified in believing certain claims without providing reasons, as long as the claims are understood as merely default entitlements which must be defended or revised when one is challenged. According to Rolin, adopting Williams's contextualism allows us to see that Longino's theory can avoid the criticism of relying on a particular set of (communally accepted) dogmatic assumptions. One is entitled to hold a claim without reason as long as one maintains a defense commitment: a commitment to defend or revise the belief if challenged by an appropriate argument.

While much less confident than Rolin on the fruitfulness of the contextualisms found within mainstream epistemology for feminism, Nancy Daukas in her contribution turns to another recent development in epistemology, virtue epistemology, and investigates its uses for feminists. Daukas argues that when properly employed, virtue epistemology can strengthen both Longino's contextual empiricism and feminist standpoint theory, and can find common ground between these feminist theories that are typically taken to be divergent. Daukas is well aware that not all virtue epistemology is compatible with feminism. However, she sees feminist promise in the agent-centered nature of virtue epistemology, and she argues that a feminist virtue epistemology can emerge when the conception of agency employed is a social one, and when the goal of virtue epistemology is understood to be the flourishing of all people.

Daukas is interested in using the resources of Helen Longino's contextual empiricism to argue for the need for agents to exercise 'oppositional agency' in their everyday knowing practices, resisting aspects of their society's epistemic practices. Longino's work stresses our reliance on background assumptions in the justification of claims, and as Daukas explains, in a society embedded with social hierarchy, some of those theoretical commitments may support a hierarchical value system. Such values must be resisted by agents who aim to exhibit *epistemic trustworthiness*, the central concept of Daukas's virtue epistemology. Agents are less epistemically trustworthy if, for example, they dismiss the testimony of certain groups of people because of the values of a social hierarchy that suggest some groups are less worthy than others. Epistemically virtuous agents however, self-consciously resist conformity with the default patterns of epistemic interaction in the exercise of oppositional epistemic agency. Whereas on its own, Longino's theory does not attend to the perspective of the epistemic agent and this need for oppositional epistemic agency, coupled with Daukas's virtue epistemology, it can. Daukas also takes up standpoint theory, rejecting versions committed to a 'women's standpoint' but taking seriously the idea that those in marginalized positions in society might experience a kind of double consciousness that offers them epistemic insight. But Daukas points out that feminists still need to make a distinction between those marginalized positions that might carry epistemic insight and those, such as a neo-Nazi viewpoint, that would not. Daukas argues that her virtue epistemology, with its notion of epistemic trustworthiness that demands open-mindedness and critical engagement, can rule out the neo-Nazi point of view while including the feminist point of view.

Feminist epistemology has also developed in concert with other fields within feminist philosophy. While there has been much cross-fertilization between feminist epistemology and feminist ethics and social and political theory, these are not the only fields from which feminist epistemologists have drawn. In this volume, Samantha Frost articulates the epistemological ramifications of recent developments in feminist metaphysics. Frost takes up the 'new materialists' such as Anne Fausto-Sterling, Elizabeth Grosz, and Karen Barad, all of whom are interested in exploring the agency of matter. Frost notes that while feminist theorists have done excellent work articulating the ways in which culture constructs the 'natural,' they have been less comfortable taking up questions of how culture is shaped by the agency of nature. The new materialists, however, focus on the interaction of nature and culture, showing how both are co-constructed: denaturalizing nature and deculturalizing culture. Frost notes that the attention these theorists give to the agency of matter means they run the danger of falling back into a troubling essentialism, turning to biology to explain gender differences and failing to thoroughly grasp the insights of constructivism that have been so important for feminist theory in developing explanations of the role of social practices in constructing gendered differences. According to Frost, however, what lies behind feminists' fears of attending to the agency of matter, organic or inorganic, is a problematic linear model of causation. Instead, the new materialists directly challenge this model of causation, with significant implications for epistemology. Frost argues that following the new materialists will lead feminist epistemologists to a more complex understanding of causation, focused on 'the interdependencies that define the contexts in which both objects and knowers exist' (78). Thus, feminist epistemologists are required to think ecologically (Code 2006), not just about objects of knowledge but also about knowers (see also McHugh this volume). According to Frost, such a reworking of the relationship between the subjects and objects of knowledge means we must also recognize additional limits to our knowledge; in addition to the standard limits of perception, we must recognize the limits that are 'intrinsic to the complexity of objects or processes themselves' (79). The new materialism calls for an 'epistemological and political humility' which could radically transform our epistemic pursuits.

Feminist epistemology has also intersected with many intellectual developments outside of philosophy, including postcolonial science studies work. Sandra Harding's essay examines the narratives of modernization, interrogating the way gender stereotypes have constituted the very contrast between modernity and tradition, delivering models of science and technology that are male supremacist. She notes that not only has the move toward modernity been masculinized, but anything premodern or traditional (and Harding notes that these are not the same) has been coded feminine. What has been missed in the narratives are the ways in which modernity has been dependent on the traditional. According to Harding, the feminist critical study of the tradition versus modernity contrast and the gendered oppositions it depends on could not be developed until postcolonial scholarship had made clear that modernization is not identical to Westernization, by showing that there are multiple forms of modernization, with each involving the incorporation of local traditional cultural features. Breaking down the narrative of modernization as a one-directional trajectory away from tradition and towards progress is crucial for feminists if they are to begin to analyze what modernization has meant for women, particularly women in 'traditional' households. Harding builds on postcolonial insights, coupling them with her previous work in feminist standpoint theory to offer a provocative methodological proposal intended to rescue scientific and technological research from its male supremacist forms.

In her 1991 landmark *Whose Science? Whose Knowledge?* Harding used the framework of a feminist standpoint theory to argue that in order to increase the objectivity of research, we needed to 'start research from women's lives.' By seriously thinking through and attending to the reality of women's lives, researchers would be accessing the epistemological advantage that standpoint theorists argued can come from marginalized social positions. Such research would contribute to what Harding calls the 'strong objectivity' of science, by revealing the cultural particularity of previously taken-for-granted assumptions that drive research. In this volume's contribution, Harding suggests that an additional methodological mandate should be that we start all research from *women's lives in households*. By doing so we will be able to see the impact of modernization and technological developments on women, as well as come to understand how such development is itself dependent on work in the household. She is quick to point out that this is not the only standpoint that should be considered, but it is an important one if we are to ensure that the knowledge produced results in the flourishing of households. Harding acknowledges that her argument takes us a long way from traditional questions in philosophy of science. But her intent is to point out just how limited those discussions have been. She demonstrates that we need to expand our questions if we are to fully understand the relationship between science and women and show how we can develop scientific and technological projects that will serve the goals of social justice.

Part II: Democracy and Diversity in Knowledge Practices

Harding's essay offers a clear reminder of why feminist epistemologists have devoted so much energy toward science as a form of knowing. Around the globe, the results of science and its technological developments are major forces in our lives, and assessing their impact on variously situated women is an important dimension of feminism. Additionally, feminist epistemologists such as Harding are well aware of the authoritative status science has enjoyed as the paradigm of sound inquiry. For both of these reasons, feminist philosophers of science have found it important to scrutinize scientific practices and the epistemologies that underwrite them. Some of the earliest feminist work on science focused on the history of exclusion and discrimination of women in science (Keller and Longino 1996). First seen as a justice or equity issue, the underrepresentation of women in science soon came to be considered by feminists as intimately connected with the androcentrism and masculine bias they were discovering in the content of science itself. Perhaps, they argued, bringing more women into science, would result in a different (and potentially better) science. To complete this argument, however, the connection between women's presence in science and the results of science needed to be made clearer. The situated knowing thesis suggested a crucial connection between social location and the perspective one brings to the epistemic (and scientific) table, though feminists still needed a philosophy of science that could prove compatible with this insight. Substantive work in this regard was undertaken in the early 1990s by such theorists as Longino, Nelson, and Harding: each argued in their own way that situated knowing was compatible with sound science when science was understood as a particular form of communal practice. Scientific communities needed *diverse* perspectives.

But the situated knowing thesis itself also needed filling out, in order to explain the nature of this connection between social location and epistemic perspective. On the ground in scientific communities, it certainly seemed far-fetched to suggest that women necessarily came to the scientific table with a women's perspective. In Carla Fehr's terminology, the link between 'situational diversity' and 'epistemic diversity' needed to be clarified. The essays in the second section of this volume take up questions of the epistemic implications, especially in science, of the underrepresentation of women in particular, and marginalized groups in general.

Kristen Intemann's contribution examines closely the calls for democracy and diversity in science. Intemann argues that feminist calls for a democratic science have much in common with the ideas of John Stuart Mill, who argued that because human beings are fallible, an epistemic community is best served when its members constitute a diversity of values and interests, who can advocate for those views in a 'free marketplace of ideas.' Dissent is taken to be important, because not only might a dissenting view be the correct one, but even if it is not, its presence encourages advocates of correct views to better understand their justification. Intemann examines how Helen Longino's and Miriam Solomon's theories, each in their own way, share some of the original Millian ideas that diversity and dissent are to be valued for their ability to contribute to good knowing in a 'free marketplace of ideas.' However, reminiscent of Daukas's arguments that feminists must be able to distinguish between the marginal points of view that are epistemically valuable and those that are not, Intemann argues that ultimately the Millian call for democracy in science is insufficient for feminist goals. The Millian call for democracy maintains a conception of value-neutrality in science that is unable to endorse or privilege any particular ethical or political values, thus placing sexist and racist values in science on a par with feminist values. Instead, Intemann argues that ethical and political values play more than an indirect and motivational role: they can be directly relevant to deciding what constitutes an empirical success. She argues for their rational scrutiny within particular contexts of research, in ways that allow us to differentiate between those values that deserve representation within the research community and those that do not.

Throughout her discussion, Intemann stresses that the demands for diversity within a scientific community do not imply maximizing the representation of different values and interests. Instead, she points out that epistemically important diversity will stem from a diversity of *experiences* rather than just interests. Gendered differences in experience, of course, have served as one of the major arguments for the inclusion of more women and underrepresented groups in the fields of science. Carla Fehr's essay looks specifically at the 'diversity that promotes excellence' theories that argue for the epistemic benefits of including diversity within scientific communities. Though 'diversity' has served as a popular and catch-all phrase for many different goals and initiatives, authors such as Fehr and others in this volume are working out very specific conceptions of diversity in an attempt to understand their epistemic implications. Like Intemann, Fehr looks to the resources of Longino's critical contextual empiricism as a way of understanding the epistemic value of diversity and dissent within a community, but Fehr finds its theoretical commitments need to be supplemented in order to provide sufficient support to convince university administrators of the need to increase gender diversity within scientific communities in order to achieve epistemic excellence. The road from diversity to excellence is more twisted than is sometimes supposed, and Fehr argues that in order to see how Longino's theoretical commitments could play out in the real world of academia and scientific research and deliver on their promise, we need a more thorough understanding of community structures and the cultural features of communities. Recognizing the different kinds of communities that scholars belong to, Fehr demonstrates that there are ways academic departments can access the benefits of diversity other than through hiring members of underrepresented groups. Because epistemic communities overlap, a formal community such as an academic department can engage in 'diversity free riding,' gaining insight through communication and interaction with informal communities without increasing the diversity in the department itself. This makes the justification for why more women and underrepresented groups should be hired more complicated. Fehr argues that to make effective use of diversity, communities need to foster cultures within which dissenting views are likely to be articulated and given uptake. Diversity free-riding off underrepresented groups on the margins fails to contribute to these kinds of changes, making it likely that greater epistemic benefits can be gained when these cultures engage in what Fehr calls 'diversity development work' that focuses on cultivating dissenting perspectives and fostering their uptake, and includes the changing of hiring practices.

In her discussion of the cultural impediments to articulating diverse perspectives and having them receive the appropriate uptake, Fehr describes some of the features of what has come to be known as the 'chilly climate' for women and minorities in STEM (science, technology, engineering and mathematics) fields. In 'What Knowers Know Well: Women, Work and the Academy,' Alison Wylie examines the evolution of the work on chilly climates in academia and the responses to it. The chilly climate research of the early 1980s sought to show how gender discrimination in academia had continued even after the passing of equal employment opportunity legislation which contributed to the hiring of more women and minorities. This research focused on the insidious form of gendered day-to-day small differences in uptake and response that accumulate and translate into significant disadvantages for women in their work-life and career trajectories. Wylie notes that whereas this early grassroots chilly climate work was met with incredulity and intense resistance, by 1999 when the highly publicized report on the status of women at the MIT School of Science came out, these same kinds of explanations appeared much more plausible to many. Wylie uses the insights of standpoint theory and Miranda Fricker's work on epistemic injustice (Fricker 2007), to explain both how grass-roots activists were able to articulate these hidden climate issues, and how the change in attitude toward gender inequities came about. In keeping with her earlier work on standpoint theory (2003), Wylie stresses that the epistemic advantage of a standpoint is always localized and contingent. Social marginality 'will confer advantage only with respect to specific epistemic problems' (163). This particular essay is an example of her putting a contextualized standpoint theory to work, arguing that in this case, feminist grassroot activists did have the advantage of their engaged standpoint which led them to formulate ideas about the reality and effects of a chilly climate.

Part III: Contexts of Oppression: Accountability in Knowing

As pointed out by Grasswick (this volume) feminist epistemologists have been particularly interested in the challenges of knowing within contexts of oppression. Their scrutiny of science as a paradigm of sound inquiry has played a major role here. Some have focused on demonstrating the ways in which scientific practices have contributed to the sustenance of oppression (for example, by offering authoritative, yet sexist explanations of the nature of women and their subordinate position in society). Others have questioned science as the dominant framework of knowing. They work instead to identify the limits of the scientific point of view and to recover other neglected and undervalued forms of knowing that may contribute to our understandings of oppression. For example, much of Lorraine Code's work has explored the possibility of taking 'knowing other people' as a model of knowing, rather than relying on the paradigm of knowing physical objects (Code 1991), and Alison Jaggar has argued for the epistemic value of attending to one's emotions, particularly in cases of oppression (Jaggar 1989). Many of the essays in the third section of this volume either explore alternate ways of knowing, or seek to reshape the idea of good scientific research by recognizing the limits of the ways science has been practiced and conceptualized. All of the contributions to this last section consider how we can know across social locations, or investigate the ways in which knowing involves us in relations of accountability with others. These authors investigate the responsibilities we have to those we try to know, or to those with whom we engage in our epistemic projects.

Nancy McHugh's essay couples John Dewey's pragmatism, and particularly his idea of 'experimental inquiry,' with feminist insights of situated knowledge to argue for the shortcomings of a scientific method that focuses exclusively on clinical trials and laboratory settings to tell us how the world is. Following Dewey, McHugh thinks science has lost its way by focusing on results from the isolated setting of the laboratory, and forgetting that in order to generate knowledge that matters to our lives, scientists and knowers must be engaged with the world. Her argument stands as a direct challenge to the recent evidence-based medicine movement that looks to randomized trials as the epistemic gold standard. Using a case study of the people living in the Aluoi Valley in Vietnam where there has been dioxin contamination from the use of Agent Orange in the Vietnam War, McHugh develops a position she calls *situated communities*, in which she argues for the need to examine the circumstances of everyday life in particular communities in order to generate scientific knowledge that is 'accurate, effective, and ethical.'

McHugh's situated communities approach challenges us to produce knowledge that matters, emphasizing that the point of producing knowledge is to make our lives better. Yet her approach also makes clear that 'our' lives are not all the same and the differences need attention. Her paper offers insight into the politics of post-Vietnam War responsibilities, demonstrating how the Aluoi Valley people lack the power to make their concerns heard in the United States, where the inconclusive results from laboratory science are used to abdicate American responsibility for the health effects experienced 'on the ground' in Vietnam. Taking situated communities seriously demands that researchers pay intimate attention to the specifics of the lives of marginalized communities.

Lorraine Code, in her contribution, shares McHugh's concerns with the responsibilities associated with knowledge and the need to attend to the specificities of people's social location. But far from focusing on scientific research of particular communities, Code focuses on our everyday interactions, exploring how one might know others across social position well enough to be able to treat them well. Code is explicit in weaving together ethical and epistemological concerns, arguing that an 'explicit, active recognition of the specificity of differences...is an epistemological prerequisite for countering inequitable social practices' (208). Yet such a recognition of differences is not always easy, particularly within a social imaginary of human sameness. To illustrate such difficulties and ultimately the limits of such knowing, Code argues that fiction and narrative forms are extremely valuable epistemological tools. Fiction and narrative forms can be more effective than other 'factually' based forms of knowing in 'showing rather than stating how 'the other' might indeed bring 'the self' up against its own limited horizons' (216). In her essay, Code turns to the characters of Nadine Gordimer's 1981 novel, *July's People* to illustrate the extreme difficulties of knowing across raced and gendered social barriers. July is the black servant of a middle-class white Johannesburg family. When political unrest forces the family to flee, July gives them sanctuary in his village, offering new circumstances from which Gordimer can explore the complexities and difficulties of her characters' knowing across their social locations. Code focuses her discussion on the difficulties Maureen, the white woman, faces in trying to know her servant July, burdened as she is by assumptions of human sameness that fail to do justice to his situation. Importantly, Code draws our attention to the limits we come up against in trying to know others well, suggesting that we need to know others well in order to treat them well ethically, but also that we must 'trim our expectations' (217). In the end, Code suggests that we take on 'the intellectual and emotional challenge of staying with indeterminancy, with ambiguity, where premature closure risks performing epistemic violence' (220).

Gaile Pohlhaus also takes up the everyday contexts in which we seek to understand others across social locations and in a move similar to Code's, she rejects the expectation of transparency of the other whom one is trying to know. In her essay, Pohlhaus argues that there are times when it is inappropriate, even offensive, to ask or expect another to understand, or follow the reasoning of a certain position. Like McHugh and Code, Pohlhaus is interested in forms of epistemic engagement that permit ethically sound relationships across social locations. Yet she notes that asking someone to follow one's reasoning, even if not demanding that they agree with you, is not always innocent. In contexts of oppression, asking the oppressed to understand the position of the privileged can in fact amount to asking them to participate in the undermining of their own agency. Opening with a case described by Patricia Williams in which people of color are asked to 'understand' racial profiling by imaginatively putting themselves in the position of whites, Pohlhaus goes on to explain how such expectations of understanding function as part of the practices of oppression, by undermining the agency of the oppressed. Pohlhaus concludes that 'refusing to understand' can be both politically and ethically productive. Her argument brings to light the need for feminist epistemologists to examine carefully how demands on our reasoning are implicated in broader social practices, including oppressive ones.

The expectations of understanding or following another's reasoning that Pohlhaus discusses are related to a broader category of epistemically-charged social interactions discussed by Heidi Grasswick: the sharing of knowledge. Grasswick's essay begins by describing the wide variety of highly localized cultural norms of sharing and withholding knowledge. Such norms set out the expectations to which we hold each other concerning when and what knowledge is shared. Grasswick considers how we might assess such cultural norms of knowledge sharing according to the ethico-political and epistemic goals of a liberatory epistemology. In spite of a recognition of the connection between power and access to knowledge, she argues against the view that more knowledge sharing always serves liberatory goals. Sometimes increased knowledge sharing will threaten those in oppressed positions, and importantly, increased knowledge sharing sometimes will prevent particular forms of knowledge from being generated, including knowledge of the specifics of oppression. Feminist and other liberatory epistemologists must consider what kinds of knowledge we need to generate in order to overcome oppression, and assess knowledge-sharing norms accordingly. Grasswick argues that the assessment of knowledge-sharing norms is complex, and such assessment must take into account the positionality of the speakers and hearers, as well as the potential for certain knowledge-sharing norms to play a role in the development of trusting relations between knowers.

Though the essays of the final section focus on the special challenges of knowing within contexts of oppression, they also indicate the wide range of social interactions that feminist epistemologists have taken up in their efforts to develop epistemological understandings that will prove adequate to the experiences of variously situated women. Such 'opening up' of epistemic terrain is a recurrent theme throughout feminist epistemology, as feminists question the boundaries and assumptions that have guided epistemic inquiry, finding new issues of epistemic relevance where none were seen before. As we see in contributions throughout this volume, their goal of understanding the power-infused nature of knowledge production and circulation and their desire to find ways of knowing well lead feminist epistemologists and philosophers of science to take up all kinds of questions, from how we can best organize our scientific or other knowledge-seeking communities, to how individuals can know well within challenging conditions of interpersonal interaction. As they take up these questions, they draw on a rich variety of sources, including other fields of feminist inquiry and other approaches in contemporary epistemology and philosophy of science.

In short, the papers collected in this volume offer an illustration of the multitude of ways feminist epistemologists and philosophers of science continue to develop their understandings of the power-infused nature of knowledge production and circulation. Taken together, they offer strong evidence of the current breadth and depth of these fields, and they clearly focus our attention toward new directions and challenges to be taken up in the future.

References

Alcoff, Linda, and Elizabeth Potter. 1993. Introduction: When feminisms intersect epistemology. In *Feminist epistemologies,* ed. Linda Alcoff, and Elizabeth Potter, 1–14. New York: Routledge.

Alcoff, Linda Martín. 2001. On judging epistemic credibility: Is social identity relevant? In *Engendering rationalities,* ed. Nancy Tuana and Sandra Morgen, 53–80. Albany: State University of New York Press.

Anderson, Elizabeth. 1995. Knowledge, human interests, and objectivity in feminist epistemology. *Philosophical Topics* 23(2): 27–58.

Anderson, Elizabeth. 2004. Uses of value judgements in science. *Hypatia* 19(1): 1–24.

Antony, Louise. 1993. Quine as feminist: The radical import of naturalized epistemology. In *A mind of one's own: Feminist essays on reason and objectivity.* ed. Louise Antony and Charlotte Witt, 185–226. Boulder: Westview Press.

Belenky, Mary Field, Blythe McVicker Clinchy, Nancy Rule Goldberger and Jill Mattuck Tarule. 1986. *Women's ways of knowing: The development of self, voice, and mind.* New York: Basic Books.

Code, Lorraine. 1981. Is the sex of the knower epistemologically significant? *Metaphilosophy* 12: 267–276.

Code, Lorraine. 1991. *What can she know? Feminist theory and construction of knowledge.* Ithaca: Cornell University Press.

Code, Lorraine. 1995. *Rhetorical spaces: Essays on gendered locations.* New York: Routledge.

Code, Lorraine. 2006. *Ecological thinking: The politics of epistemic location.* Oxford: Oxford University Press.

Collins, Patricia. 1990. *Black feminist thought: Knowledge, consciousness, and the politics of empowerment*. New York: Routledge.

Fricker, Miranda. 2007. *Epistemic injustice: Power and ethics in knowing*. Oxford: Oxford University Press.

Grasswick, Heidi. 2004. Individuals-in-communities: The search for a feminist model of epistemic subjects. *Hypatia* 19(3): 85–120.

Grasswick, Heidi. 2008. *Feminist social epistemology*. From http://plato.stanford.edu/archives/fall2008/entries/feminist-social-epistemology/

Haraway, Donna. 1988. Situated knowledges: The science question in feminism and the privilege of partial perspective. *Feminist Studies* 14: 575–599.

Haraway, Donna. 1991. *Simians, cyborgs, and women: The reinvention of nature*. New York: Routledge.

Harding, Sandra. 1982. Is gender a variable in conceptions of rationality: A survey of issues. *Dialectica* 36(2–3): 225–242.

Harding, Sandra. 1993. Rethinking standpoint epistemology: What is strong objectivity? In *Feminist epistemologies*, ed. Linda Alcoff and Elizabeth Potter, 49–82. New York: Routledge.

Harding, Sandra. 1998. *Is science multicultural? Postcolonialisms, feminisms, and epistemologies*. Bloomington: Indiana University Press.

Harding, Sandra and Merrill Hintikka, eds. 1983. *Discovering reality: Feminist perspectives on epistemology, metaphysics, methodology, and philosophy of science*. Dordrecht: D. Reidel.

Hartsock, Nancy. 1983. The feminist standpoint: Developing the ground for a specifically feminist historical materialism. In *Discovering reality,* ed. Sandra Harding and Merrill Hintikka, 283–310. Dordrecht: D. Reidel.

Jaggar, Alison. 1989. Love and knowledge: Emotion in feminist epistemology. *Inquiry* 32: 151–176.

Keller, Evelyn Fox and Helen Longino. 1996. Introduction. In *Feminism and science,* ed. Evelyn Fox Keller and Helen Longino, 1–14. Oxford: Oxford University Press.

Longino, Helen. 1990. *Science as social knowledge: Values and objectivity in scientific inquiry.* Princeton: Princeton University Press.

Longino, Helen. 2002. *The fate of knowledge*. Princeton: Princeton University Press.

Nelson, Lynn Hankinson. 1990. *Who knows: From Quine to a feminist empiricism*. Philadelphia: Temple University Press.

Nelson, Lynn Hankinson. 1993. Epistemological communities. In *Feminist epistemologies,* ed. Linda Alcoff and Elizabeth Potter,121–159. New York: Routledge.

Scheman, Naomi. 2001. Epistemology resuscitated. In *Engendering rationalities,* ed. Nancy Tuana and Sandra Morgen, 23–52. Albany: State University of New York Press.

Schmitt, Frederick. 1994. Socializing epistemology: An introduction. In *Socializing epistemology: The social dimensions of knowledge,* ed. Frederick Schmitt, 1–27. Lanham: Rowman & Littlefield.

Sullivan, Shannon, and Nancy Tuana. eds. 2007. *Race and epistemologies of ignorance*. Albany: State University of New York Press.

Tuana, Nancy. 2001. Introduction. In *Engendering rationalities*, ed. Nancy Tuana and Sandra Morgen, 1–20. Albany: State University of New York.

Tuana, Nancy and Shannon Sullivan. 2006. Introduction: Feminist epistemologies of ignorance. *Hypatia* 21(3): 1–3.

Wylie, Alison. 2003. Why standpoint matters. In *Science and other cultures: Issues in philosophies of science and technology,* ed. Robert Figueroa and Sandra Harding, 26–48. New York: Routledge.

Part I
Intersections: Feminism, Epistemology, and Science Studies

Chapter 1
The Marginalization of Feminist Epistemology and What That Reveals About Epistemology 'Proper'

Phyllis Rooney

Abstract Though feminist epistemology has been in place for a quarter century, it still remains marginalized, if not invisible, in 'mainstream' epistemology. An implicit, if not explicit, assumption that feminist epistemology is not epistemology 'proper' regularly underwrites this marginalization. The construction of feminist work as 'other' to epistemology 'proper' reflects the legacy of a philosophical history of sexism and racism more than it reflects a uniform coherent project or area of inquiry that has been in place under the rubric 'epistemology.' Specific epistemological as well as political insights into the development of epistemology (of knowledge about knowledge) are available when we critically examine the relationship between feminist epistemology and mainstream epistemology. These epistemological or, in many cases, *metaepistemological* insights merit particular attention and development at this time. The proliferation of different approaches or directions in epistemology in recent decades provides rich ground for advantageous feminist intervention. Such intervention is necessary for the recovery of epistemology as a central philosophical discipline attuned to worlds of moral and political complexity.

Keywords Marginalization of feminist epistemology • Epistemology 'proper' • Feminist as other • Metaepistemology • Feminist metaepistemic advantage

I thought how unpleasant it is to be locked out;
and I thought how it is worse perhaps to be locked in.

– Virginia Woolf, A Room Of One's Own

P. Rooney (✉)
Department of Philosophy, Oakland University, Rochester, MI 48309-4401, USA
e-mail: rooney@oakland.edu

H.E. Grasswick (ed.), *Feminist Epistemology and Philosophy of Science: Power in Knowledge*, DOI 10.1007/978-1-4020-6835-5_1,

1.1 A View from the Margins

It happened again just in this past year (2007), something that those of us who work in feminist epistemology had hoped to consign to the first 5 or 10 years in the development of the field. I was at a conference talking with two 'mainstream' epistemologists when I mentioned that my main area of research is feminist epistemology. Both looked somewhat puzzled and one of them asked, 'What is feminist epistemology?'

On the face of it, this question seems quite innocuous, an expression of interest in feminist epistemology. It is a question that we who work in the area frequently address in our teaching or in explaining what we do to people who don't work in epistemology. But situated in this professional context the question points to something of a problem: it underscores the persisting marginalization of feminist epistemology within the field of epistemology.

Feminist epistemology has been around for a quarter of a century now and it has not been hidden. Even a quick search on the Internet yields introductory essays and bibliographies that give a very good sense of the field.[1] My two discussants, both within a decade of graduate school, both claiming epistemology as their main area of interest in philosophy, had, it seems, never encountered feminist epistemology in any meaningful way in their classes or readings or professional contacts. If they had heard of the area (it's difficult to know how they could not), their interest hadn't been piqued enough to do a minimal search that would soon give a good indication of the variety of questions, interests, topics, and debates that those who work in the area engage, a variety that speaks against the kind of simple answer, the circumscribed description of feminist epistemology that my discussants seemed to expect on this occasion. (Another feminist epistemologist with whom I discussed this phenomenon of the irksome question in epistemology contexts said that she felt like carrying around a list of key readings in feminist epistemology to give to such questioners, asking them to come back after they had read them for a more meaningful conversation about feminist epistemology!)

The marginalization of feminist epistemology within the field as a whole is illustrated in other ways. Those of us who have wanted to incorporate feminist epistemology into our general epistemology classes have been hard-pressed to find epistemology texts that do that. The vast majority of these texts (which demarcate

[1] Among such readily accessed Internet essays are Elizabeth Anderson's 'Feminist Epistemology and Philosophy of Science' and Heidi Grasswick's 'Feminist Social Epistemology,' both in the *Stanford Encyclopedia of Philosophy* (Anderson 2009 [first published in 2000], Grasswick 2008), and Marianne Janack's 'Feminist Epistemology' in the *Internet Encyclopedia of Philosophy* (Janack 2004). The earliest works in what is now identified as feminist epistemology/philosophy of science include papers by four prominent pioneers in the field: Sandra Harding (1980), Lorraine Code (1981), Helen Longino (1981), and Evelyn Fox Keller (1982). A significant number of the papers in Harding and Hintikka's *Discovering Reality* (1983) deal with feminist epistemological questions. These works were soon followed by Genevieve Lloyd's *Man of Reason* (1993 [1984, first edition]) and the development of feminist epistemology was well under way.

what, for now, I take to be 'the mainstream') still give no indication that there is even such a thing as feminist epistemology, though they now regularly give some attention to other relatively new directions or approaches in epistemology – naturalized epistemology, virtue epistemology, or neo-pragmatist epistemology, for instance. The situation is somewhat better with philosophy of science texts, which have begun to include papers or discussions in feminist philosophy of science, though that inclusion is still relatively minimal. Feminist philosophy of science and feminist science studies have been very prominent within feminist epistemology (broadly construed), and this accounts for some, but not all, of this discrepancy in textual inclusion.[2] Yet feminist epistemologists have also drawn significant attention to other areas of human knowledge: moral knowledge, historical knowledge, or 'everyday' personal, social, cultural, or political knowledge. In addition, feminist epistemologists engage in debates about general understandings and philosophical conceptions of *evidence, reason, objectivity, justification, and knowledge* – the same core topics that mainstream epistemologists examine.

I will attend to feminist epistemology in this broader sense, particularly since feminist epistemology's relationship to mainstream epistemology (as that is related to, but also differentiated from, philosophy of science) is a central focus in my paper. There are, I will argue, specific epistemological as well as political insights into the development of epistemology (of knowledge about knowledge) that are available with the critical examination of this relationship that I undertake. These epistemological or, in many cases, *metaepistemological* insights merit particular attention and development at this time. Drawing in part on a standpoint epistemological perspective, I will argue that certain (meta)epistemic advantages accrue to feminist epistemology's marginal status, facilitating, in effect, specific insights about epistemology that are not otherwise available. But first, we need to reflect further on the contours of this marginality.

If my two discussants had previously heard of feminist epistemology it's quite likely that their disinterest rather than their interest would have been piqued. There has been a persistent refrain in mainstream epistemology circles that feminist epistemology is not epistemology 'proper,' and thus not something with which epistemologists need concern themselves. This refrain has ranged from hostile to dismissive to limited acknowledgement. On the more hostile side, feminist epistemologists are dogmatic ideologues, driven by 'political correctness' and 'agendas'

[2] By the mid-1980s important works by, among others, Ruth Bleier (1984), Anne Fausto-Sterling (1992 [1985]), Keller (1985), and Harding (1986) provided significant impetus to the development of feminist philosophy of science and feminist science studies (which also includes historical and sociological examinations of the sciences and technology). Also see Tuana (1989) for an important collection of essays on feminism and science that were published in *Hypatia* in 1987 and 1988. By the 1980s many mainstream philosophers of science were incorporating historical and sociological studies of real-world scientific developments into their philosophical projects, and this helped facilitate some mainstream recognition of the philosophical significance of feminist work on science.

rather than by the traditional norms of apolitical, value-free, rational investigation that have long been the hallmarks of good inquiry – including in epistemology itself. Feminist epistemology can, therefore, be dismissed, or, at best, recognized as part of political philosophy.[3] On the limited acknowledgement side, feminist epistemology is recognized as epistemology but only in a restricted sense. It is a form of applied epistemology, for example, something that might be applicable in contexts where gender roles or practices are epistemically significant. Or it is recognized as a type of epistemology but it is not new: it is a form of, or subarea within, naturalized epistemology, social epistemology, pragmatist epistemology, or virtue epistemology – where these projects have achieved some recognition within the mainstream. (The relationships between feminist epistemology and these other approaches or directions in epistemology are important and I will return to these later.)

I was introduced to such hostility and dismissal quite soon after I began to develop interests in feminist epistemology. At an institution where I formerly taught I mentioned to a colleague that I was interested in the feminist question with reason and rationality. Without even asking what this question might involve, he simply quipped as he moved on, 'Oh, so you [meaning those of us with interests in this question] are going to tell us that there is a male logic and a female logic!' And that was the end of that conversation, which was quite unlike any other conversation I'd had or witnessed with this particular philosopher. Such a reaction was not isolated, I soon learned, even if others were a bit more subtle. I've heard about numerous similar experiences from others who work in this area. Though quite painful to live through and hear about still, these experiences have given us a sharper awareness and understanding of the ways in which the subtle or not-so-subtle operations of epistemic authority and credibility work within epistemology – and within philosophy more generally. These dismissals and exclusions are problematic for epistemological as well as political reasons. In particular, specific *epistemic* problems are evident in documented efforts to summarize, dismiss, or denigrate feminist epistemology, and these, as we will see, are quite revealing of unexamined assumptions about epistemology 'proper.'

My third discussant (if one could call this exchange a discussion!) illustrates a not uncommon phenomenon, epistemically confounding as it often is. There are philosophers with minimal knowledge of feminist epistemology who nonetheless have little hesitancy about deciding what it is. Their assumptions seriously strain epistemic norms of accurate description. For example, feminist epistemology has sometimes been described as involving theories of knowledge about 'women's

[3] Stressing the political connection with feminist epistemology, Susan Haack has remarked, 'The rubric "feminist epistemology" is incongruous on its face, in somewhat the way of, say, "Republican epistemology"' (2003, 8). Similar moves to deem feminist epistemology a non-starter, or to make it disappear, are evident in paper titles such as 'Why Feminist Epistemology Isn't' (Richards 1996) and 'Feminist Epistemology as Folk Psychology' (Klee 2003).

ways of knowing.'[4] The idea of 'women's/feminine ways of knowing' has indeed surfaced in feminist epistemology, but the primary focus has been on how problematic the idea is. Among other things, it involves generalizations about women (across different races, classes, and cultures, for example) that have been the focus of significant critical scrutiny in the past three decades of feminist theorizing.[5] But the ready acceptance of this particular inaccuracy has not been innocuous. It has, one suspects, prompted the hostility that some women in philosophy have shown to the very idea of feminist epistemology. *If* feminist epistemology is promoting a 'women's ways of knowing,' they can well argue, then it is getting too close to lending support to one of the most persistent sexist refrains in the history of Western philosophy – the idea that women have different *and inferior* ways of reasoning and knowing, something that feminists surely ought to be challenging.

Other problematic modes of critique of feminist epistemology are evident in *Scrutinizing Feminist Epistemology* (Pinnick et al. 2003), a significant number of the essays in which adopt a disparaging attitude toward the field. In their reviews of this volume both Elizabeth Potter and Elizabeth Anderson draw attention to critical assessments of the field as a whole. Most of the authors are animated not by the impulse 'to make it better [but] to make it go away' (Potter 2004, 7); they aim 'to show that the entire enterprise [of feminist epistemology] is a failure' (Anderson 2006).[6] The move to disparage a whole field of inquiry is, in itself, disquieting as well as politically suspect, but especially so when standard *epistemic* norms of respectful philosophical discussion and argumentation – the very norms we teach in our introductory philosophy, epistemology, and logic classes – are set aside. I noted above that general characterizations of feminist epistemology can strain

---

[4] For example, Noretta Koertge begins an essay as follows: 'Feminist epistemology consists of theories of knowledge created *by* women, *about* women's modes of knowing, *for* the purpose of liberating women. By any reasonable standard, it should have expired in 1994' (1996, 413). Though he does not mention feminist epistemology directly, Michael Williams seems to have it clearly (and dismissingly) in mind when he writes, 'cultural relativism sometimes leads to the embrace of "standpoint epistemology," according to which ethnic, class, gender, or other "cultural" differences are associated with distinct "ways of knowing"' (2001, 220).

[5] See Lorraine Code (1991, 251–262) for a critical examination of the use of 'women's ways of knowing' as a sociological or epistemological category. In particular, in reference to a particular sociological study of 'women's ways of knowing,' she is concerned that, 'Essentialist assumptions about "women" are mirrored [...] in essentialist assumptions about knowledge, experience, and authority' (260).

[6] Anderson notes, however, that unlike most of the other essays in this volume, those by Sharon Crasnow and Janet Kourany 'are models of respectful, intellectually serious critical scholarship.' (Anderson 2006). The idea that feminist epistemology can be accepted or rejected as a whole is also evident in a special issue of *The Monist* devoted to the topic: 'Feminist Epistemology – For and Against' (*The Monist*, vol. 77, no. 4, 1994). This 'astonishing topic,' Lynn Hankinson Nelson has remarked, 'suggested that whether one is "for" or "against" "feminist epistemology" is a matter of subscribing to one of two clearly delineated, complete, and mutually exclusive sets of tenets... [which] badly mischaracterized much of the work at the intersections of feminism, epistemology, and philosophy of science' (Nelson 1995, 32).

norms of accurate description. Straw man attacks, or, as Potter remarks, 'unrecognizable caricatures of feminist scholars' (2004, 7) and their work make a frequent appearance in these critiques. Discrepancies bordering on inconsistencies also surface. Anderson notes, '[these critics of feminist epistemology tell] us that it's wrong to think that all women think alike. But apparently [these same critics think] that it's ok to think that all feminist epistemologists think alike' (2006). Hasty generalizations tend to accompany such notions. The views of one or two feminist epistemologists/philosophers of science are taken to be the views of most or all feminist epistemologists.[7] These 'representative' views are, in turn, often taken from the work of a particular scholar without adequate attention to the full context and development of her work. Sandra Harding seems to have drawn particular attention in this regard. Accounts of her positions are regularly drawn from a limited reading of her work, or her early positions are taken as static and representative when, as is the case with most epistemologists, her positions have developed and changed, often in response to respectful criticism from others who work in feminist epistemology. A recent essay titled 'The Failure of Feminist Epistemology' (Shelton 2006) cites feminist works which were all published in the 1980s, save two, both published in 1991. In effect, a decade and a half of very active feminist scholarship is simply ignored. Harding's 1986 book is cited (and there are references to two papers in her 1987 co-edited volume) but not one of her six authored or edited volumes published from 1987 to 2006 is mentioned.

What is going on here? Why is it that critics of feminist epistemology (of the field as a whole especially) regularly set aside standard philosophical and epistemic norms of careful research, reading, and reasoning in their critiques? Even if specific works by feminist epistemologists have errors or poor arguments (which are best illuminated by careful reading and reasoning in any case), why does it seem acceptable to these critics to then disparage the field as a whole? Suppose that we were to take the principle operating here and apply it more broadly, that is, decide that any area of philosophy that has produced poor work should be deemed a failure and banished from the philosophical map. We would, in effect, soon have to close down all of our philosophy departments! So why is feminist epistemology singled out in this regard? (Certainly other areas of feminist philosophy have met with similar resistance, but not quite to the extent that feminist epistemology has.) We cannot simply confine this less-than-respectful attitude toward feminist epistemology to

[7] One of the most egregious examples of this tendency to morph feminist epistemologists into one comes from Ellen Klein's description of feminist critical analyses of reason and objectivity: '"Reason" does not deliver to us "a single objective truth" [1] because "objectivity, the 'ostensibly noninvolved stance'," [2] "is the male epistemological stance…we see a male-created truth and reality, a male point of view, a male-defined objectivity" [3].' (Klein 1996, 18) As Klein indicates in footnotes, the three quotes in this single sentence are sentence fragments that come from three different authors discussing reason or objectivity in quite different contexts: Genevieve Lloyd in her analysis of the historical philosophical 'man of reason' (Lloyd 1993 [1984]); Catharine MacKinnon with her primary focus on objectivity in legal and political arenas (MacKinnon 1982); and Ruth Bleier (1984) in a discussion of objectivity in science.

individual critics: as personal anecdotes and publications continue to reveal, there has been a kind of professional acceptance, if not endorsement, of this attitude.[8]

The most immediate explanation of these problematic critiques of feminist epistemology is that there are lingering forms of sexism – perhaps in some cases misogyny – at work. The fact that this is an area of epistemology that has been developed primarily by women seems to trigger assumptions to the effect that the work is likely be of an inferior sort and, consequently, doesn't merit the careful research and reading accorded other areas of epistemology developed primarily by men. In his paper, 'Feminist Epistemology as Whipping-Girl,' Mark Owen Webb maintains that 'Feminist epistemology has become one of the whipping-girls of choice for overtly and covertly sexist elements in philosophy'(Webb 2002, 49). In this case the discipline simply reflects and reinforces forms epistemic injustice evident in the larger culture.[9] But claiming sexism as an explanation is only a first step. The sources, forms, and workings of sexism in epistemology, as elsewhere, are not all as readily apparent as we might initially think they are – or indeed hoped they were. Gender is a factor, but not in some obvious ways. Not all of those who contribute to the development of feminist epistemology are female, and not all of

[8] One might note, for example, that these problematic critiques of feminist work also pass through the hands of philosophy referees and editors, seemingly without comment. In addition to dismissive exchanges such as those noted earlier, quite telling indicators of professional resistance come from reports by feminist scholars (I've heard two more in the past year) who say that they did not go 'public' with their interests in feminist philosophy until after they had job security with tenure. Sally Haslanger (2008) provides a very helpful analysis of 'the ideology and culture of philosophy' that still sustains discrimination against women and minorities, a culture that, she notes, has also contributed to the hostile reception that feminist philosophy and feminist philosophers have often received. In her recent examination of feminist work in philosophy of science, Sarah Richardson (2010) also emphasizes connections between the marginalization of feminist thought and the marginalized status of women in philosophy. In addition to noting explicit marginalization (exclusions from elite publications and faculty positions), Richardson comments on the more implicit 'subtle everyday' forms of marginalization 'through discursive and disciplinary constructions that exclude, other, and delegitimize gender as a properly philosophical topic and feminist thought as a properly philosophical occupation' (2010, 351). These persisting problems of exclusion and marginalization still figure among the recurring topics in feminist philosophy discussion lists. All of this reflects poorly on a profession that, on the surface at least, promotes intellectual diversity and freedom.

[9] In her examination of the backlash against feminist philosophy, Cressida Heyes has noted that claims that feminist work is narrow, biased, and dogmatically ideological 'mesh neatly with sexist beliefs that have long and dishonorable histories: this culture commonly understands women as excessively concerned with the parochial and personal, incapable of seeing the "big picture," and [...] unable to exercise our rationality to attain intellectual objectivity' (Heyes 1999, 37). In addition to Webb's, other papers in Superson and Cudd (2002) also examine philosophy's particular backlash against feminism. For an important analysis of the concept of *epistemic injustice* see Fricker (2007). The marginalization of feminist epistemology can also be explored in a comparison with the documented patterns of exclusion and invisibility that women in many science disciplines still experience. Alison Wylie examines these patterns and the specific forms of epistemic injustice they reveal (Wylie, this volume). Carla Fehr also addresses the epistemic problems linked to the diminished levels of intellectual authority granted women in the sciences (Fehr, this volume).

those who disparage it are male. In addition, sexist assumptions and attitudes are not always a matter of consciously available, articulable beliefs. Many of these critics might claim that they support equal opportunity and respect for women in philosophy – though the hostility and invective of some would lead one to seriously doubt that in some cases.

A deeper understanding of the workings of sexist or gender-limited attitudes and practices in this situation requires insights developed specifically in feminist work, the very insights that many of these critics (with their relatively superficial reading and understanding of that work) clearly aim to keep at arm's length. As feminist theorizing more generally continues to illuminate, sexism and problematic gendering are significantly a matter of background social and cultural institutions and practices. These include linguistic practices, cognitive practices of attention or inattention, and power-inflected epistemic practices that confer or withhold credibility, for instance. Such practices frame individual and community attitudes and behavior, and they do so in ways that may not be visible as sexist or gender-inflected without specific *feminist* intervention. Indeed, the task of *making gender visible* has been taken to be constitutive of a range of feminist epistemological projects.[10] Thus, for example, making visible the impact of women's epistemic disenfranchisement, that is, their dismissal as serious reasoners and knowers in a variety of knowledge areas and disciplines as well as in philosophy, continues to be a defining project in feminist epistemology. However, making gender visible has also required the development of nuanced understandings of gender.[11] Among these we can include the textual, metaphorical construction of *gender* in traditional philosophical theorizing. Something of a displacement of *gender* as the primary or only focus of attention must also be included here, given important feminist work on the intersection of gender with other epistemically salient social/status divisions such as race and class. These enhanced understandings provide additional insights into the problematic marginalization of feminist epistemology, and some are further developments of Simone de Beauvoir's key insight into the philosophical construction of *woman* as 'Other.'

In her paper 'The Feminist as Other' Susan Bordo reflects on the marginal status accorded feminist cultural critique and feminist ethical theory within their respective mainstream theoretical arenas. Her explanatory framework also applies to the situation with feminist epistemology, though most immediately to the less hostile, limited acknowledgement forms of its marginalization:

> ...feminists [are construed] as engaging in a specialized critique, one that cannot be ignored, perhaps, but one whose implications are contained, self-limiting, and of insufficient general consequence to amount to a new knowledge of 'the way culture operates.'

[10] Helen Longino, for example, has proposed as 'a bottom line requirement of feminist knowers on cognitive standards [across a range of knowledge/epistemology projects]: that they reveal or prevent the disappearing of gender' (1994, 481).

[11] Sally Haslanger (2002, esp. pp. 87–91) provides a helpful examination of feminist theoretical work on the concept of *gender*. For feminist epistemological work, in particular, Haslanger emphasizes an understanding of *gender* as something that is fundamentally constituted by systems of social relations and norms.

> One does 'gender' *or* one engages in criticism of broad significance; pick one. [...] this construction is not merely an annoying bit of residual sexism but a powerful conceptual map that keeps feminist scholarship, no matter how broad its concerns, located in the region of what Simone de Beauvoir called the 'Other' ...feminist theory swims upstream against powerful currents whenever it threatens to assume the mantle of *general* cultural critique, rather than simply advocate for the greater inclusion or representation of women and their 'differences' (Bordo 1998, 297, 306–07).

The regular 'misreading' of feminist work, Bordo argues, ensures that 'the insights of feminist philosophy are "kept in their place," where they make no claim on "philosophy proper"' (308).

The 'powerful conceptual map' that Bordo refers to is quite familiar to feminist philosophers. However, where the traditional construction of woman as 'Other' remains unrecognized and unexamined – as it does still in non-feminist philosophy circles – the conceptual map it has engendered remains largely intact. And where that map remains intact, feminists' concern with women and gender (along with the common assumption that only women are gendered) is sufficient to slot feminists themselves and their work into the place of the 'Other' – even when their concern involves challenging this historical construction.

Bordo's explanation of the persisting marginalization of feminist work lends itself to further expansion in connection with feminist epistemology. For example, critics of feminist epistemology often use wording such as 'feminist epistemologists maintain...' or 'the feminists say...' where, apart from problems with hasty generalizations, it is quite clear that the authors want to distance themselves from 'the feminists.' They present feminists as 'other,' just as, historically, male philosophers regularly discussed women as other to themselves and their anticipated audience. Yet women were sometimes assigned their special circumscribed place; so too mainstream responses to feminist epistemologists' work sometimes accord it its theoretically circumscribed place. Misogynistic refrains of hostility, of battling, of warding off women or 'womanly' characteristics also informed the historical construction of woman as Other – not least when ideals of pure reason and knowledge were under discussion. Women, 'woman' or 'the feminine' regularly served as the literal embodiment of, or the metaphorical representation of body, emotion, subjectivity, particularity, and disorder, the *other* in relation to which (or in contrast to which) the realm of the truly rational, the truly philosophical was regularly conceptualized.[12] This imaginary or metaphorical underpinning thus insinuated the view

[12] In her book, *The Philosophical Imaginary*, Michèle Le Doeuff sums up Western philosophy's 'imaginary portrait of "woman"' as follows: 'a power of disorder, a being of night, a twilight beauty, a dark continent, a sphinx of dissolution, an abyss of the unintelligible, a voice of underworld gods, an inner enemy who alters and perverts without visible signs of combat, a place where all forms dissolve' (1989, 113). For specific examinations of the metaphorical construction of 'woman' as *Other* to 'the man of reason' see Rooney (1991) and Lloyd (2002). This displacement of woman as 'Other' was often confounded (and rendered 'invisible') by the related displacement of metaphor as 'feminine' decoration or embellishment of language, as that which is *other to* the linguistic place of clarity and purity in language – the place of the core content of philosophy (Rooney 2002).

that following the path of reason and true knowledge required a kind of continual warding off, a battling against threatening, confounding, or bewitching 'feminine' elements. The additional hostility that feminist epistemology garners thus likely reflects the lingering impact of these misogynistic aspects of philosophical theorizing. Feminist epistemologists don't generally endorse the feminine other of traditional conceptions of reason and knowledge, since it is largely a caricatured construction of a sexist and misogynistic cultural imaginary – a point that was clearly made and well-taken a quarter of a century ago.[13] However, they *do* stand ground against the sexism and misogyny that engendered that construction, and they question ideals of reason and knowledge that have implicitly imported aspects of that construction. But this distinction is quite lost among critics of feminist epistemology who simply assimilate feminist perspectives with 'the other' of reason and knowledge. Feminist epistemology thus emerges as a hostile principle, as a threat to 'pure' epistemology.

Some of the epistemically confounding criticisms of feminist work now come into clearer focus. Despite the fact that feminist epistemologists present a range of different and nuanced arguments about a variety of epistemological topics, those differences are morphed and assimilated when filtered through the specter of the threatening feminine Other. And despite the fact that feminist epistemologists are significantly engaged in constructively developing accounts of reason and objectivity that pay attention to a greater range of reasoning and knowing situations and contexts than many traditional conceptions did, that work is still regularly framed as an attack or 'assault' on reason and objectivity, as something hostile to the very ground of epistemology 'proper.'

Women, 'woman' or 'the feminine' are not the only real or imaginary constructions to occupy the realm of the Other, however. Otherness, Bordo notes, has many faces and is reflected also in the (mainstream) marginalization of philosophical work on race: 'Every time black authors are quoted only for their views on race – expertise about "general" topics being reserved for white males who are imagined to be without race and gender – the Otherness of the black is perpetuated' (1998, 298). Racial and other cultural forms of otherness also made their mark in the history of philosophy. Charles Mills has documented the facility with which philosophers of the pre-Enlightenment and Enlightenment discussed 'barbarians… men who are like beasts' (Grotius), 'savage people' (Hobbes), 'savages' (Rousseau – even if some of his savages were 'noble'), or the fundamental 'difference between [the black and white] races of man' (Kant) (Mills 1997, 64–71). Along with 'woman,' the categories of 'the savage' and 'the primitive' carried epistemological weight – they too were regularly invoked to mark that which is beyond the realm

[13] Though this point was made elsewhere, Genevieve Lloyd made it quite clearly in the concluding remarks of her *Man of Reason*: 'The affirmation of the value and importance of "the feminine" cannot of itself be expected to shake the underlying normative structures, for, ironically, it will occur in a space already prepared for it by the intellectual tradition it seeks to reject… What has happened has been not a simple exclusion of women, but a constitution of femininity through that exclusion' (1993, [1984], 105–06).

of the truly rational. (Debates about whether or how 'primitive people' are rational cannot even be consigned to the distant past in epistemology.)

Mills has argued that centuries of racial injustice and white epistemic authority were sustained by what he calls an 'epistemology of ignorance.' He notes that political theories (social contract theories supporting white privilege, for instance) typically require epistemological commitments about what counts as credible experience and genuine knowledge about the world. By excluding the experiences and knowledge of those not counted among the theories' ideal moral and political agents, these theories thus sustain systematic ignorance, not only about the social realities of those 'others,' but, just as significantly, about the ways in which the social realities of the included are constructed and privileged by the lives and work of those others. According to Mills, centuries of white racism have thus prescribed:

> an inverted epistemology, an epistemology of ignorance [which is] a particular pattern of localized and global cognitive dysfunctions (which are psychologically and socially functional), producing the ironic outcome that whites will in general be unable to understand the world they themselves have made…[this involves] a cognitive model that precludes self-transparency and genuine understanding of social realities (Mills 1997, 18).

Among recent developments in feminist epistemology are many that address the ways in which patterns of race and gender subordination and exclusion reflect and reproduce patterns or forms of ignorance.[14] Thus, here and elsewhere, the epistemological concerns raised by liberatory movements addressing gender injustice intersect with those addressing race or other group-based forms of injustice. Where it incorporates these intersections feminist epistemology is no longer focused exclusively on *gender*. (It seldom was to the extent that its critics often assume – Harding's work is a notable case in point.) I understand 'feminist epistemology' to encompass these theoretical intersections and expansions, though some now prefer the term 'liberatory epistemology(ies).' Whatever term is used, however, these liberatory perspectives in epistemology share common concern with making visible the forms of ignorance systematically produced and reinforced by mainstream perspectives that still insist – explicitly or otherwise – that particular groups of knowers, particular forms of knowledge, understanding, and insight, or particular topics and questions about human knowledge (including questions about connections between human knowledge and human justice) are beyond the pale of epistemology 'proper.' As we will next explore, the mainstream's insistence here is the kind of thing that can come back to haunt, particularly when reflected back from the margins, from the place of the Other. The idea that there is such a thing as a clearly demarcated epistemology 'proper' turns out to be the product of the same philosophical imaginary that created the idea of its Other.

---

[14] The 2004 conference at Penn State University, 'Ethics and Epistemologies of Ignorance,' foregrounded these important connections between feminist epistemology and philosophy of race, and between epistemology and moral and political philosophy. The conference directors, Shannon Sullivan and Nancy Tuana, have published two volumes of papers from the conference: *Feminist Epistemologies of Ignorance*, a special issue of *Hypatia* (Tuana and Shannon 2006), and *Race and Epistemologies of Ignorance* (Sullivan and Tuana 2007).

1.2 Feminist Metaepistemic Advantage

Being on the margins is not all bad – especially when one has good company there! Some epistemic advantages may also accrue to this location, a key insight in (feminist) standpoint epistemology particularly. We need not assume here what Alison Wylie calls the 'thesis of *automatic epistemic privilege*... [the claim] that those who occupy particular standpoints (usually subdominant, oppressed, marginal standpoints) automatically know more, or know better, by virtue of their social, political location' (2004, 341). As Wylie notes, this has been, at best, a controversial thesis in standpoint epistemology and it is not clear that any theorist has endorsed it in this general form. (It is, however, the kind of general thesis that critics of feminist epistemology are likely to attribute to it.) The kinds of standpoints that merit particular epistemological attention are, as Wylie and others stress, those that are achieved by a critical consciousness attuned to 'the nature of our social location and the difference it makes epistemically' (Wylie 2004, 344). The epistemic import of any such critical consciousness thus depends on specifics of the subdominant or marginal location and the particular forms of marginality it represents.

Despite differences in specifics there are recurring themes in arguments for standpoint-informed epistemic *advantage* that also pertain to arguments for feminist metaepistemic advantage. Prominent among these are claims of straightforward experiential or empirical advantage: the lived experience of marginalization can enable one to see and understand things that are quite 'invisible' to those not marginalized. (For instance, I doubt that I would have seen so clearly some of the workings of epistemic authority and credibility within epistemology – and professional philosophy more generally – had I not experienced them first-hand by coming up sharply against them when I simply expressed positive interest in feminist epistemology.) Empirical advantage is sometimes spelled out in terms of the 'double consciousness' of the 'outsider within,' as Patricia Hill Collins has done in her exploration of Black women's status in sociology. As academicians Black women have acquired a certain 'within' or 'insider' status, Collins argues, but when they find their lived experiences devalued or negated in sociological paradigms they may also become quite conscious of their 'outsider' status there, and 'their difference sensitizes them to patterns that may be more difficult for established sociological insiders to see' (1991, 53). Feminist epistemologists have also acquired something of an outsider within status. To the extent that we have the time and resources to read, teach, or write feminist epistemology we are likely to have attained some type of insider status in philosophy or in other academic disciplines. Our outsider status is made known to us, however, when we confront the various forms of marginalization noted above. In addition, many of us have come to this work after developing more traditional epistemological interests – in effect, we were 'proper' epistemologists before we became 'improper'! We can thus mine the epistemic benefits of this particular double consciousness, of what Collins calls the 'the creative tension of outsider within status' (53).

As suggested in the previous section and elucidated further below, the marginalization of feminist work affords specific insights into the limited understandings of epistemology that are conveyed in the endorsement of a non-feminist or contra-feminist 'real' or 'true' or 'pure' epistemology, or an epistemology 'proper' – even if these terms are not used explicitly.[15] Yet, as we will see, the unity or uniformity that these terms suggest stands in marked contrast to the proliferation of approaches and directions in epistemology developed during these past decades, not to mention historically. This proliferation has ushered in something of an age of metaepistemology, insofar as it invites us to be more reflective (and reflexive) about the different approaches we might adopt in philosophical thinking about knowledge and related epistemic concepts. This invitation is also advanced by (recent) futuristic speculations about where epistemology is going or ought to go. I will now explore this metaepistemological terrain, paying particular attention to the ways in which it proves to be rich ground for feminist metaepistemic advantage.

In my paper so far the term 'mainstream epistemology' has seemed to indicate a relatively uniform, circumscribed body of inquiry. But that uniformity has been largely constituted by its practitioners' quite uniform neglect, dismissal, or inability to engage and incorporate feminist work in epistemology. We have noted that when mainstream dismissals assume or posit a (non-feminist) 'real' epistemology they, in part, follow the contours of an entrenched conceptual map linked to a philosophical history of sexism and racism. Thus, such dismissals illustrate the very significance of feminist epistemology. They also suggest that the notion of epistemology 'proper' is largely a defensive front that is marshaled against the threat of the very idea of 'feminist epistemology.' Let us explore further how this suggestion pans out.

Elisabeth Lloyd has examined a similar stance of unity or uniformity in connection with mainstream reactions to feminist work on *objectivity*. When feminist epistemologists and philosophers of science argue for the importance of paying attention to sex and gender in examining the ways in which scientific knowledge is produced and validated, they are portrayed, Lloyd notes, as: 'playing "out-of-bounds" in terms of mainstream understandings of the problems of epistemology and philosophy of science; feminist work can, therefore, be safely ignored, set aside, or characterized as of interest only in marginal cases' (Lloyd 1995, 352). She maintains that this marginalization has its source in a specific 'philosophical folk story about objectivity,' in which...

> 'objectivity' [is taken] as a sort of beautiful primitive, self-evident in its value, and all-powerful in its revelatory power... a unified front is implicitly presented against feminist epistemologists: 'objectivity' is of utmost clarity and importance to everyone except the feminists, who are caricatured as disregarding it in order to further their political agendas. (375)

Lloyd argues that this philosophical folk story about objectivity is just that. She examines in some detail different conceptions of *objectivity* that are in broad use in

[15] Suppositions to the effect that feminist epistemology is not epistemology 'proper' are often conveyed by the marginalizing gestures noted above. Webb describes how he was 'met with puzzlement' (including from fellow graduate students) when he pursued interests in feminist epistemology, and he was sometimes asked, 'When are you going to get back to real epistemology?' (2002, 51).

debates in contemporary analytic philosophy. These conceptions, she argues, incorporate substantial recognition of the significance of social practices and social standards in good inquiry (a central focus in *social epistemology* particularly), and they do so in a way that cannot, without argument, preclude examinations of social practices linked to sex and gender. When we take into account recent developments in (mainstream) work on objectivity, she continues, it is not the term 'feminist epistemology' that emerges as an oxymoron but, instead, terms such as 'value-free inquiry,' 'disinterested knowledge,' and 'pure epistemology' (374).

An analogous (and, of course, related) argument can be made about a philosophical folk story about *epistemology* that surfaces in mainstream reactions to feminist work, where *epistemology* also emerges as kind of 'beautiful primitive… as transparent, simple, stable, and clear in its meaning' (Lloyd 1995, 375). When we push beyond initial characterizations of epistemology (as something like 'the philosophical study of the nature, sources, and limits of knowledge') we find anything but general agreement about what, more precisely, its core topics, questions, methods, and directions are or should be. This has been the case historically but, quite specifically for our purposes here, this has been the case during these past decades – these same decades in which feminist epistemology has been ignored or deemed out-of-bounds by mainstream perspectives.

By the end of the twentieth century, in fact, it looked like the worst of times and the best of times for epistemology. Some had notably proclaimed the 'end' of epistemology. Richard Rorty figured quite prominently with such claims, though his main focus was the modern project of establishing secure foundations for knowledge – with knowledge understood in terms of mental representations that mirror truths in the world. W. V. O. Quine also proposed something like an end to epistemology with his argument for 'epistemology naturalized,' for the replacement of epistemology (as primarily a philosophical project of conceptual analysis) by the cognitive scientific study of knowing and knowledge as natural phenomena.[16] Many deemed a core project in standard analytic epistemology, the analysis of *knowledge* as justified true belief (JTB analyses), as futile, given the seemingly endless production of Gettier-like counterexamples to such analyses – those 'countless and wonderfully rococo counterexamples' (Bishop and Trout 2005, 702).

Yet in the last few decades of the twentieth century a whole range of new epistemologies, new directions or approaches in epistemology, or newly reworked versions of older epistemological orientations also emerged. In addition to naturalized epistemology, we now have social epistemology, virtue epistemology, pragmatist or neo-pragmatist epistemology, and, of course, feminist epistemology. New work on the epistemic status of moral judgments and beliefs gave some prominence to moral epistemology. Late twentieth century developments in continental and postmodern epistemology can also be included here, even if some of this work is positioned as deflationary with respect to the 'traditional' project of epistemology. Some projects in feminist epistemology especially (by Linda Martín

[16]See especially Rorty (1979) and Quine (1969).

Alcoff, Susan Bordo, and Lorraine Code, for instance) connect the analytic and continental traditions in constructive and original ways, and thus contribute new topics and directions in epistemology.

These worst and best of times mark an interesting time for epistemology, certainly, but they also signal a significant shift into *meta*epistemological terrain. In pronouncing the end or limited viability of particular ways of doing epistemology, 'end' claims invite serious consideration of other ways of doing it (as Rorty did in his engagement with pragmatist epistemology and Quine did with his proposal for naturalized epistemology). Many have questioned whether *knowledge* ought to be the constitutive or core concept of epistemology: some have argued that *epistemic justification* is the more tractable core concept; *understanding* is also a proposed alternative (Elgin 2006). In addition, when presented with a range of new epistemologies or new directions in epistemology, we are encouraged to ask metaepistemological questions such as the following: How does one choose among these different directions or approaches in epistemology? What are the goals of epistemology anyway – goals that would help guide one's choices here? What constitutes epistemological progress? (Many epistemologists make individual choices by engaging the opportunities or directions in epistemology that are available to them, or that they find interesting – or that their thesis/dissertation advisers did. They may proceed with little more than a hand-waving dismissal of other questions and approaches as uninteresting, or as not 'real' epistemology – a move that simply begs the question in the best of these metaepistemological times.) In a chapter titled 'Epistemology's End,' Catherine Elgin draws attention to the metaepistemological considerations that she thinks attend basic disagreements in or about epistemology:

> To view [epistemological theories] as supplying alternative answers to the same questions is an oversimplification. For they embody disagreements about what the real questions are and what counts as answering them....To understand a philosophical position and evaluate it fairly requires understanding the network of commitments that constitute it; for these commitments organize its domain, frame its problems, and supply standards for the solution of those problems (1996, 3).

Feminist epistemology's development in constructive conversation with a variety of approaches or directions in epistemology merits particular attention in this discussion. Code's early work, *Epistemic Responsibility* (1987), though not as explicitly feminist as many of her later works, set the stage for the ongoing intersection of her work in feminist epistemology with virtue epistemology. Her title concept ranks centrally among the intellectual and epistemic virtues that virtue epistemologists examine and promote – others are truthfulness, open-mindedness, curiosity, intellectual integrity, epistemic trust, and intellectual autonomy and courage. (For the purposes of this paper, it is it is worth noting that just about all of the non-feminist virtue epistemologists who make reference to her early book make no mention of Code's later feminism-inspired work.) Miranda Fricker's recent work, *Epistemic Injustice* (2007), also exemplifies important connections between feminist epistemology, virtue epistemology, and social epistemology. She examines the epistemic significance of power-inflected social relations such as gender, race, and class, particularly as they effect unequal distributions of epistemic trust, credibility, and authority. Many identify feminist epistemology as a form of social epistemology, where the latter is broadly characterized as encompassing the epistemological study of the ways in which social relations and institutions shape knowledge practices and achievements. Though Heidi Grasswick thinks that 'by far the majority of work in feminist epistemology is best understood as a form of social epistemology,' she also notes that feminist epistemology predates social epistemology as the latter is now understood (Grasswick 2008). Thus, instead of thinking of feminist epistemology as simply a subarea or form of social epistemology, we might think of it as a significant inspiration and resource in the development of social epistemology, something that cannot be ignored by (mainstream) social epistemologists.

There are, however, feminist epistemological projects that develop important connections with other – though arguably related – directions or perspectives in epistemology. Many have emphasized specific links with pragmatist epistemology.[17] In promoting understandings of knowers as engaged inquiring actors in the world, and in drawing connections between practices of inquiry and broader social and political practices and concerns, pragmatism facilitates the incorporation of political awareness into epistemological reflections, and is thus consonant with many feminist projects. Pursuing a somewhat different orientation, many of the authors in Nelson and Nelson (2003) develop feminist epistemological perspectives in both easy and uneasy conversation with Quine-inspired naturalized epistemology. I argue there, however, that an 'uneasy alliance' between these two areas or directions in epistemology speaks against an easy subsumption of feminist epistemology as a part of naturalized epistemology – as the latter is commonly understood (Rooney 2003). A significant naturalist dimension is also evident in ongoing feminist epistemological engagement with feminist research projects across a range of areas and disciplines of knowledge – notably, though not exclusively, in the natural and social sciences.[18]

---

[17] Lisa Heldke (1989) argued for significant similarities between the epistemological projects of John Dewey and Keller. A special issue of *Transactions of the C.S. Peirce Society: A Quarterly Journal in American Philosophy* (vol. 27, no. 4, Fall 1991) was devoted to 'Pragmatism and Feminism.' Charlene Haddock Seigfried edited a special issue of *Hypatia* (vol. 8, no. 2, Spring 1993) on 'Feminism and Pragmatism,' and developed specific connections between feminist philosophy and the work of William James and Dewey (Seigfried 1996). I have examined notable links between feminist epistemology and pragmatist epistemology in Rooney (1993). Sharyn Clough (2003) also examines connections between feminism and pragmatism as she develops 'a pragmatist approach to feminist science studies.'

[18] Interdisciplinary work continues be significant in feminist epistemology, broadly construed. For example, the program for the first FEMMSS (Feminist Epistemologies, Methodologies, Metaphysics, and Science Studies) conference in Fall 2004 listed participants from the following disciplines, in addition to philosophy and women's studies: physics, education and information sciences, english, psychology, law, political science, science and technology studies, economics, sociology, nursing, mathematics and computer sciences. http://depts.washington.edu/femmss. Also see Alcoff and Potter (1993) and Tuana and Morgen (2001) – important collections of essays in feminist epistemology that engage with a variety of approaches in epistemology. The introductions to these volumes also provide helpful overviews of topics and methods in feminist epistemology.

The boundaries between or among these different directions or perspectives in epistemology are neither rigid nor static. Specific differences or connections among them depend on how narrowly or how broadly one characterizes any given perspective. Some, for instance, understand social epistemology (insofar as it incorporates empirical studies of social practices and institutions) as an extension of naturalized epistemology. These different perspectives also lend themselves to various hybrid epistemological viewpoints, as connections with and through feminist work make especially clear.

Yet some boundaries between not-specifically-feminist projects and feminist epistemology stubbornly persist in familiar ways. While mainstream proponents of 'new' epistemological approaches regularly contrast their work with 'traditional' approaches, their work still reflects the marginalization inspired by that very tradition. More specifically, conversations between feminist and mainstream projects have, to date, been notably one-way. Quite typical in this regard is a recent analysis of intellectual virtues that incorporates an endorsement of virtue epistemology as a 'regulative epistemology' (Roberts and Wood 2007). The authors maintain that 'the triviality of standard epistemology's examples' and the 'cottage industry' that sprang up with Gettier-informed analyses of knowledge contributed to analytic epistemology's becoming 'increasingly ingrown, epicyclical, and irrelevant to broader philosophical and human concerns' (2007, 5–8). They argue that virtue epistemology 'holds enormous promise for the recovery of epistemology as a philosophical discipline with broad human importance… [where epistemology] connects with ethical and political issues' (6, 9). Such a 'recovery of epistemology' has been central to many projects in feminist epistemology for some decades, yet again (and this, by now, *is* a broken record) feminist work is nowhere mentioned in this text.[19]

More generally, other than, at best, limited recognition of feminist work as an example or a subarea within their respective developments, mainstream advocates of these new perspectives do not engage feminist work in any substantive way – indeed most seem quite unaware of its existence. If open-mindedness as openness to new or different perspectives is an intellectual or epistemic virtue (and I certainly think it is), then openness to readily-available new or different epistemological perspectives and directions is surely a metaepistemic virtue, and it is one that feminist epistemologists are in a position to claim to their advantage. In particular, by failing to avail themselves of the reflexive awareness and critique of their own epistemological assumptions and practices that more thoughtful attention to feminist work would surely bring, mainstream proponents of these other perspectives put themselves at a distinct disadvantage in the metaepistemological terrain mapped out by these different perspectives in epistemology.

[19] There is mention of Code's early work (1987) but, again, no mention of her later explicitly-feminist work in epistemology. In a much more inclusive vein, Laura Ruetsche (2004) makes constructive use of a model of Aristotelian virtue in her examination of the concept of *warrant* in feminist epistemology.

Feminist metaepistemic advantage can also be assessed in connection with recent futuristic thinking in philosophy, some of which is prompted by the kind of drawing back and looking at the longer view that the turn of a century – or in this case a millennium – inspires.[20] Most significant for my discussion is Stephen Hetherington's edited volume *Epistemology Futures* (2006). In his introduction, Hetherington lists some key metaepistemological questions that frame the essays in the volume. These include back-to-the-drawing-board questions about what the purpose or goal of epistemology is, what cognitive or epistemic phenomena it should study, what core concepts it should examine ('maybe other epistemic concepts... [besides] knowledge, evidence, warrant...would be more penetrating and apt'), what methods it should use, and what should count as epistemological progress or achievement (2006, 1–9).

How does feminist epistemology figure into Hetherington's epistemology futures? At first glance, not at all. In none of the 13 essays in this volume is feminist epistemology mentioned or referenced, though, taken together, the essays engage a range of directions or approaches in epistemology – in standard analytical, naturalist, pragmatist, and virtue epistemology. According to this text, not only is feminist epistemology nonexistent in the present but it doesn't figure into any epistemology future either.[21]

On second glace, however, the volume, as well as specific remarks by Hetherington, establish both the legitimacy and the necessity of feminist epistemology. For a start, critics of feminist epistemology who assume or claim that it is not epistemology 'proper' stand on shaky ground when, as is evident in this volume and elsewhere, what constitutes epistemology 'proper' is very much open to debate. In addition, Hetherington frames the metaepistemological explorations in his volume in a way that (unwittingly it seems) directs attention to the significance of feminist or other 'outsider' perspectives in epistemology. He wonders whether...

> ...our current grasp of epistemological possibilities is itself more limited than we realize.... How good are we at judging epistemological proposals without reflecting entrenched yet narrow or misleading central concepts, standards, methods, questions, and so on? How good are we at improving upon those, even at imagining *new* central concepts, standards, methods, questions, and the like? ...This process [of moving into an improved epistemological future] can stagnate, as we assume that some proposals are irrelevant or mistaken, simply because of how 'implausible' they can currently strike us as being. Bare assessments of implausibility tend to give voice merely to our professional training...but what is entrenched need not be true. Nor need it be able fair-mindedly to assess fundamental challenges or alternatives to itself (2006, 5).

The absence of any mention of feminist epistemology in his volume establishes Hetherington's concerns as quite real and justified. As we have seen, mainstream

[20] Brian Leiter's edited volume, *The Future for Philosophy* (2004) foregrounds this futuristic trend. As Leiter notes: 'Meta-philosophical questions, i.e. questions about what philosophy is, its proper concerns, methods, and limitations, and its rightful ambitions are inevitably on the table in any consideration of philosophy's future... Philosophy today – especially, though not only, in the English-speaking countries – is not a monolith, but a pluralism of methods and topics' (2004, 1).

[21] Hendricks and Pritchard's 2008 volume, *New Waves in Epistemology*, also presents a range of new directions in the field, but, again, feminist epistemology is not recognized among them.

assessments of 'implausible' feminist work often have less to do with feminist work than with the inability of mainstream perspectives 'fair-mindedly to assess fundamental challenges or alternatives' to 'entrenched' understandings of what epistemology is, and, more importantly, what epistemology can or ought to be. But such assessments also establish the necessity of feminist work for moving beyond those same 'narrow or misleading' understandings and moving toward 'an improved epistemological future.'

I do not claim that my examinations above exhaust the possibilities for advantageous feminist metaepistemological intervention and insight. In addition, the interventions outlined intersect and connect in ways that suggest developments that go beyond the scope of this paper. (I examine some of these specific metaepistemic advantages in more detail in Rooney (forthcoming)). They do, however, help to establish specific findings about feminist epistemology and its relationship to a supposed epistemology 'proper.' First, the effort to contain feminist epistemology (as one distinct, circumscribed project, or as a subarea within a more 'mainstream' area or approach in epistemology) is misguided, and has been for some time. Second, the positing of a contrasted epistemology 'proper' reveals more about the lingering effects of philosophy's history of sexism and racism than it reveals about some supposed unified, coherent area of inquiry in philosophy. Third, feminist epistemology (as encompassing a range of epistemological projects informed and linked by efforts to uncover the political and epistemological fallout of the epistemic disenfranchisement of women and other 'others') proceeds in fruitful conversation with a range of approaches or directions in epistemology, and it is unique in this kind of epistemological flexibility. Fourth, in part because of this flexibility, feminist epistemology provides fruitful ground for new metaepistemological reflections about how epistemology is defined, about what its core concepts, questions, and directions are, or, more to the point, ought to be in an improved epistemology future. These reflections inspire nothing less than a recovery of epistemology as a central philosophical discipline attuned to worlds of moral and political complexity, including worlds that have been informed by Western epistemology's own moral and political history.[22]

I don't recall how I responded to my two discussants (mentioned at the beginning) who wanted to know what feminist epistemology is. I may well have responded to their puzzlement with an equally puzzled look of my own, a kind of counter-puzzlement! To the extent that this paper is an answer to their question I suspect it contains a lot more than they were bargaining for, and not just because of its length. Then again, if they were to read it (that's the catch, of course), I think that doing so would make them better epistemologists, and not just because they would know some additional things about what some fellow epistemologists are up to.

[22] In her paper 'How is Epistemology Political?' Alcoff has argued that epistemology cannot but be political in quite specific ways. Among other things, it has distinct discursive effects in philosophy and in broader social and political arenas – it can '[influence] whose arguments are considered plausible enough to be given consideration …[it can] authorize or disauthorize certain kinds of voices, certain kinds of discourses' (Alcoff 1993, 69, 73).

Acknowledgments I would like to thank Laura Kramer for helpful comments on an early draft of this paper and Heidi Grasswick for her encouragement and support with its development.

References

Alcoff, Linda. 1993. How is epistemology political? In *Radical philosophy: Tradition, counter-tradition, politics*, ed. Roger S. Gottlieb, 65–85. Philadelphia: Temple University Press.

Alcoff, Linda, and Elizabeth Potter. 1993. *Feminist epistemologies*. New York: Routledge.

Anderson, Elizabeth. 2006. How not to criticize feminist epistemology: A review of *scrutinizing feminist epistemology*. Elizabeth Anderson's philosophy home page. Jan. 19, 2006. http://www-personal.umich.edu/~eandersn/hownotreview.html A shorter version of this review appeared in *Metascience* 13 (2004): 395–399.

Anderson, Elizabeth. 2009. Feminist epistemology and philosophy of science. *The Stanford encyclopedia of philosophy (Winter 2009 edition)*, Edward N. Zalta, ed. http://plato.stanford.edu/archives/win2009/entries/feminism-epistemology/.

Bishop, Michael, and J.D. Trout. 2005. The pathologies of standard analytic epistemology. *Nous* 39(4): 696–714.

Bleier, Ruth. 1984. *Science and gender: A critique of biology and its theories on women*. Elmsford: Pergamon Press.

Bordo, Susan. 1998. The feminist as other. In *Philosophy in a feminist voice: Critiques and reconstructions*, ed. Janet A. Kourany, 296–312. Princeton: Princeton University Press.

Clough, Sharyn. 2003. *Beyond epistemology: A pragmatist approach to feminist science studies*. Lanham: Rowman & Littlefield.

Code, Lorraine. 1981. Is the sex of the knower epistemologically significant? *Metaphilosophy* 12(3–4): 267–276.

Code, Lorraine. 1987. *Epistemic responsibility*. Hanover: University Press of New England.

Code, Lorraine. 1991. *What can she know? Feminist theory and the construction of knowledge*. Ithaca: Cornell University Press.

Collins Patricia, Hill. 1991. Learning from the outsider within: The sociological significance of black feminist thought. In *Beyond methodology: Feminist scholarship as lived research*, ed. Mary Margaret Fonow and Judith A. Cook, 35–59. Bloomington: Indiana University Press.

Elgin, Catherine Z. 1996. *Considered judgment*. Princeton: Princeton University Press.

Elgin, Catherine Z. 2006. From knowledge to understanding. In *Epistemology futures*, ed. Stephen Hetherington, 199–215. Oxford: Oxford University Press.

Fausto-Sterling, Anne. 1992. *Myths of gender: Biological theories of women and men*, 2nd ed. New York: Basic Books. 1985.

Fricker, Miranda. 2007. *Epistemic injustice: Power and the ethics of knowing*. Oxford: Oxford University Press.

Grasswick, Heidi. 2008. Feminist social epistemology. *The Stanford encyclopedia of philosophy (Fall 2008 edition)*, Edward N. Zalta ed. http://plato.stanford.edu/archives/fall2008/entries/feminist-social-epistemology/.

Haack, Susan. 2003. Knowledge and propaganda: Reflections of an old feminist. In *Scrutinizing feminist epistemology: An examination of gender in science*, ed. Cassandra L. Pinnick et al., 7–19. Piscataway: Rutgers University Press.

Harding, Sandra. 1980. The norms of social inquiry and masculine experience. In *PSA 1980: Proceedings of the Biennial meeting of the Philosophy of Science Association*, vol. 2, ed. Peter D. Asquith and Ronald N. Giere, 305–324. East Lansing: Philosophy of Science Association.

Harding, Sandra. 1986. *The science question in feminism*. Ithaca: Cornell University Press.

Harding, Sandra, and Merrill B. Hintikka (eds.). 1983. *Discovering reality: Feminist perspectives on epistemology, metaphysics, methodology, and philosophy of science*. Dordrecht: D. Reidel Publishing Company.

Haslanger, Sally. 2002. On being objective and being objectified. In *A mind of one's own: Feminist essays on reason and objectivity*, 2nd ed, ed. Louise M. Antony and Charlotte E. Witt, 209–253. Boulder: Westview Press.

Haslanger, Sally. 2008. Changing the ideology and culture of philosophy: Not by reason (Alone). *Hypatia* 23(2): 210–223.

Heldke, Lisa. 1989. John Dewey and Evelyn Fox Keller: A shared epistemological tradition. In *Feminism & science*, ed. Nancy Tuana, 104–115. Bloomington: Indiana University Press.

Hendricks, Vincent F., and Duncan Pritchard (eds.). 2008. *New waves in epistemology*. New York: Palgrave Macmillan.

Hetherington, Stephen (ed.). 2006. *Epistemology futures*. Oxford: Oxford University Press.

Heyes, Cressida J. 1999. The backlash against feminist scholars and scholarship: Introduction. *American Philosophical Association Newsletter on Feminism and Philosophy* 99(1): 36–40.

Janack, Marianne. 2004. Feminist epistemology. *Internet encyclopedia of philosophy*. www.iep.utm.edu/fem-epis/.

Keller, Evelyn Fox. 1982. Feminism and science. *Signs* 7(3): 589–602.

Keller, Evelyn Fox. 1985. *Reflections on gender and science*. New Haven: Yale University Press.

Klee, Robert. 2003. Feminist epistemology as folk psychology. In *Scrutinizing feminist epistemology: An examination of gender in science*, ed. Cassandra L. Pinnick et al., 31–44. Piscataway: Rutgers University Press.

Klein, Ellen R. 1996. *Feminism under fire*. Amherst: Prometheus Books.

Koertge, Noretta. 1996. Feminist epistemology: Stalking an un-dead horse. In *The flight from science and reason*, ed. Paul R. Gross, Norman Levitt, and Martin W. Lewis, 413–419. New York: The New York Academy of Sciences.

Le Doeuff, Michèle. 1989. *The philosophical imaginary* (trans: Colin Gordon). Stanford: Stanford University Press.

Leiter, Brian (ed.). 2004. *The future for philosophy*. Oxford: Oxford University Press.

Lloyd, Genevieve. 1993. *The man of reason: 'Male' and 'Female' in Western philosophy*. Minneapolis: University of Minnesota Press. 1984.

Lloyd, Elisabeth A. 1995. Objectivity and the double standard for feminist epistemologies. *Synthese* 104: 351–381.

Lloyd, Genevieve. 2002. Maleness, metaphor, and the "crisis" of reason. In *A mind of one's own: Feminist essays on reason and objectivity*, 2nd ed, ed. Louise M. Antony and Charlotte E. Witt, 73–89. Boulder: Westview Press.

Longino, Helen. 1981. Scientific objectivity and feminist theorizing. *Liberal Education* 67(3): 33–41.

Longino, Helen. 1994. In search of feminist epistemology. *The Monist* 77(4): 472–485.

MacKinnon, Catharine A. 1982. Feminism, marxism, method, and the state: An agenda for theory. *Signs* 7(3): 515–544.

Mills, Charles. 1997. *The racial contract*. Ithaca: Cornell University Press.

Nelson, Lynn Hankinson. 1995. The very idea of feminist epistemology. *Hypatia* 10(3): 31–49.

Nelson, Lynn Hankinson, and Jack Nelson (eds.). 2003. *Feminist interpretations of W. V. Quine*. University Park: Penn State Press.

Pinnick, Cassandra L., Noretta Koertge, and Robert F. Almeder (eds.). 2003. *Scrutinizing feminist epistemology: An examination of gender in science*. New Brunswick: Rutgers University Press.

Potter, Elizabeth. 2004. The science war front. *The Women's Review of Books* 22(1, October): 7–8.

Quine, W.V.O. 1969. Epistemology naturalized. In *Ontological relativity and other essays*, 69–90. New York: Columbia University Press.

Richards, Janet Radcliffe. 1996. Why feminist epistemology isn't. In *The flight from science and reason*, ed. Paul R. Gross, Norman Levitt, and Martin W. Lewis, 385–412. New York: The New York Academy of Sciences.

Richardson, Sarah S. 2010. Feminist philosophy of science: History, contributions, and challenges. *Synthese* 177:337–362.

Roberts, Robert C., and W. Jay Wood. 2007. *Intellectual virtues: An essay in regulative epistemology*. Oxford: Oxford University Press.

Rooney, Phyllis. 1991. Gendered reason: Sex metaphor and conceptions of reason. *Hypatia* 6(2): 77–103.

Rooney, Phyllis. 1993. Feminist-pragmatist revisionings of reason, knowledge, and philosophy. *Hypatia* 8(2): 15–37.

Rooney, Phyllis. 2002. Philosophy, language, and wizardry. In *Feminist interpretations of Ludwig Wittgenstein*, ed. Naomi Scheman and Peg O'Connor, 25–47. University Park: Penn State Press.

Rooney, Phyllis. 2003. Feminist epistemology and naturalized epistemology: An uneasy alliance. In *Feminist interpretations of W.V. Quine*, ed. Nelson Lynn Hankinson and Jack Nelson, 205–239. University Park: The Penn State University Press.

Rooney, Phyllis. Forthcoming. What is distinctive about feminist epistemology at 25? In *Out From the Shadows: Analytical Feminist Contributions to Traditional Philosophy*, ed. Sharon Crasnow and Anita Superson. Oxford: Oxford University Press.

Rorty, Richard. 1979. *Philosophy and the mirror of nature*. Princeton: Princeton University Press.

Ruetsche, Laura. 2004. Virtue and contingent history: Possibilities for feminist epistemology. *Hypatia* 19(1): 73–101.

Seigfried, Charlene Haddock. 1996. *Pragmatism and feminism: Reweaving the social fabric*. Chicago: University of Chicago Press.

Shelton, Jim D. 2006. The failure of feminist epistemology. *Academic Questions* 19(2): 82–92.

Tuana, Nancy and Shannon Sullivan (eds.). 2007. *Race and epistemologies of ignorance*. Albany: State University of New York Press.

Superson, Anita M., and Ann E. Cudd (eds.). 2002. *Theorizing backlash: Philosophical reflections on the resistance to feminism*. Lanham: Rowman & Littlefield.

Tuana, Nancy (ed.). 1989. *Feminism & science*. Bloomington: Indiana University Press.

Tuana, Nancy, and Sandra Morgen (eds.). 2001. *Engendering rationalities*. Albany: State University of New York Press.

Tuana, Nancy and Shannon Sullivan eds. 2006. *Feminist epistemologies of ignorance*. Special issue of *Hypatia* 21(3, Summer 2006).

Webb, Mark Owen. 2002. Feminist epistemology as whipping-girl. In *Theorizing backlash: Philosophical reflections on the resistance to feminism*, ed. Anita M. Superson and Ann E. Cudd, 49–65. Lanham: Rowman & Littlefield.

Williams, Michael. 2001. *Problems of knowledge: A critical introduction to epistemology*. Oxford: Oxford University Press.

Wylie, Alison. 2004. Why standpoint matters. In *The Feminist standpoint theory reader: Intellectual & political controversies*, ed. Sandra Harding, 339–351. New York: Routledge.

Chapter 2
Contextualism in Feminist Epistemology and Philosophy of Science

Kristina Rolin

Abstract I introduce a contextualist theory of epistemic justification in order to defend Helen Longino's contextual empiricism against three criticisms. The critics claim that contextual empiricism (1) implies dogmatism with respect to standards of argumentation, (2) lacks naturalistic justification, and (3) implies relativism with respect to moral and social values. I argue that the three criticisms fail. If we understand contextual empiricism as a contextualist theory of epistemic justification, standards of argumentation do not need to be adopted dogmatically, Longino's social account of objectivity is justified in virtue of advancing epistemic responsibility, and her account of objectivity does not imply that any moral and social values are acceptable in scientific debates.

Keywords Contextualism • Feminist epistemology • Helen Longino • Skepticism • Michael Williams

2.1 Introduction

During the last three decades feminist epistemology and philosophy of science have played a central role in the increasingly popular turn towards studying scientific knowledge in its social and cultural context. While feminist philosophers have joined the contextual turn in philosophy of science, they have also raised the concern that the critical edge in feminist work is lost by suggesting that the standards of scientific knowledge are relative to context. Among feminist philosophers there is an urgency to reflect on how the contextual turn is taken unless we want to admit that sexist or androcentric research is after all good research in *some* contexts. In this chapter my aim is to offer a critical overview of these debates as they concern Helen Longino's influential and controversial theory of scientific knowledge: contextual empiricism.

K. Rolin (✉)
Aalto University School of Economics, P.O. Box 21210, 00076 Aalto, Finland
e-mail: kristina.rolin@aalto.fi

H.E. Grasswick (ed.), *Feminist Epistemology and Philosophy of Science: Power in Knowledge*, DOI 10.1007/978-1-4020-6835-5_2,

Longino considers contextual empiricism to be a form of empiricism in that it treats experience as the basis of knowledge claims in the sciences (1990, 219). Whereas her focus on philosophy of science distinguishes her view from other 'contextualisms' in contemporary epistemology, the 'contextual' component of contextual empiricism distinguishes her view from other philosophies of science. In Longino's work, the term 'contextual' refers to three notions of context, the context of particular background assumptions, the context of scientific communities, and the social and cultural context of science. The first notion of context is employed in her well-known argument that epistemic justification is relative to background assumptions because such assumptions are needed to establish the relevance of empirical evidence to a hypothesis or a theory (1990, 43). The second notion of context is employed in her analysis of objectivity, in which she argues that the objectivity of scientific knowledge is a function of a community's practice rather than an individual scientist's observations and reasoning (1990, 74). The third notion of context is employed in her analysis of the role of values in science, in which she argues that values belonging to the social and cultural context of science can enter into epistemic justification via background assumptions (1990, 83). Longino's approach to philosophy of science is innovative in that she combines the three notions of context when she argues that we should adopt a social account of objectivity because values belonging to the social and cultural context of science can legitimately have an impact on the context of background assumptions. In Longino's contextual empiricism, the contextual nature of scientific knowledge is not merely a source of epistemic problems; it is also a solution to them.[1]

Longino's contextual empiricism is of interest to feminist epistemologists and philosophers of science because it invites us to rethink the role of moral and social values in epistemic justification. In the traditional view, moral and social values are not allowed to play a role in epistemic justification. Longino challenges this view by arguing that no method of scientific inquiry can guarantee that an accepted hypothesis or theory is fully value-free (1990, 12). Instead of embracing an unfeasible ideal, philosophers of science should acknowledge that the influence of moral and social values on scientists' choice of background assumptions is not necessarily a sign of 'bad science' (1990, 83). Longino argues that background assumptions are necessary in scientific reasoning because an observed state of affairs in itself does not tell for what hypothesis or theory it can be taken as evidence (1990, 40–43, see also 2002, 127). As she explains, 'a state of affairs will only be taken to be evidence that something else is the case in light of some background belief or assumption asserting a connection between the two' (1990, 44). While background assumptions do not always 'encode social values,' they sometimes do and they can do so in ways that elude an individual scientist (1990, 216).

[1] In *The Fate of Knowledge* (2002), Longino suggests that *critical* contextual empiricism is an even more appropriate term for her theory than mere contextual empiricism because it emphasizes the social aspect in contextual empiricism (208).

Given Longino's analysis of the role of background assumptions in scientific reasoning, feminist philosophers and scientists can criticize particular assumptions as sexist or androcentric instead of rejecting a whole research or research program as biased (see e.g., Fausto-Sterling 2000; Lloyd 2005). Such a partial and nevertheless piercing critique is more likely to make an impact on research than an attempt to rebuke a whole research program. Additionally, Longino's theory allows us to see how feminist values can play not only a critical but also a positive role in science insofar as they direct scientists to look for new evidence or to interpret evidence in new ways (see e.g., Anderson 2004; Wylie 2002, especially chapter 14). Moreover, Longino's contextual empiricism draws attention to the social structure and dynamics of scientific communities. It enables feminist philosophers to argue that women's underrepresentation in some areas of science is problematic not merely from the point of view of social justice; it is an epistemic problem insofar as it impedes criticism from diverse perspectives (Rolin 2004, see also Fehr and Wylie in this collection).

In spite of the positive uptake of contextual empiricism among feminist philosophers, it has also been criticized. In this chapter I defend Longino's contextual empiricism against three objections. One objection is that it collapses into dogmatism with respect to standards of argumentation. The critics suggest that dogmatism is implicit in the view that objectivity is a function of a community's practice. Another objection is that Longino's social account of objectivity lacks naturalistic justification. The critics claim that Longino should present case studies or other kind of empirical evidence in support of the norms she recommends for a knowledge-seeking community. Yet another objection is that Longino's social account of objectivity implies a kind of relativism with respect to moral and social values. The critics argue that Longino's social account of objectivity is not sufficiently normative because it does not explicitly set constraints on the kinds of moral and social values that are allowed to enter into epistemic justification.

While in Sect. 2.2 I explain Longino's contextual empiricism and in Sect. 2.3 the three criticisms, in Sect. 2.4 I take a detour to what many feminist philosophers perceive as mainstream epistemology. I introduce Michael Williams's contextualist theory of epistemic justification (henceforth contextualism) with two aims, the aim of preparing ground for my defense of Longino's contextual empiricism and the aim of showing that a juxtaposition of feminist epistemology to mainstream epistemology is misleading in some cases. Even though Longino and Williams approach epistemology with different interests, the feminist interest in developing a normative theory of values in science and the more traditional interest in solving the problem of skepticism, they end up sharing many views about epistemic justification. In Sect. 2.5 I explain how Williams's contextualism enables me to develop a response to the three criticisms.[2]

---

[2] Longino discusses David Annis's (1978) and Stewart Cohen's (1987) contextualist theories of epistemic justification but not Williams's (2001) contextualism (see Longino 2002, 104–106).

2.2 Helen Longino's Contextual Empiricism

The three criticisms I take up raise objections to Longino's view of objectivity, that is, her claim that scientific knowledge is objective to the degree that a relevant scientific community satisfies the four norms of public venues, uptake of criticism, public standards, and tempered equality of intellectual authority (Longino 2002, 129–131, see also 1990, 76–81). I call this claim the thesis of social objectivity. Longino argues that the four norms are aptly included in the thesis of social objectivity because they each facilitate 'transformative criticism' (1990, 76). The norm of public venues facilitates transformative criticism by requiring that criticism of scientific research be given the same or nearly the same weight as original research (2002, 129). The norm of uptake facilitates transformative criticism by requiring that each party to a critical exchange is willing to revise its views instead of merely 'tolerating dissent' (2002, 129–130). The norm of public standards facilitates transformative criticism by requiring that criticism appeals to at least some standards of argumentation publicly recognized in a relevant scientific community (2002, 130–131). The norm of tempered equality of intellectual authority facilitates transformative criticism in two ways, by disqualifying those communities where certain perspectives dominate because of the political, social, or economic power of their adherents (1990, 78), and by making room for a diversity of perspectives which is likely to generate critical perspectives in a scientific community (2002, 131).

Longino argues that we should accept the thesis of social objectivity because it provides a solution to the problem of epistemic relativism that is generated by her contextual understanding of epistemic justification. According to Longino, epistemic justification is relative to a context of background assumptions (henceforth the thesis of contextual evidence) (Longino 1990, 44). The thesis of contextual evidence implies that moral and social values can enter into epistemic justification indirectly, by influencing what background assumptions scientists rely on in their evidential reasoning. As Longino explains, 'contextual values, interests and value-laden assumptions *can* constrain scientific practice in such a way as to affect the results of inquiry and do so without violating constitutive rules of science' (1990, 83). Moreover, she does not believe that it is possible to eliminate all influence of moral and social values on epistemic justification 'without seriously truncating the explanatory ambitions of the sciences' (1990, 223). As she explains, 'the reliance on assumptions directly encoding contextual values is not by itself grounds for rejecting the work as science' (1990, 128). The thesis of contextual evidence generates the problem of epistemic relativism because in some cases scientists can appeal to different background assumptions that seem to be equally plausible (1990, 61, see also 2002, 127). As Longino explains, 'Without some absolute or nonarbitrary means of determining acceptable or correct background assumptions there seems no way to block the influence of subjective preferences' (1990, 61).[3]

[3] By *contextual* values Longino means value judgments concerning what is morally acceptable or praiseworthy, or what is a desirable social order (1990, 4). By *constitutive* values she means values

In response to this problem, Longino argues that a community practice constrained by the four norms advances objectivity because it forces scientists to examine critically the background assumptions that facilitate evidential reasoning as well as the moral and social values that may have motivated the choice of certain background assumptions (1990, 73). Without such a community practice, many ungrounded or even false assumptions may pass without criticism. As Longino explains, 'As long as background beliefs can be articulated and subjected to criticism from the scientific community, they can be defended, modified, or abandoned in response to such criticism' (1990, 73–74). And she adds that 'As long as this kind of response is possible, the incorporation of hypotheses into the canon of scientific knowledge can be independent of any individual's subjective preferences' (1990, 74).

Additionally, in *The Fate of Knowledge* (2002) Longino argues that we should accept the thesis of social objectivity not only because a well-designed community practice is likely to increase the objectivity of scientific knowledge but also because scientific knowledge is *necessarily* an outcome of a community practice (134). According to Longino, scientific observation is 'dialogical' in the sense that observation reports are ordered and organized in accordance with social conventions (100). Also, scientific reasoning is dialogical in the sense that it is part of 'a practice of challenge and response' (103). As both observation and reasoning are distributed in a community, there has to be a method of bringing individual contributions into interaction with each other in order for these contributions to count as scientific knowledge. No individual scientist can produce scientific knowledge independently of other scientists, no matter how rigorous and open-minded she is in her inquiry.

2.3 Three Objections to Longino's Contextual Empiricism

I argue that the three criticisms of Longino which I outline below deserve to be taken seriously because they satisfy the minimum standards that criticisms need to meet: accuracy and perspective (see also Anderson 2005). By the standard of accuracy I mean the principle that a critique must represent relevant works accurately and by the standard of perspective that a critique should make its presuppositions explicit and defend them if needed. All three of the criticisms I focus on satisfy these two criteria.[4]

that are generated from an understanding of the goals of science (1990, 4). So whereas such values as truth, consistency, and explanatory power are constitutive values in science, moral and social values such as equality and justice are contextual values in science. In her essay 'Gender, Politics, and the Theoretical Virtues' (1995), Longino suggests that the distinction between constitutive and contextual values is not as clear-cut as she had assumed in *Science as Social Knowledge* (1990).

[4] See Anderson (1995) and Rolin (2002) for a response to Haack's (2003) critique and Rolin (2006) for a response to Pinnick's (2003) critique of feminist epistemology in *Scrutinizing Feminist Epistemology*.

2.3.1 Dogmatism with Respect to Standards of Argumentation

Sharon Crasnow argues that we should not accept the thesis of social objectivity because it implies dogmatism with respect to a community's standards of argumentation (2003, 136, see also Clough 1998, 91; Haack 1996, 80; Smith 2004, 145; Pinnick 2005, 112). As she sees it, the problem of dogmatism is inherent in Longino's narrow understanding of relativism and objectivity. Longino equates relativism with subjectivism, that is, the view that epistemic justification is relative to an individual scientist's 'subjective preference' (1990, 61, see also 74). Crasnow acknowledges that Longino's social account of objectivity protects scientific knowledge from subjectivism because it requires that knowledge claims be justified in a community practice (2003, 138). As Longino herself explains, 'The role of background assumptions in evidential reasoning is grounds for unbridled relativism only in the context of an individualist conception of scientific method and scientific knowledge' (1990, 216). Yet, Crasnow argues that Longino fails to address relativism with respect to a community's practice because on Longino's account scientific knowledge can be objective *only* in relation to such a practice. In Crasnow's view, the social account of objectivity merely moves relativism from one level to another, from the level of individual preferences to the level of 'rules governing the practice of science' (2003, 138). As Crasnow explains, the social account of objectivity 'does not take us beyond intersubjectivity' (2003, 136).

Reflecting on the criteria for serious critique, there is no doubt that Crasnow's critique is based on a relatively accurate presentation of Longino's position. Given Longino's social account of objectivity, two scientific communities can have different and conflicting standards of argumentation even though they are required to have some overlapping standards of argumentation (such as empirical adequacy).[5] While Longino's social account of objectivity recommends that scientific communities live up to the norms of public venues, uptake of criticism, public standards, and tempered equality of intellectual authority, the third norm requires merely that standards of argumentation are publicly recognized in the community and participants in a dialogue share at least some terms, principles of inference, and aims (2002, 130). Thus, it is not surprising that Crasnow finds the thesis of social objectivity problematic on the basis that it does not offer any grounds to resolve disputes about standards of argumentation (2003, 136). In her view, this means that such standards would have to be adopted dogmatically in the respective communities. As Crasnow explains, 'we do not ultimately have a way of adjudicating between competing claims from competing cultures or even

[5] Like Longino, many philosophers argue that philosophy of science should account for the persistence of the diversity of standards throughout the history of science (see e.g., Kuhn 1970; Rolin 2010; Solomon 2001).

groups within cultures because standards themselves are socially constructed' (2003, 138).[6]

Likewise, Crasnow's critique makes its perspective explicit. Crasnow shares with Longino the view that a feminist philosophy of science should be concerned with developing a normative account of values in science as well as a conception of objectivity (see Crasnow 2003, 130–131). However, when Crasnow argues that Longino's social account of objectivity is not *objective enough*, she presupposes that we can have a more objective account of objectivity. I will return to this assumption in Sect. 2.5.

2.3.2 Lack of Naturalistic Justification

Miriam Solomon and Alan Richardson (2005) argue that we should not accept the thesis of social objectivity because Longino does not provide empirical evidence in its support. They suggest that in order for the thesis of social objectivity to be plausible, some case studies or some other kind of empirical evidence are needed to support the claim that the four norms *actually* promote the epistemic goals of science, either truth or empirical success (2005, 213). In their view, it is not sufficient to suggest, as Longino does, that the four norms would facilitate transformative criticism if they were realized. As Solomon and Richardson explain, 'Longino rests with presenting her standards as intuitively reasonable' (Solomon and Richardson 2005, 216, see also Solomon 2001, 143–145). In their view, this is unsatisfactory because '"intuition" has a poor track record in philosophy of science' (Solomon and Richardson 2005, 213).

Solomon's and Richardson's critique also presents Longino's position in a relatively accurate way. Their reading of Longino's position is given support by Longino's own account of her method: 'The argument I have offered for these conditions depends on an analysis of the relation between cognitive aspirations and cognitive resources and on an intuitive distinction between knowledge and opinion that I take to be shared. To the extent the intuition is shared and correctly articulated in the conditions, the analysis specifies in normative terms the meaning of a normative concept. Those who reject the conditions have a different concept of knowledge, or perhaps, a concept of something else' (2002, 174).

---

[6] Philip Kitcher (1994) argues that Longino's contextual empiricism collapses into relativism with respect to *truth* because Longino identifies truth with consensus belief in communities that follow certain types of procedures (1994, 132, note 26). As K. Brad Wray points out, Kitcher's criticism is based on the mistaken premise that Longino identifies truth with consensus belief in a community (Wray 1999, 545). This premise is false because Longino does not accept the view that knowledge entails truth (1990, 93). In Longino's view, acceptance of a theory or a hypothesis involves a belief in its empirical adequacy (1990, 94).

Also, Solomon and Richardson make their perspective explicit. As they explain, 'Most philosophers of science are, either obviously or ultimately, consequentialists, typically giving a justification of their account of good scientific methods in terms of conduciveness to scientific success and/or truth (however qualified)' (2005, 212–213). In their view, 'It is unusual for a philosopher of science to give an account of good scientific methods without some kind of consequentialist justification of the account' (2005, 213). Moreover, they themselves discuss historical studies of science (see also Solomon 2001), and argue that if philosophers insisted upon adherence to Longino's four norms during the Scientific Revolution, philosophers would have to discard nearly all early modern science as bad science (2005, 215). Thus, they suggest that Longino's contextual empiricism fails to live up to the standards of naturalized philosophy of science even though Longino herself claims to be a kind of naturalist (see Longino 2002, 10). However, Solomon and Richardson take it for granted that empirical success and truth are the epistemic goals of science that philosophers of science should appeal to when they assess normative views such as the thesis of social objectivity. In Sect. 2.5 I return to this assumption.[7]

2.3.3 Relativism with Respect to Moral and Social Values

Kristen Intemann (2008) and Janet Kourany (2005) argue that we should not accept the thesis of social objectivity as it is because it implies a kind of relativism with respect to moral and social values. As Intemann (2008) sees it, feminist values do not have a special status in Longino's contextual empiricism even though contextual empiricism is meant to address feminist concerns about values in science (see Longino 1990, 10). On Longino's social account of objectivity, feminist values are epistemically valuable only insofar as they play an instrumental role in science, that is, they help scientists identify problematic background assumptions. According to Intemann, the thesis of social objectivity is based on the assumption that inclusive

[7] In a more recent paper Solomon (2006) cites evidence in support of the claim that the practice of rational deliberation recommended by Longino does not always lead to epistemic success. Irving Janis's work suggests that a group that deliberates with the aim of reaching a consensus is vulnerable to the so called 'groupthink' phenomenon (Solomon 2006, 31). Such a phenomenon takes place when peer pressure and pressure from those in authority leads dissenting individuals to change their minds or not to share their knowledge of contrary evidence. As a result, the outcome of group deliberation is biased in the sense that it does not reflect all the information that individual group members have. I would argue that even if there is such a phenomenon as 'groupthink' it is less likely to take place in scientific communities than in research teams because scientific communities are socially more dispersed than research teams. Insofar as equality of intellectual authority is respected scientific communities, as Longino recommends it be, scientific communities are unlikely to be subject to 'groupthink.' Ideally, such communities are open to outside criticism. As Alison Wylie (2006) argues, evidence of the 'groupthink' phenomenon does not support the hypothesis that the aggregation of opinion is epistemically superior to rational deliberation. It shows merely that under certain circumstances rational deliberation may fail (2006, 44).

scientific communities, where some scientists have feminist values, are more likely to screen out sexist background assumptions than other communities. While Intemann does not reject the assumption, she argues that the thesis of social objectivity has counter-intuitive consequences. In her view, it implies that 'all contextual values are equally beneficial in contributing to a diverse research community' (2008, 1070). As she explains, 'feminist political commitments will be no more important in contributing to diversity than anti-feminist commitments' (2008, 1070).

In Kourany's (2005) view, Longino is right to suggest that objectivity should not be equated with the traditional ideal of value-free science, the view that epistemic justification should be fully free of moral and social values (see Longino 1990, 6). However, Kourany argues that Longino's ideal of 'social value management' (Longino 2002, 50) is not sufficiently normative to count as a *feminist* philosophy of science. As Kourany explains, 'According to this ideal, all social values should be welcomed into science and all social values, and the science they engender, should be subjected to criticism' (2005, 296). In Kourany's view, it is not sufficient to recommend that scientific communities be inclusive of feminist scientists or that all background assumptions be subjected to criticism. A feminist philosophy of science should recommend the ideal of 'socially responsible science' that directs all scientists to include only specific social values in science (2005, 297).

Thus, the kind of relativism that Intemann and Kourany object to is the view that all moral and social values are equally legitimate in scientific debates. As Intemann explains, on Longino's social account of objectivity it is the *diversity* of the moral and social values represented and not the content of any particular value judgment that increases objectivity (2008, 1070).

Intemann's and Kourany's critiques are accurate in that the thesis of social objectivity requires that knowledge claims be justified in a well-designed community practice but does not require that epistemic justification is fully free of moral and social values. Given the thesis of social objectivity, a background assumption can legitimately encode moral and social values insofar as no one has raised an appropriate objection to the assumption. The thesis of social objectivity does not *explicitly* set any constraints on the kinds of moral and social values that are allowed to enter into epistemic justification via background assumptions. Thus, it is not surprising that some feminist critics such as Intemann and Kourany find Longino's position too evenhanded vis-à-vis moral and social values.

Intemann's and Kourany's critiques satisfy also the standard of perspective. They are both interested in developing not only a normative theory of values in science but more specifically – a *feminist* one. While they do not recommend that we abandon Longino's social account of objectivity, they suggest that it be revised so that it privileges 'egalitarian social values' (Kourany 2005, 297; Intemann 2008, 1067). In Sect. 2.5 I argue that Longino's thesis of social objectivity is not as neutral with respect to feminist values as Intemann and Kourany take it to be.

To summarize, the three objections aim to challenge different aspects of the thesis of social objectivity. Crasnow objects to the view that the objectivity of scientific knowledge is a function of a community practice. She suggests that objectivity should be based on some other feature of scientific knowledge than its being an outcome of a community practice. This other feature of scientific knowledge should be something that takes us 'beyond intersubjectivity.' Solomon and Richardson question the view that the four norms of public venues, uptake of criticism, public standards, and tempered equality of intellectual authority rather than some other norms should be included in a normative account of a community practice. They suggest that some case studies or other kind of empirical evidence are needed to support the claim that these four norms actually advance empirical success or truth. Kourany and Intemann challenge the assumption that the four norms are sufficient for a feminist account of values in science. They suggest that the thesis of social objectivity be revised so that it privileges feminist values over anti-feminist ones. While I have argued that the three criticisms deserve attention, in the remainder of the paper I show that Longino's contextual empiricism will stand up to the challenge if we understand it as a version of contextualism.

2.4 A Contextualist Theory of Epistemic Justification

In order to prepare ground for my defense of Longino's contextual empiricism against the three above objections, I introduce a contextualist theory of epistemic justification as it has been developed by Williams (2001). As Williams's main argument in support of his version of contextualism is that it provides a solution to the problem of skepticism, I begin my discussion with skepticism and then move on to explain Williams's contextualism and its relation to Longino's contextual empiricism. While many feminist philosophers perceive the problem of skepticism as remote from feminist interests in epistemology and philosophy of science (see e.g., Clough 2003), I suggest that an engagement with the problem can bring about interesting lessons for feminists. Most importantly, it can help feminist philosophers see that some attacks against feminist epistemology are launched from epistemic positions that are untenable. Additionally, William's contextualism is of interest to feminist epistemology because unlike some other contextualists, he does not grant that skeptical arguments are sound in some 'high standard' contexts while denying that they are sound in 'ordinary' contexts (see e.g., DeRose 2009, 42–43). Williams's contextualism is meant to disarm the skeptic in all contexts.

Skepticism is the view that knowledge is impossible because no one can have justified beliefs (Williams 2001, 59). The problem of skepticism is best understood by introducing two famous (or notorious, if you like) arguments for skepticism, a classical and a modern one. The classical argument appeals to the premise that any attempt to provide an epistemic justification for a claim leads one to face three discouraging prospects (Williams 2001, 62). One prospect is that one will embark on an infinite regress by being forced to provide reasons for reasons *ad infinitum*. An infinite regress undermines the very possibility of epistemic justification. Another prospect is that one will bring the regress to an end at some point by claiming to know some beliefs without justification. However, this move makes one vulnerable to the charge of dogmatism. Yet another prospect is that one repeats something one has already justified before. But this move makes one guilty of circular reasoning. On the basis of these three prospects, the skeptic concludes that no claim is ever justified to the slightest degree (Williams 2001, 63).

The modern argument aims to compel me to accept the conclusion that I do not know a claim such as 'I have two hands' even though this claim seems to me to be a simple common sense truth. The argument includes the following two premises: (1) I do not know that I am not a brain in a vat, and (2) if I do not know that I am not a brain in a vat, then I do not know that I have two hands. The conclusion is that (3) I do not know that I have two hands. Thus, the modern argument for skepticism appeals to an imagined brain-in-a-vat scenario which is designed so that there is nothing in my inner experience that enables me to tell whether I am part of a brain-in-a-vat scenario or a common sense scenario where I have two hands (Williams 2001, 187–188).

Williams argues that contextualism provides a solution to the problem of skepticism because it enables one to make a diagnosis of what is wrong in the two arguments and it explains how one can have justified beliefs (2001, 146). According to Williams, the classical argument is based on the *false* assumption that the claimant has merely three alternatives: infinite regress, dogmatism, or reasoning in a circle (2001, 151). Contextualism provides a fourth alternative: the claimant can be justified in believing that p even though she does not provide reasons in support of p. Her believing that p can be justified in virtue of her belief having the status of a default entitlement. As Williams explains, in contextualism epistemic justification is thought to have a default and challenge structure (2001, 25). This means that an entitlement to one's belief is the default position but one has a duty to defend or revise one's belief as soon as it is challenged with an appropriate argument. An appropriate argument can cite reasonable and relevant error-possibilities (2001, 149). For example, the challenger can provide evidence in support of the claim that the claimant's belief is false or that the claimant's belief has been acquired in an unreliable way. Williams argues that contextualism enables the claimant to avoid both infinite regress and circular reasoning because the chain of reasons is brought to an end by beliefs that function as default entitlements (2001, 151). Moreover, contextualism enables the claimant to avoid dogmatism because beliefs that function as default entitlements are justified beliefs; they are not adopted dogmatically (2001, 151).

Yet another way to make a diagnosis of the classical argument is to say that the skeptic is mistaken in the assumption that she always has a right to demand that others present evidence in support of their beliefs (Williams 2001, 150). Williams argues that we have to abandon this assumption if we want to avoid the classical argument's conclusion (2001, 147–148). As soon as we abandon the assumption, the burden of proof is shifted to the skeptic. The skeptic has a duty to explain why others should doubt their beliefs in the first place (2001, 151). However, the modern argument for skepticism is more challenging than the classical one because the modern skeptic does not simply assume that she always has a right to demand that the claimant presents reasons for her beliefs. Instead, the skeptic provides an argument that gives the claimant a reason to doubt her common sense beliefs. Thus, the skeptic seems to be successful in her attempt to undermine the default status of my common sense beliefs such as 'I have two hands' (2001, 186).

With respect to the modern argument for skepticism, Williams adopts a strategy similar to the one above in that he aims to analyze the argument's underlying assumptions and shift the burden of proof to the skeptic by questioning the seemingly self-evident nature of the skeptic's assumptions. According to Williams, the modern argument for skepticism is based on two problematic assumptions (2001, 189–190). One assumption is that we can distinguish between the so called 'experiential knowledge' which is based on our inner experience and the so called 'knowledge of the external world.' Another assumption is that 'experiential knowledge' is epistemically privileged in relation to 'knowledge of the external world.' Both of these assumptions are implicit in the skeptic's argument that I do not know that I am not a brain in a vat because there is nothing in my inner experience that enables me to tell whether I am part of the brain-in-a-vat scenario or a common sense scenario. According to Williams, the skeptic should explain why I should accept the distinction between 'experiential knowledge' and 'knowledge of the external world' in the first place and why I should privilege the former to the latter. As long as the skeptic does not provide such an explanation, I am justified in believing that I am not a brain in a vat (2001, 188).

In summary, Williams argues that we should accept contextualism because it provides a solution to – or more aptly, a *diagnosis* of – the problem of skepticism both in its classical and modern form. Let me make explicit the central ideas in Williams's contextualism.

Williams's notion of default entitlement will turn out to be significant when I address the charge of dogmatism raised against Longino's contextual empiricism. Williams's contextualism includes the view that epistemic justification takes place in a context of assumptions, some of which are justified in virtue of functioning as default entitlements and others in other ways (2001, 226–227). Default entitlements are not plain assumptions since they are adopted with a defense commitment. A defense commitment means that one accepts a duty to defend or revise one's belief when it is challenged with an appropriate argument. As Williams explains, default entitlements can be articulated and challenged but only by a recontextualization of inquiry which involves assumptions of its own (2001, 227). So, even if one can challenge an assumption that functions as a default entitlement, it does not follow that one can transcend the contextual nature of epistemic justification. When one accepts such a challenge, one's inquiry is moved into another context of assumptions; it does not become less contextual. According to Williams, recontextualization of inquiry can go on indefinitely (2001, 227).

Williams's contextualism includes a notion of epistemic responsibility that will turn out to be significant when I address the charge that Longino's thesis of social objectivity lacks naturalistic justification. In Williams's view, epistemic justification is a matter of epistemic responsibility and not merely a matter of having adequate grounds for a claim. In order to understand why epistemic responsibility is not the same thing as having adequate grounds, it is necessary to make a distinction between two different conceptions of epistemic justification. When we ask for epistemic justification, we can ask under what conditions a person is justified in believing a particular proposition, or we can ask under what conditions the proposition she believes is justified (Williams 2001, 21). Whereas the first question inquires whether a belief has been responsibly formed or is responsibly held, the second question inquires whether a belief has adequate grounds (2001, 22). Traditionally, it has been assumed that one is epistemically responsible in believing that p *only if* one's belief that p is based on adequate grounds (2001, 24). Williams argues that we should give up this assumption because it enables the skeptic to pull us into a dialectical situation where she has an upper hand. In Williams's contextualism, a person can be epistemically responsible in believing that p *even if* she does *not* have adequate grounds for p (2001, 25). A person believing that p is epistemically responsible if she adopts a defense commitment with respect to her belief that p (2001, 25).[8]

Williams's contextualism can be used to defend Longino's contextual empiricism against her critics because Williams and Longino share the view that epistemic justification is relative to a context of background assumptions (the thesis of contextual evidence). While Williams arrives at this view on the basis of his diagnosis of skepticism, Longino endorses it on the basis of her reflections on scientific reasoning. Moreover, like Longino Williams grants a special status to empirical evidence. In Williams's view, observation reports are paradigmatic cases of beliefs that function as default entitlements in many contexts of inquiry. Scientists are justified in believing in them as long as no one has provided reasons to suspect that they are false or that they have been produced in an unreliable way (Williams 2001, 175).

Despite these similarities there is an obvious difference between the two positions. While Williams's contextualism focuses on the epistemic responsibility of *individuals*, Longino's contextual empiricism focuses on the features of epistemically ideal *communities*. The explanation for the difference in their focus may be that Williams' research interests are in general epistemology, not in philosophy of science. Scientific knowledge is usually identified as knowledge that satisfies standards of argumentation set by relevant scientific communities. Thus, the epistemic justification of *scientific* knowledge is not dependent on just *any* context. As Longino explains, contextual empiricism is 'contextual' in the sense that it highlights the context of scientific communities (1990, 219). Keeping the limitations of Williams's contextualism in mind, I argue that it nevertheless offers conceptual tools that enable me to defend Longino's contextual empiricism against its criticisms.

2.5 Defending Contextual Empiricism

2.5.1 Dogmatism with Respect to Standards of Argumentation

Contrary to Crasnow (2003), I argue that Longino's contextual empiricism does not imply dogmatism with respect to a community's standards of argumentation because it involves a contextualist understanding of such standards. In Williams's

[8] Williams stresses that both conceptions of epistemic justification, justification as epistemic responsibility and as having adequate grounds, are necessary elements in an adequate theory of knowledge (2001, 23).

contextualism, standards of argumentation are understood to be justified in virtue of functioning as default entitlements, or in some other way.[9] To treat standards as default entitlements means that they are adopted with a defense commitment. A defense commitment implies a duty to defend, revise or abandon a standard when it is challenged with an appropriate argument. Insofar as standards are adopted with a defense commitment, they are not adopted dogmatically, that is, without justification. Indeed, Williams-style contextualist understanding of standards is implicit in *The Fate of Knowledge* where Longino writes that 'standards are not a static set but may themselves be criticized and transformed, in reference to other standards, goals, or values held temporarily constant' (2002, 131).

Moreover, it is false to claim, as Crasnow does, that Longino's contextual empiricism does not provide grounds for resolving disputes about standards of argumentation (2003, 136). Certainly, a standard can be articulated and challenged. Doing so does not require that the challenger appeals to a set of meta-standards. Indeed, the question of whether there are meta-standards does not even rise in contextualism because an appropriate challenge requires that the challenger appeal to another context of inquiry where some assumptions function as default entitlements. It is also important to notice that leaving behind one context of inquiry and moving into another one does not require that one leaves behind all the assumptions that constitute the former context of inquiry. In order for a challenge to count as appropriate it has to appeal to some cross-contextual assumptions. Cross-contextual assumptions may include empirical evidence or standards other than the contested one. Thus, Longino's contextual empiricism does provide grounds for resolving a dispute about a standard of argumentation.

Also, an analysis of contextualism reveals that Crasnow's criticism of contextual empiricism is based on a fundamental misunderstanding of what epistemic justification is about. In Williams's contextualism, epistemic justification is a matter of epistemic responsibility and not merely of having adequate grounds for a claim (2001, 23). A person is epistemically responsible in believing that p if she provides sufficient evidence in support of p (or some other kind of argument) or if she adopts a defense commitment with respect to p. In either case her being epistemically responsible is dependent on the intersubjective features of her situation. What counts as sufficient evidence depends on what kind of challenges she has to respond to. Similarly, whether she is justified in believing that p as a default entitlement depends on whether potential challengers are able to raise appropriate objections to p. An appropriate challenge to an assumption that functions as a default entitlement is expected to provide reasons to suspect that the assumption is false or that it has been acquired in an unreliable way. Thus, epistemic values such as empirical adequacy and truth are built into the default and challenge structure of epistemic justification. Crasnow is right to point out that Longino's social account of objectivity 'does not take us beyond intersubjectivity' (2003, 138). But this is not an argument against Longino's contextual empiricism because the point in epistemic justification

[9] For example, the standard of consistency can be given a justification on the grounds that it is derived from the epistemic value of truth.

is not to 'take us beyond intersubjectivity.' The point is to bring epistemic values such as empirical adequacy and truth into 'intersubjectivity' that is understood to have a default and challenge structure.

To summarize, the key difference between Williams-style contextualism and the kind of relativism that Crasnow objects to is in the epistemic status we assign to standards of argumentation. Whereas relativism is the view that epistemic justification is relative to standards that function as *plain* assumptions, contextualism is the view that epistemic justification is relative to standards that function as default entitlements. The difference between default entitlements and plain assumptions is that the former are adopted with a defense commitment whereas the latter are adopted dogmatically. As Williams explains, both a relativist and a contextualist can hold the view that epistemic justification is relative to some context, but only a relativist would hold the view that contexts are 'frameworks of ultimate commitments' (2001, 224–225). In contextualism, no context includes ultimate commitments which are beyond criticism (2001, 226–227).

2.5.2 *Lack of Naturalistic Justification*

While I am not convinced of the objection that Crasnow advances against Longino's social account of objectivity, I admit that philosophers can ask questions about the justification of Longino's account. This is what Solomon and Richardson (2005) do. Contrary to Solomon and Richardson, I argue that we do not need to take their demand for a naturalistic justification at face value if we can provide another kind of epistemic justification for the four norms in Longino's thesis of social objectivity.

I argue that Williams's notion of epistemic responsibility provides such a justification. In order to understand how the notion of epistemic responsibility can provide a justification for the thesis of social objectivity, it is necessary to notice that Solomon and Richardson have a narrow view of the epistemic goals of science. In their view, the epistemic goals of science are empirical success and in some cases also truth. While I do not deny that the epistemic goals of science include such goals as empirical success and truth – or to be more precise, significant truth (Anderson 1995, 37), I argue that yet another epistemic goal of science is *epistemic justification*.

The goal of epistemic justification helps us understand why we can observe what Jutta Schickore (2008) calls 'the mismatch between what scientists do and what they state they did when they communicate their findings in their publications' (323). Scientists are engaged in a number of material activities whereby they 'represent and intervene in' natural processes (Hacking 1983). Scientists design and perform experiments, develop instruments to make observations and measurements, interpret data, and construct models and simulations. Scientists publish their findings in a venue that is designed for the purpose of communicating research results to a relevant scientific community. Yet, conference presentations and journal papers do not give an exhaustive, detailed account of how scientists have obtained their results. The goal of epistemic justification helps us understand why there is a mismatch between 'doing science' and 'writing science.' While scientists undertake many activities with the purpose of producing evidence, 'writing science' constitutes a break from these other activities since it is done with an eye to the epistemic justification of knowledge claims. As Schickore explains, scientists 'present their findings embedded in a web of arguments and reasons, thereby changing the order and justification of their research steps' (2008, 323).

As soon as we recognize that epistemic justification is one epistemic goal of science along with empirical success and truth, we can re-evaluate the thesis of social objectivity. While it may be difficult to provide empirical evidence in support of the hypothesis that Longino's four norms promote empirical success or truth (see also Rolin 2009, 74), it is not difficult to argue that they serve epistemic justification, and that provides merit in itself.

The four norms are required by epistemic responsibility as it is understood in Williams's contextualism. Public venues are required because epistemic responsibility demands that those presenting a challenge to a view have a hearing in a relevant community. Uptake of criticism is required because epistemic responsibility demands that an appropriate challenge receives a response. Indeed, this second norm is parallel to Williams's notion of a defense commitment because uptake of criticism means that at least some scientists in a relevant community have a duty to defend a community's view (or revise it) when it is challenged in an appropriate way. Publicly recognized standards are necessary because epistemic responsibility requires that both the challenger and the claimant are aware of the standards that determine what counts as an appropriate challenge. Tempered equality of intellectual authority is required because epistemic responsibility demands that an appropriate challenge be taken seriously independently of who presents it.

2.5.3 Relativism with Respect to Moral and Social Values

So far I have argued that Longino's thesis of social objectivity can be given an epistemic justification that is an alternative to the naturalistic justification demanded by Solomon and Richardson. However, one can still ask whether the thesis of social objectivity meets the *feminist* expectation that it screens out anti-feminist values in science. Against Intemann and Kourany I argue that the thesis of social objectivity does not imply the view that all moral and social values are equally legitimate in scientific debates.

While the thesis of social objectivity does not prevent moral and social values from *entering* into scientific debates, not just any moral and social values are likely to *survive* scientific debates. In Williams's contextualism, value judgments are subject to the default and challenge structure of justification in the same way as scientific theories, hypotheses, pieces of empirical evidence, and standards of argumentation. This means that some value judgments are likely to lose their justification because they will be met with an appropriate challenge. Scientific communities are not isolated islands of culture; they are embedded in liberal democratic societies where debates about moral and social values are common. Moreover, feminist moral philosophy provides plenty of resources for a scientist who finds it necessary to argue for feminist values or against anti-feminist values. If moral and social values are allowed to enter into scientific debates, then certainly feminist moral theories and other argumentative resources are allowed to enter into scientific debates. Therefore, it is unlikely that just anything will pass with respect to moral and social values in a scientific community that fulfills to a high degree the four norms in Longino's thesis of social objectivity.

Longino's (2002) discussion of creationism serves as an illustration of how the thesis of social objectivity can function to discredit certain value judgments, such as the view that scientific theories should be consistent with a religion. Longino argues that creationists do not qualify as responsible challengers to scientific communities because they fail to meet all the norms included in the thesis of social objectivity except the first one (2002, 158). They fail to meet the second norm of uptake because they do not respond to the criticisms of the theory of intelligent design. They fail to meet the third norm of public standards because they appeal to a standard not publicly recognized in scientific communities, the view that scientific theories should be consistent with a religion. Also, they do not respond to the criticism of this standard. Moreover, they fail to respect the fourth norm of tempered equality because they do not treat non-believers as fully equals to their community of believers. Thus, the value judgment requiring consistency between science and religion is discredited as a piece of dogmatism (see also Longino 2002, 159).

Longino's thesis of social objectivity can ground the criticism of creationism because the four norms are meant to be mutually binding. Not only scientific communities but also other communities are expected to respect them insofar as they make claims to scientific knowledge.

2.6 Conclusion

Finally, I am in a position to explain why the contextualist diagnosis of skepticism is of interest to feminist epistemology and philosophy of science. One answer is that the diagnosis points towards a conception of epistemic responsibility that is relevant to feminist epistemology and philosophy of science. The diagnosis reveals that the skeptic's arguments are based on a problematic conception of epistemic responsibility, the view that a person is epistemically responsible in believing that p only if she has adequate grounds for p. An alternative conception of epistemic responsibility is the view that a person is epistemically responsible in believing that p if she presents sufficient evidence in support of p – or at least adopts a defense commitment with respect to p. Given the alternative conception, a person can be epistemically responsible in believing that p even if she does not have adequate grounds for p.

This alternative conception of epistemic responsibility plays an important role in my defense of Longino's contextual empiricism. I have argued that scientists can be epistemically responsible even when epistemic justification is understood to be relative to background assumptions and standards of argumentation. Scientists are epistemically responsible insofar as they adopt a defense commitment in relation to relevant background assumptions and standards of argumentation. They do not have to adopt a dogmatic attitude towards such assumptions and standards. Moreover, I have argued that the four norms of public venues, uptake of criticism, public standards, and tempered equality of intellectual authority are themselves justified because they are required by this alternative sense of epistemic responsibility. Also, I have argued that scientists can be epistemically responsible even when their background assumptions encode moral and social values. They are epistemically responsible insofar as they either provide arguments in support of their moral and social values or adopt a defense commitment in relation to them.

The contextualist diagnosis of skepticism is of interest to feminist epistemology and philosophy of science also because it offers a clear picture of the options we have. One option is that we continue to develop feminist epistemology and philosophy of science along the lines indicated by contextualism. This involves such things as developing a more refined account of epistemic responsibility than what I have presented here, for example, by analyzing what counts as an appropriate challenge in actual scientific debates, how the burden of proof shifts in these debates, and how relations of power influence these practices. Another option is to reject the contextualist conception of epistemic responsibility, and consequently, the arguments I have presented in defense of contextual empiricism. The latter option may seem to be attractive for those philosophers who think that contextualism comes too close to relativism in approving of the view that epistemic justification is relative to a context of assumptions. However, before one makes up one's mind about the latter option, it is necessary to understand what it involves. In the pursuit of more fundamental epistemic principles than the ones provided by contextualism, one may fall prey to a view which is hardly more attractive than contextualism: skepticism.

Acknowledgments I am grateful to Heidi Grasswick for detailed comments on an earlier version of the manuscript. A version of the manuscript was presented in the Second Conference of the Society for Analytical Feminism at the University of Kentucky in 2008. I wish to thank Sharon Crasnow, Nancy Daukas, Phyllis Rooney, and Miriam Solomon for their critical feedback.

References

Anderson, Elizabeth. 1995. Knowledge, human interests, and objectivity. *Philosophical Topics* 23(2): 27–58.

Anderson, Elizabeth. 2004. Uses of value judgments in science: A general argument, with lessons from a case study of feminist research on divorce. *Hypatia* 19(1): 1–24.

Anderson, Elizabeth. 2005. How not to criticize feminist epistemology: Review of Pinnick, Koertge, and Almeder. http://www-personal.umich.edu/~eandersn/hownotreview.html. Accessed 5 May 2005.

Annis, David. 1978. A contextualist theory of epistemic justification. *American Philosophical Quarterly* 15(3): 213–219.

Clough, Sharyn. 1998. A hasty retreat from evidence: The recalcitrance of relativism in feminist epistemology. *Hypatia* 13(4): 88–111.

Clough, Sharyn. 2003. *Beyond epistemology: A pragmatist approach to feminist science studies.* Lanham/Oxford: Rowman & Littlefield Publishers.

Cohen, Stewart. 1987. Knowledge, context, and social standards. *Synthese* 73(1): 3–26.

Crasnow, Sharon. 2003. Can science be objective? Feminism, relativism, and objectivity. In *Scrutinizing feminist epistemology: An examination of gender in science*, ed. Cassandra L. Pinnick, Noretta Koertge, and Robert F. Almeder, 130–141. New Brunswick/London: Rutgers University Press.

DeRose, Keith. 2009. *The case for contextualism.* Oxford/New York: Oxford University Press.

Fausto-Sterling, Anne. 2000. *Sexing the body: Gender politics and the construction of sexuality.* New York: Basic Books.

Haack, Susan. 1996. Science as social? – Yes and No. In *Feminism, science, and the philosophy of science*, ed. Lynn H. Nelson and Jack Nelson, 79–93. Dordrecht: Kluwer.

Haack, Susan. 2003. Knowledge and propaganda: Reflections of an old feminist. In *Scrutinizing feminist epistemology: An examination of gender in science*, ed. Cassandra L. Pinnick, Noretta Koertge, and Robert F. Almeder, 7–19. New Brunswick/London: Rutgers University Press.

Hacking, Ian. 1983. *Representing and intervening: Introductory topics in the philosophy of natural science.* Cambridge: Cambridge University Press.

Intemann, Kristen. 2008. Increasing the number of feminist scientists: Why feminist aims are not served by the underdetermination thesis. *Science & Education* 17(10): 1065–1079.

Kitcher, Philip. 1994. Contrasting conceptions of social epistemology. In *Socializing epistemology*, ed. Frederick F. Schmitt, 111–134. Lanham/London: Rowman & Littlefield Publishers.

Kourany, Janet. 2005. A feminist primer for philosophers of science. In *Philosophy – Science – Scientific philosophy*, ed. Christian Nimtz and Ansgar Beckermann, 287–305. Paderborn: Mentis.

Kuhn, Thomas. 1970. *The structure of scientific revolutions*, 2nd ed. Chicago: The University of Chicago Press.

Lloyd, Elizabeth. 2005. *The case of the female orgasm: Bias in the science of evolution.* Cambridge: Harvard University Press.

Longino, Helen. 1990. *Science as social knowledge.* Princeton: Princeton University Press.

Longino, Helen. 1995. Gender, politics, and the theoretical virtues. *Synthese* 104(3): 383–397.

Longino, Helen. 2002. *The fate of knowledge.* Princeton: Princeton University Press.

Pinnick, Cassandra. 2003. Feminist epistemology: Implications for the philosophy of science. In *Scrutinizing feminist epistemology: An examination of gender in science*, ed. Cassandra L. Pinnick, Noretta Koertge, and Robert F. Almeder, 20–30. New Brunswick/London: Rutgers University Press.

Pinnick, Cassandra. 2005. The failed feminist challenge to "fundamental epistemology". *Science & Education* 14: 103–116.

Rolin, Kristina. 2002. Is "science as social" a feminist insight? *Social Epistemology* 16(3): 233–249.

Rolin, Kristina. 2004. Why gender is a relevant factor in the social epistemology of scientific inquiry. *Philosophy of Science* 71(5): 880–891.

Rolin, Kristina. 2006. The bias paradox in feminist standpoint epistemology. *Episteme* 3(1–2): 125–136.

Rolin, Kristina. 2009. Scientific knowledge: A stakeholder theory. In *The social sciences and democracy*, ed. J. van Bouwel, 62–80. Hampshire: Palgrave Macmillan.

Rolin, Kristina. 2010. Diversity and dissent in the social sciences: The case of organization studies. *Philosophy of the Social Sciences*, published online september 8.

Schickore, Jutta. 2008. Doing science, writing science. *Philosophy of Science* 75(3): 323–343.

Smith, Tara. 2004. "Social" objectivity and the objectivity of values. In *Science, values, and objectivity*, ed. Peter Machamer and Gereon Wolters, 143–171. Pittsburgh: University of Pittsburgh Press.

Solomon, Miriam. 2001. *Social empiricism*. Cambridge: The MIT Press.
Solomon, Miriam. 2006. Groupthink versus the wisdom of crowds: The social epistemology of deliberation and dissent. *The Southern Journal of Philosophy* 44: 28–42.
Solomon, Miriam, and Alan Richardson. 2005. A critical context for Longino's critical contextual empiricism. *Studies in the History and Philosophy of Science* 36: 211–222.
Williams, Michael. 2001. *Problems of knowledge: A critical introduction to epistemology*. Oxford/New York: Oxford University Press.
Wray, K. Brad. 1999. A defense of Longino's social epistemology. *Philosophy of Science* 66(3): 538–552.
Wylie, Alison. 2002. *Thinking from things: Essays in the philosophy of archaeology*. Berkeley/Los Angeles/London: University of California Press.
Wylie, Alison. 2006. Socially naturalized norms of epistemic rationality: Aggregation and deliberation. *The Southern Journal of Philosophy* 44: 43–48.

Chapter 3
Altogether Now: A Virtue-Theoretic Approach to Pluralism in Feminist Epistemology

Nancy Daukas

Abstract In this paper I develop and support a feminist virtue epistemology and bring it into conversation with feminist contextual empiricism and feminist standpoint theory. The virtue theory I develop is centered on the virtue of epistemic trustworthiness, which foregrounds the social/political character of knowledge practices and products, and the differences between epistemic agencies that perpetuate, on the one hand, and displace, on the other hand, normative patterns of unjust epistemic discrimination. I argue that my view answers important questions regarding epistemic agency which both contextual empiricism and standpoint theory leave open, but need to have answered. Feminist virtue epistemology thus emerges as providing an integrative framework for pluralism in feminist epistemology that illuminates connections among theories through engagement with the lived experiences, aspirations, and epistemic work of feminist epistemic agents.

Keywords Contextual empiricism • Feminist epistemology • Standpoint theory • Testimonial justice • Trustworthiness • Virtue epistemology

I've argued elsewhere that a feminist virtue epistemology provides a useful analysis of the politics of testimony – that is, of the normatively entrenched patterns in practices of testimonial exchange whereby epistemic authority is granted, or withheld, along lines of gender, race, class, and so on.[1] In this paper I further develop a

[1] Daukas (2006), in which I discuss 'epistemic exclusion on the basis of social location', which should be compared to Miranda Fricker's 'testimonial injustice', in (2003) and (2007). I sketched my virtue-theoretic approach to 'epistemic exclusion' before encountering Fricker's work, but am pleased to find congruities in our approaches, and I've benefited greatly from reflection on her work as I further develop my own. See also Karen Jones' discussion of testimonial injustice in (2002), which is firmly and effectively grounded in a finely detailed case study. Finally, compare 'epistemic exclusion' and 'testimonial injustice' to 'epistemic discrimination' in Code (1991, 57). I am indebted to Code's work in much of my thinking on the issues discussed in this paper.

N. Daukas (✉)
Department of Philosophy, Guilford College, 5800 West Friendly Avenue, Greensboro, NC 27410, USA
e-mail: ndaukas@guilford.edu

H.E. Grasswick (ed.), *Feminist Epistemology and Philosophy of Science: Power in Knowledge*, DOI 10.1007/978-1-4020-6835-5_3,

feminist virtue epistemology centered on the virtue of epistemic trustworthiness, and argue that it supplements and strengthens, both descriptively and prescriptively, two prominent approaches in feminist epistemology which are typically seen, if not as rivals, at least as divergent – feminist standpoint epistemology and feminist contextual empiricism – thereby providing them common ground. The three theories – feminist virtue epistemology, feminist standpoint epistemology and feminist contextual empiricism – thus fall into place as differently focused but mutually supporting theoretical approaches within a pluralistic feminist epistemology.

3.1 Virtue Epistemology: A Brief Introduction

With roots in Aristotle, virtue epistemology received little explicit attention or development in mainstream Anglo-American epistemology until late in the twentieth century, when it emerged as an alternative to the seemingly intractable internalism/externalism, and foundationalism/coherentism disputes, and interminable, increasingly arcane, disembodied analyses of '*S* knows that *p*.[2] It is often understood as the epistemic analogue of virtue ethics, as its approach is agent-centered instead of principles-centered. Where a principles-centered epistemology asks questions such as, 'what criteria must a belief satisfy in order to constitute knowledge?' and 'what are the necessary and sufficient conditions for a belief to be justified?', an agent-centered epistemology asks questions such as, 'what does it mean to be a knower?', or, 'what is involved in being an excellent epistemic agent?'. It responds in terms of epistemic virtues,[3] understood as states that dispose an agent to fulfill (or skillfully strive to fulfill) epistemic goals such as maximizing true beliefs and minimizing false beliefs, acquiring significant knowledge, or developing significant understanding. The appeal to virtue may replace, or perhaps flesh out, the justification condition of the standard analysis of knowledge as justified-true-belief; or a virtue epistemology may instead be understood as a 'successor epistemology'.

Some virtue epistemologies (such as reliabilism) are 'externalist': they define epistemic virtues as 'faculties' or 'mechanisms' (such as perception, memory, and inferential abilities) that reliably produce true beliefs, with no awareness or

[2] Here I'm thinking particularly of Code's *Epistemic Responsibility* (1987) and Zagzebski's *Virtues of the Mind* (1996). In (1994), Code characterizes her (1987) as an alternative to what she calls '"S knows that p" epistemology', with its implicit conception of a generic, disembodied, disinterested (masculine) subject of knowledge, and limitation to artificially simplistic propositional knowledge. Zagzebski (1996, Part I) argues that virtue epistemology promises to cut through dead-end disputes (e.g., internalism vs externalism) and (other) confusions and controversies regarding the concept of justification which have bogged down traditional epistemology. But no doubt the recent interest in virtue epistemology has various sources. Greco and Turri (2010) see it as resisting the Quinean turn from normativity to descriptive naturalism in epistemology.

[3] For present purposes we can think of 'epistemic virtues' as a broad category that includes what might be called 'intellectual virtues' and 'doxastic virtues'.

epistemic motivation on the part of the agent assumed or required.[4] Since I am interested in exploring epistemic agency, I will set aside any purely externalist approach. Other, 'responsibilist', virtue epistemologies interpret virtues as enduring dispositional character states or 'habits of mind' required for responsible epistemic agency, which self-reflective agents deliberately develop over time, motivated by an explicit desire for acquiring significant knowledge and understanding.[5] Responsibilist accounts are thoroughly internalist if they do not require epistemic dispositions to reliably tend toward success in order to qualify as virtues.[6] Others, including my own, do require (a tendency toward) success, and thus incorporate an externalist component.[7] A 'pure'(or 'strong') virtue epistemology derives the concepts 'knowledge' and 'justification' from virtue-based concepts, so that, roughly, a justified belief is defined as what a virtuous epistemic agent would believe under the circumstances.[8] A 'weak' virtue epistemology takes 'what a virtuous agent would believe under the circumstances' (which, on my view, may be different for differently 'situated' virtuous agents) as the criterion for *identifying* justified beliefs, without defining justification in terms of virtue. Mine is a weak responsibilist/reliabilist hybrid, which takes epistemic trustworthiness to be the primary epistemic virtue around which others are organized. Its analysis and development of epistemic trustworthiness distinguishes it from other virtue epistemologies, and grounds its feminism.

3.2 Feminist Virtue Epistemology

Virtue epistemology need not, but may, be feminist[9]; it all depends on how one conceives of epistemic agents and their virtues. Although I don't intend to take on the question what makes an epistemology feminist, and although there are many and diverse feminist epistemologies,[10] it seems safe to characterize them collectively as grounded in the awareness that gender, and more generally, the 'social

[4] For example, Sosa (1991).

[5] Code (1987) breaks new ground as the first responsbilist view.

[6] For example, Montmarquet (1993).

[7] Code's responsibilism in (1987) is 'mixed' (although not cast in the 'internalism/externalism' vocabulary), as is Zagzebski's (1996).

[8] Zagzebski's is a 'strong' theory, as is M. Fricker's in (2007).

[9] I consider Braatan (1990), Daukas (2006), Fricker (2003) and (2007), and Ruetsche (2004) to be work in feminist virtue epistemology (although Fricker doesn't identify her view as explicitly feminist). Code's (1987) precedes her explicitly feminist work, but it lends itself to feminist interpretation and is clearly foundational for feminist work in virtue theory.

[10] Code's entry on 'Feminist Epistemology' in Jagger and Young (1998), and Anderson's (2009) entry in the *Stanford Encyclopedia of Philosophy*, both offer comprehensive and insightful overviews and analyses of feminist epistemology that convey the field's common ground while capturing its diversity.

location' or 'situatedness' of the subject of knowledge, matters to knowledge practices and products in contextually variable ways; and that knowledge practices and products, along with theorizing about them, matter socially and politically. Therefore the 'social location' of the knowing subject matters to feminism and epistemology, in ways that may vary with different contexts of inquiry, and in ways that have powerful social/political consequences. From this we can distill the following three commitments shared by feminist epistemologies: (1) to engage the point of view of epistemic agency, conceiving of agents in their particularity, as embodied and politically 'situated'; (2) to attend to the particularity of contexts in which epistemic questions arise and are pursued, in a way that especially illuminates social structures and their politics; and (3) to contribute to feminist political goals, by providing both a descriptive analysis of the epistemic aspects of oppression, and regulative guidance for transforming them.

Responsibilist virtue epistemology (and from here on, when I say 'virtue epistemology' I mean 'responsibilist virtue epistemology') is at least poised to fulfill the first commitment insofar as it is agent-centered. I'll say much more about that soon. Further, although virtue epistemology may consider general normative principles to be useful as 'rules of thumb,' it tends to see them as artificial, because context- and agent-insensitive. Typically, a virtue theory in ethics or epistemology holds that what one should do, or believe, in a given situation depends on any number of situational variables. Part of what makes an agent virtuous is her ability to discern what the particular context requires of her morally and epistemically, which may be different for different agents. Hence a virtue epistemology is poised to fulfill the second feminist commitment (to context-sensitivity) above, while also remaining attuned to agential factors, including agential differences. This by itself – a theoretical structure that accommodates both agential and contextual variability while remaining a (generalized) epistemological theory (no easy trick!) – goes a long way toward recommending a feminist virtue epistemology.

Finally, consider the third, political commitment of a feminist epistemology. Traditionally, virtue ethics understands the virtuous agent as living a goal-directed life with the ultimate *telos* of human flourishing. When paired with virtue ethics, virtue epistemology defines and enables the epistemic aspects of that flourishing. Whether or not a particular virtue epistemology meets the third feminist commitment, then, depends on how flourishing is defined. So long as it is defined consistently with liberatory values, and desired for all persons, virtue epistemology prescribes the development of habits of mind that contribute to feminist political goals, and therefore meets the third feminist commitment noted above.

I will say more about what virtue epistemology offers feminism as the arguments of this paper unfold. But I also want to acknowledge that virtue theories may be antithetical to feminism. Although agent-centered, a virtue theory may assume a traditional masculinist conception of agency – an atomistic individualism on which agents are understood to be (artificially) independent, self-interested, and rationally self-governed (where 'rationally' is understood in opposition to 'emotionally', and 'emotion' is understood to be disorderly, in need of control and containment, and characteristically feminine). And a virtue theory's interpretation of human flourishing may presuppose a patriarchal value system and so reinforce the power structures which feminists seek to displace.

We can effectively align virtue theory with feminist goals by turning away from a traditional, individualistic conception of the self, instead embracing a relational or social conception, on which individual well-being and community well-being are interdependent.[11] 'Doing well' for an individual then entails doing well in mutually self-defining engagement with others, which is impossible if not through mutually beneficial, non-hierarchical relationships. And when relationships and community are defined non-hierarchically, the habits and dispositions that maintain structures of domination are understood to be vicious, while those that tend toward fulfillment of liberatory goals are understood to be virtuous.

Along with being more feminist-friendly, the resulting picture of selves aligns well with the social character of knowledge.[12] By this I mean, minimally, that epistemic products, practices, and norms are social: knowledge and understanding are produced collaboratively and shared by epistemic communities. We develop epistemic skills and dispositions together, through participation in social practices whose regulative norms are culturally inscribed and socially maintained. Solo epistemic endeavors are only superficially solo, and successes are interdependent. Doing well as an epistemic agent therefore implies doing well as a member of an epistemic community.

This is not to deny that individuals may pursue their own epistemic projects, critically assess the assertions of other community members or critically reject components of their inherited cultural traditions that are still widely, perhaps normatively, accepted by others in the community. The point is simply that the ground on which one stands when undertaking such critique in part constitutes the very inherited framework under critical scrutiny. So, for example, I have learned the value of identifying and evaluating unstated assumptions, and the means by which to proceed in such a project, through participation in the very cultural tradition some of whose unstated assumptions I now wish to identify and question. I proceed by articulating how my experience challenges established normative conceptions of social reality, and in this process work critically from, with, and against the work of others with whom I share an epistemic culture.

At this point it should be clear that a virtue theory can meet the three general commitments of feminist epistemology mentioned above. Regardless of what more we may want from feminist epistemology, or what different forms a feminist epistemology may take, it should be clear that a feminist virtue epistemology

[11] There is a substantial feminist literature that critiques the traditional conception of 'the autonomous self' and explores views of the self as social or relational. See Moody-Adams (1998) for an overview of work from the 1980s and early 1990s; see also Meyers (1997) and Mackenzie and Stoljar (2000).

[12] For the most part, feminist epistemologies are social epistemologies, that is, they see epistemic communities, as opposed to isolated Cartesian subjects, as primary knowledge producers. See Code (1991), Nelson (1993), and Potter (1993). For a recent discussion of feminist epistemology as social epistemology, see Grasswick (2008).

has something to offer. In the next section I develop a feminist socialized virtue epistemology, and then argue that it offers more than one might initially expect.

3.3 A Feminist Virtue Epistemology Centered on Epistemic Trustworthiness

The virtue theory I propose is both character-based and truth-connected (and therefore an internalist (responsibilist)/externalist (reliabilist) hybrid). It holds epistemic agents accountable for their beliefs and responsible for the character of their epistemic interactions with others: this is its responsibilism. And it requires epistemic virtues to reliably produce (significant) true beliefs and understanding, thereby holding agents accountable to how things are, rather than to how a particular doctrine or ideology says things are; this is its externalism. It is centered on 'epistemic trustworthiness', by which I refer to the complex character state that grounds dispositions to represent oneself as a more or less credible 'informant' or 'testifier' (relative to the given context and subject matter), and to judge others' credibility as 'informants' or 'testifiers.'[13] Hence the degree to which epistemic agents are epistemically trustworthy significantly determines the ways in which epistemic practices play out, which I take to be fundamentally, pervasively, implicitly or explicitly testimonial. And epistemic trustworthiness is therefore crucial for projects that aim to set right the wrongs of testimonial politics in a white-supremacist patriarchy.

Given its social and moral importance, and given that we typically think of moral character when we use the vocabulary of trust, one might assume that the epistemic trustworthiness I have in mind is really a moral virtue expressed in epistemic contexts. But I think it is important to theoretically isolate a thoroughly *epistemic* dimension of, or thread within, trustworthiness. This allows us to foreground and analyze the core role of *epistemic* self-critique and agent-assessment in responsible epistemic agency. To that end, let us assume that, unless otherwise indicated, the epistemic agents under discussion here are morally trustworthy – that is, they are disposed to be honest and forthcoming, to keep promises, to intend to be egalitarian, and to approach others with a general attitude of good will. So imagine that you and I are morally trustworthy, that you care about *p*-related inquiries, and that I assert that *p*. In order for you to take what I say to have epistemic value for your *p*-related epistemic interests, you must see me as *worth listening to*, particularly with respect to *p*-type matters. You must experience my testimony that *p*, in the given circumstances, as providing you with a reason to consider that *p* is a real possibility. In sum: you must see me as credible, and so respect me as an *epistemic* agent, with respect to *p*. Given my assumed honesty, I am worthy of that epistemic

[13] Compare this 'other-directed' aspect of epistemic trustworthiness for a recipient of testimony with M. Fricker's 'virtue of testimonial justice.' See esp. (2007), chapter 4. Part of my goal here is to explore the interdependence of 'other-directed' and 'self-directed' epistemic beliefs and virtues.

respect only if (other things equal)[14] I possess the epistemic skills, abilities, and attitudes required for successful *p*-related inquiries.

Other analyses of trustworthiness tend to end here[15]: they take trustworthiness to include sincerity, as its moral component, and competence, as its epistemic component. But 'competence' is vague, and easily interpreted as primarily first-order. If so, then this standard analysis of trustworthiness neglects a crucial, second-order epistemic component. To see this, assume not only that I am honest, but also, that I possess the first-order epistemic competence required for *p*-related inquiry (such as, for example, *p*-related perceptual acuity and inferential skills). I still may or may not deserve *p*-related epistemic respect: it all depends on whether I really am credible regarding *p*, that is, whether I am epistemically trustworthy in my *p*-related testimonial performance. I am epistemically trustworthy as a 'testifier' so long as I generally claim to have *p*-related knowledge in, and only in, circumstances in which it is reasonable for me to believe I have that knowledge. That is, I deserve *p*-related epistemic respect only to the extent that my *epistemic self-presentation* (with respect to *p*) is generally sound. Since we are assuming honesty, whether or not my epistemic self-presentation is sound turns on the integrity and accuracy of my *epistemic self-knowledge*. If I frequently take my *p*-related views to be more epistemically secure than they are, due to a tendency toward epistemic overconfidence regarding *p*, and I behave in a way that conveys that undue epistemic confidence to you, I thereby behave in an epistemically irresponsible manner toward you: I authorize you to accept claims that I am not qualified to authorize you to accept. I am undesirable as an epistemic partner; I lack credibility, at least when it comes to *p*. Note that I am also a poor partner in epistemic inquiry, in a different way, if I frequently take my *p*-related views to be less epistemically secure than they are, due to a tendency toward excessive epistemic diffidence, and so fail to authorize you to accept views that I am qualified to authorize you to accept. For either kind of reason, I am not epistemically trustworthy regarding *p* under the circumstances *because of my second-order epistemic self-assessment* and its behavioral expression, even though I possess the first-order epistemic competence required for successful *p*-related inquiry.

Here we see the necessity of epistemic self-critique to responsible epistemic agency. Developing reliably accurate dispositions for epistemic self-critique is no easy task. It requires the integrated efforts of other, simpler virtues. Most obviously, being epistemically trustworthy requires having the right degree of confidence with respect to one's salient beliefs under relevant kinds of circumstances (where the 'right degree' is the degree warranted by one's actual epistemic status *vis-à-vis* the subject matter under the given (saliently defined) circumstances). That confidence

[14]That is, there are no prudential or practical constraints in the situation that override epistemic constraints on the warranted assertability of knowledge claims.

[15] See, for example, Bernard Williams (2002), which analyzes trustworthiness in terms of sincerity and accuracy; I take such a view of trustworthiness to be implicit in M. Fricker's discussion in (2007) (see, e.g., pp.76–79). See also Lehrer (2006) (e.g., p. 157). My discussion of epistemic trustworthiness here builds on my discussion in (2006).

must be tempered with the appropriate degree of humility, and yet empowered. At a minimum, appropriately tempered epistemic self-confidence requires a finely-tuned ability to discern what features of different contexts are most salient to a given inquiry, what is required epistemically for a given inquiry in light of its context and goals, the degree to which one is, and is not, equipped to meet those requirements (and how), and how the salient features of one's epistemic situation compare to, contrast with, or complement those of others.

This, in turn, requires a well-developed sense of one's own, and others', contextually relevant epistemic strengths and weaknesses – a sense which neither under- nor over-estimates one's relevant epistemic competence. That 'sense' requires attunement to the relevant attitudes of the appropriate others, which requires (frequently tacit) context-sensitive judgments regarding who those appropriate others are. All of this discernment regarding self and others develops, and constantly evolves, through engagement with others in politically-permeated epistemic activities and practices, partly through 'reading' and assessing others' epistemic attitudes toward oneself and one another.

The emerging picture is this: epistemic trustworthiness emerges from a complex of testimonially related dispositions. It is possible only when an agent is attuned to her own, and particular others', epistemic strengths and weaknesses relative to particular contexts and projects. That attunement, along with a desire for significant truth and understanding, disposes the agent to find the right balance (which is sure to be different for different agents) of open-mindedness and charitability with critical astuteness; of courage (as when, with good reason, we take seriously a view that our community rejects, or an agent whom our group tends to shun epistemically) with carefulness; humility with confidence, and so on. The epistemic self- and other-attunement which all of this requires is socially learned and continually evolves. How a particular agent develops it depends (in part) on her unique epistemic history, which is thoroughly social and permeated by consequences of her social location. When all the pieces come together constructively, the epistemic agent works from dispositions which enable her to unmask and transform epistemically unsound (and socially unjust) practices of epistemic exclusion (i.e., testimonial injustice).

That epistemic trustworthiness is necessary for testimonial justice becomes more obvious when we consider that we don't know many of the individuals with whom we interact epistemically, and whom we must therefore trust epistemically to some extent, at least provisionally, in order to function productively as epistemic agents. For this reason, participating in epistemic practices often requires extending to one another what I elsewhere call an *epistemic principle of charity* (Daukas 2006), or a presumption that all participants are epistemically trustworthy to a 'baseline' degree. That is, in order to function epistemically, we presume that our partners in inquiry are credible as 'testifiers' to some threshold degree; and we presume that we are entitled to expect the same in return. Such mutual epistemic respect is an enabling condition, and a default expectation, of inquiry among relative strangers. Under such conditions, if you and I find that our respective views conflict, then unless the relevant experience and knowledge-base of one of us clearly surpasses that of the other, each of us should be willing to take a critical stance toward our own views in light of the other's, and should expect the other to reciprocate.

Brief as it is, I think this sketch shows that the socialized virtue epistemology I have in mind meets the three commitments of a feminist virtue epistemology stated above. It conceives of epistemic agency, and the development and functioning of epistemic excellence, in a way that incorporates the particularity of epistemic history and character, and the particularity of context and situation. Because epistemic trustworthiness (or its absence) underwrites epistemic interaction, and epistemic interaction is (implicitly or explicitly) integral to all forms of practical, social, political, theoretical, and moral decision making and therefore pivotal for concerns regarding social justice, developing this type of approach in epistemology, both descriptively and prescriptively, contributes to feminist political goals.

As mentioned earlier, I don't see this feminist virtue epistemology as competing with other feminist epistemologies; rather, it builds on work of feminist epistemologists,[16] and complements at least two influential approaches in a way that provides a framework for reconceiving their relation as collaborative.

3.4 Contextualisms

One such view is Helen Longino's 'contextual empiricism', which provides an account of the production of scientific and 'everyday' knowledge that reveals the normal role of contextual values in inquiry.[17] As I see it, Longino develops a feminist (descriptive and prescriptive) interpretation of the political/epistemic consequences of the Quine-Duhem thesis regarding the relation between data and theory, and Kuhnian insights into the framework-relativity of 'normal science'. Her basic picture is this: a given inquiry is methodologically and conceptually guided by a framework of theoretical and evaluative background commitments. Some of those commitments are specific to the project, others, to the type or field or subfield of inquiry, and still others permeate the culture, remain largely tacit, and are inherited and maintained through socialization. Such a framework guides an inquiry or research programme insofar as it determines what kinds of questions it makes sense to ask, what kinds of hypotheses should be taken seriously, how data are to be interpreted, what data constitute adequate evidence for a given hypothesis, and so on.

Since the framework of a given inquiry often includes social values embedded in an inquiry's 'home culture', Longino's approach challenges the traditional distinction between the 'context of discovery' and the 'context of justification', showing that social values (of the 'context of discovery') play a role in the justificatory methods and products of normal science. Hence 'normal' practices of inquiry provide various opportunities for political commitments and social values to influence epistemic practices and products. In a hierarchical society, commitments that (purport to) naturalize, and thereby perpetuate, unjust power relations therefore may influence, determine, or otherwise constitutively contribute to the practice and

[16] Especially Code (1987), (1991), and (1995) and Fricker (2003).

[17] Developed as an epistemology of science in Longino (1990) and more generally in (2002).

content of science – not in spite of, but by means of, its normal methodological structure.[18] It follows that we can do 'feminist science' by unmasking the influence of patriarchal assumptions and values on the community's epistemic work, replacing them with deliberately chosen liberatory values and assumptions, and reinterpreting empirical data and revising theory accordingly.[19]

Longino associates her contextual empiricism with the epistemological contextualism of recent popularity in mainstream epistemology, developed, for example, by Keith DeRose, Stewart Cohen, Michael Williams, and David Lewis.[20] But it's not clear that her view is contextualist in the mainstream sense, and, as I am about to argue, Longino's feminist goals recommend distancing contextual empiricism from the mainstream view. Doing so may require that some aspects of Longino's view be further, or differently, developed.

The contextualism of mainstream epistemology was initially developed as a strategy for insulating skeptical arguments from 'ordinary' knowledge claims. It asserts that the truth-conditions on knowledge attributions vary contextually, so that '*S* knows that *p*' may be true when asserted in one context, but false when asserted in another, for the same *S* and *p*. The variation in contextual standards is usually characterized in terms of levels of rigor: 'the bar is raised' in some contexts and 'lowered' in others; the context in which Cartesian skepticism is true maintains the highest standards of all, but since 'everyday' standards are 'lower', skepticism is false in those 'everyday' contexts and thus does not threaten 'ordinary' (including scientific) knowledge claims. This form of contextualism identifies its foil as *invariantism*, the view that the truth-conditions on knowledge claims are stable, and so do not vary contextually. Let us call this type of contextualism 'semantic contextualism'.

As does semantic contextualism, so Longino's contextual empiricism implies that the goals, purposes, and methods of particular inquiries are context-relative, so that different contexts of inquiry might require different sorts and degrees of epistemic labor in order to arrive at epistemically acceptable results. But as I see it, the common ground should, and may, stop here. Consider Longino's analysis of knowledge in (2002): 'S knows that *p*' is true if (i) S accepts that *p*, (ii) *p* is true, and (iii) 'S's response to contextually appropriate criticism of *p* or of S's accepting *p* is or would be epistemically acceptable in C,' where 'C' refers to the relevant epistemic community (2002, 138). The social acceptability criterion (iii) stands alongside two other stable acceptability conditions taken for granted at this point in the text, namely, logical coherence and support by empirical evidence (although

[18] By now feminist philosophers of science have produced a number of case studies to illustrate this. Along with examples in Longino (1990) see Bleier (1984), Fausto-Sterling (1992, 2000), Hubbard (1990), Keller (1984), Longino and Doell (1983), Martin (1991), Spanier (1995), Biology and Gender Study Group (1988), to mention only a sample.

[19] The arguments in (1983) and (1989) develop this implication for 'doing science as a feminist.' In (1989), Longino argues 'for the deliberate and active choice of an interpretive model and for the legitimacy of basing that choice on political considerations' in specific cases (54) (such as the case illustrated in detail in (1983)).

[20] See (2002, 104–106). Examples of what I will call 'semantic contextualism' include Cohen (1987), DeRose (1999), Lewis (1979), Williams (1996).

presumably these criteria may be interpreted or enacted differently by different communities). This appears to be an *invariantist* analysis of knowledge, insofar as it specifies a stable, uniform set of necessary and sufficient conditions for the truth of '*S* knows that *P*'.

We can think of condition (iii), Longino's social epistemic acceptability criterion, as fulfilling the role of the justification condition of the traditional analysis of knowledge as justified-true-belief. Unlike traditional forms of justification, Longino's ties justification to context, critical social engagement, and epistemic community. But note that not any form of social criticism will do, nor will any community do: a community is an epistemic community only if it is 'epistemically productive', and its critical/testimonial interactions are governed by epistemic norms that guard against political hijacking or other intrusions of social power (2002, 128–134). Although meeting that condition requires different epistemic activities in different contexts, it is nevertheless a stable condition: a statement is justified just so long as it is logically coherent, supported by empirical evidence with sound reasoning, and meets the standards of effective criticism for that context of inquiry (which must include the just distribution of epistemic authority) as (otherwise) determined by the relevant epistemic community.

In contrast, semantic contextualism is consistent with the possibility that in some contexts, knowledge claims will be true in the absence of empirical support (although of course empirical adequacy would be required by all scientific contexts), and/or in the absence of the critical assessment of an epistemic community. Some contexts may require community assessment with epistemic authority explicitly being granted only hierarchically, in accordance with a patriarchal power structure; some may require consistency with religious doctrine. And so on. In short: Longino's normative acceptability conditions are stable, and have regulatory force cross-contextually; contextual variability therefore must be limited to the particular ways stable norms are expressed. This is a *practical* variability in how knowledge is acquired in different contexts, not a semantic/theoretical variability in what 'knowledge' means in different contexts.

Another difference: unlike mainstream contextualists, Longino does not characterize contextual variability in terms of degrees of rigor required by context-relative standards for knowledge. For it is apparently because the contextual differences in mainstream contextualism are a matter of degree – a matter of how 'high' or stringent the context demands its epistemic standards to be – that the truth of a given knowledge claim is context-relative, and that it therefore does not 'convey' from context to context. Although Longino doesn't explicitly rule it out, there's no indication that she understand contextual variability in terms of 'higher' and lower' standards. (As is so often the case: the mainstream view interprets difference hierarchically; the feminist view allows difference to be non-hierarchical).

As feminist epistemologists, we should embrace an invariantist interpretation of Longino's account of the context-sensitivity of epistemic practices, and steer clear of semantic contextualism. One reason is that the latter deprives the critical theorist of a 'place to stand' from which to logically engage with knowledge claims made in contexts other than its own (and this is the source of its appeal if our primary goal is to avoid skepticism). If '*S* know that *p*' has different truth-conditions in different contexts, then analyses that take place within particular contexts of feminist epistemology do not logically connect to, and so have no critical bearing on, knowledge claims made in the many contexts of everyday and scientific life which the feminist epistemologist seeks to understand. Further, semantic contextualism implies that when a feminist (or any other) epistemologist critiques a particular theory of knowledge, her critique may simply illustrate how the conventions defining 'knowledge' in the critic's 'home' context of inquiry differ from those of the 'target' context of inquiry. So semantic contextualism precludes the possibility of a vantage point from which to argue that one theory offers a more accurate, more insightful, more empirically adequate analysis of epistemic practices and attitudes than does another.

Finally, mainstream semantic contextualism fails to engage the point of view of epistemic agency. To see why, imagine going through the day in a semantically contextualist world: in one context you know that *p*; you shift to another context, and don't know that *p*, but that doesn't present a conflict or a problem, because the kind of knowledge you had in the one context is different from ('less than') what you claimed to have in the other. What you come to know in one context therefore remains logically disconnected from what you learn in another (so, for example, one cannot correctly say 'I know that *p* and *q*', if one acquired the knowledge of *p* in a different context from the context in which one came to know that *q*.). This picture is artificially fragmented, and completely at odds with the lived experience of a dedicated epistemic agent engaged in the ongoing project of constructing a comprehensive body of knowledge and understanding.[21] And it is inconsistent with the goals of a feminist agent who struggles to create a potentially transformative understanding of things that can serve as a basis for political action. She works from an enduring set of liberatory values, in opposition to mainstream entrenched patriarchal/white supremacist values, which continually inform and transform her theoretical commitments and unify her various epistemic pursuits, despite their localized, context-specific challenges.

3.5 Feminist Contextual Empiricism, with Virtue

Earlier in this paper I identified three commitments shared by feminist epistemologies. It is obvious that Longino's contextual empiricism is context-sensitive, and effectively directed toward feminist political goals. It engages with the point of view of agency to an extent: it acknowledge the importance of agency in knowledge production by emphasizing the role of critical exchanges within a diversely constituted epistemic community. And it requires that the epistemic norms guiding those exchanges run counter to the traditional norms by which epistemic authority is granted and withheld along lines of power and social position. But its engagement

[21] My critique of mainstream epistemic contextualism here draws on my discussion in Daukas (2002).

with the point of view of agency itself – with the experiences, cognitive possibilities and challenges of differently situated subjects of knowledge/agents of inquiry – remains abstract and undeveloped. This is where the value of a partnership with feminist virtue epistemology is most evident.

The values and value-serving contextual commitments that shape our (diverse) ways of seeing and understanding things operate both at the first-order level, shaping how we 'see' and 'manage' objects of inquiry, and also at the second-order level, guiding the ways in which we present ourselves, and 'read' others, as epistemic agents engaged in testimonial exchange. Among the culturally inherited commitments at play are those that perpetuate testimonial injustices – those that support normalized patterns of testimonial exchange whereby epistemic authority is granted, or not, along lines of social privilege and power. Given their historical persistence and ubiquity, it is reasonable to locate those commitments in the central scaffolding of the framework of 'the western worldview' (or in 'bedrock', to shift to a Wittgensteinian metaphor). They take the form of a tacit social metaphysics that links 'successful' white males to 'Reason', constructed as the creative source of knowledge, culture, and moral order, which ('rightfully') opposes and controls the 'lower', affective, disorderly influences of body and emotion, associated with 'other', inferior humans.[22]

The consequences for testimonial practices are clear: in the absence of a deliberately counter-cultural, oppositional framework of inquiry, the testimony of members of the privileged class *just is,* inherently, *prima facie* credible, while that of members of subordinate classes *just is,* inherently, *prima facie* suspicious, and therefore 'justifiably' ignored, greeted with skepticism, or dismissed or trivialized, particularly when it questions, conflicts with, or even simply differs from, that of a member of the 'Reasoning' class. (This is borne out by a number of empirical studies, such as those in which the same essay is graded differently depending on the assumed gender of the author).[23] Longino's acceptability conditions counteract the continued epistemic influence of that social ideology by requiring that the community whose critical assessment determines the acceptability of a belief must be diversely populated, and must distribute epistemic authority or credibility among community members justly.

The latter is crucial: the premise of this entire discussion is that a diverse epistemic community that does not self-consciously aspire to overcome the influences of a white-supremacist patriarchy is likely to continually reproduce the patterns in testimonial exchange largely responsible for epistemic injustices. But how is that crucial condition for the just distribution of epistemic authority to be

[22] Cf. Lloyd (1984), Code (1991), esp. chapter 2, Plumwood (1993), esp. chapter 2, and Rooney (1991). Compare my discussion of the role of this traditional social metaphysics in testimonial discrimination to Fricker's discussion of the role of 'identity stereotypes' in (2007). I discuss the difference in Daukas (2006).

[23] See Paludi and Strayer (1985). For references to a number of studies regarding gender bias in academic evaluation, see Wylie, Jakobson and Fosado (2007). Many thanks to Sharon Crasnow for pointing me to these materials.

met? Without an answer to this question, Longino's contextual empiricism provides a descriptive account of how an unjust power structure and its distorted social metaphysics are perpetuated through normal epistemic practices, but no effective prescriptive account of what to do about it.

This is where feminist virtue epistemology steps in: for diversity within an epistemic community to pay off epistemically, members of subordinated groups must be empowered *as* epistemic agents. For this, the community, as a collective, must understand, value, and (critically) retain the epistemic/perspectival differences enabled by members' diverse social locations, and those members must be appropriately epistemically self-confident and courageous, if they are socially marginalized, and appropriately humble and courageous, if they are socially privileged. They must be disposed to grant epistemic authority to those who may otherwise be epistemically overlooked or even shunned, and to critically withhold it when it is usually granted without question. And this requires social self-awareness and epistemic self-understanding, which, as we have seen, is interdependent with epistemic attunement to others. That is, the social diversity Longino requires of an epistemic community can deliver its intended epistemic benefits only if we assume that its members are (jointly, collaboratively) epistemically trustworthy.[24]

By pairing contextual empiricism with a feminist virtue epistemology, then, we can imaginatively put Longino's prescriptive epistemology into practice. We can also solve residual features of the problem of fragmentation sketched in the previous section. The challenges posed by contextual empiricism require responsible inquirers who make explicit choices about the values and theoretical commitments that will guide them as they formulate questions, pursue and interpret data, and extend, or not, the epistemic principle of charity to potential epistemic partners. As a particular agent navigates from context to context, she may shift foreground and background commitments and reprioritize epistemic values. How does she do this? What provides the normative location, so to speak, from which she makes these choices? And what provides the logical space in which to integrate the results (or questioning, doubting, etc.) of inquiries from different contexts? (For example, how does one move among contexts of working as a feminist epistemologist, a biologist, a teacher, a parent, and living as an environmentally and socially responsible citizen, allowing the epistemic work of each context to enrich that of the others, and integrating all around the *telos* of feminist political goals?)

The answer now seems obvious: it is the epistemic character and cognitive habits of the socially attuned and engaged epistemic agent – in short, her epistemic agency itself – that unifies her various epistemic projects, experiences, and values. The epistemically trustworthy agent has developed higher-order, critically reflexive attitudes that constitute a continually evolving, self-aware interpretive metaframework. That metaframework provides the normative and conceptual space from

[24] Remember: This is not to say that these epistemic agents need not also be morally trustworthy. Dispositions they need, that seem to me to stem from trustworthiness that is both moral and epistemic, include being disposed to value collaborative epistemic work over individual recognition, and truth and significant understanding over proving oneself right.

which to map relations among localized epistemic projects and interactions in a way that enriches her understanding of the particularity of each, while integrating the products of distinct inquiries into a complex, relatively coherent (yet not 'unified') view of the world and her place in it.

The emerging constellation of virtue epistemology, contextual empiricism, and feminist values looks roughly like this: contextual empiricism provides a theoretical framework for analyzing the structure of epistemic practices; virtue epistemology provides a framework for understanding how those structured practices translate into the activities and attitudes of epistemic agents in interaction with one another, seeking to live epistemically productive, integrated epistemic lives. A feminist lens reveals how, through participation in testimonial practices, the individual knower may serve as an agent of entrenched social/political hierarchical values. Feminist virtue epistemology characterizes the kinds of agency mutually required and supported by transformative epistemic practices. A virtuous epistemic agent has developed a critical acuity specifically attuned to recognizing and challenging the values and theoretical commitments embedded in epistemic practices and products that maintain a culture of domination and oppression. She explicitly identifies and intentionally embraces contextually salient, liberatory/oppositional values to guide her as she participates in local inquiries, and more generally, strives to transform the standard practices of the community.

This bring us into dialogue with standpoint theory.

3.6 Feminist Standpoint Epistemologies

Turning to standpoint theory in a discussion of feminist contextual empiricism might seem inadvisable, especially given that Longino explicitly rejects it (1990, 188). But what she rejects is any standpoint theory that either presumes that there is such a thing as a single 'woman's standpoint', which is clearly false; or that acknowledges the multiplicity of women's perspectives and yet privileges one of them (typically that of the white, straight, educated middle-class woman) over others, which is clearly unacceptable. However, feminist standpoint epistemologies have evolved beyond those trends.[25] Approaches that emphasize that standpoints are diversely located, diversely empowered epistemic achievements avoid those mistakes and complement the contextual empiricism discussed here.[26]

---

[25] Longino also objects that standpoint theory is circular: it needs itself to identify the groups whose perspectives 'count' as potential standpoints (1990, 12). Although I am not convinced that this is a vicious circularity, the issue is moot if we appeal to epistemic virtues to distinguish mere marginalization from standpoint, as I argue below.

[26] I'm thinking especially of views that identify marginalized standpoints by reference to 'double consciousness' (or 'binocular vision' or 'the outsider within'). See, for example, Collins (1986) and (1991), Frye (1983, 147–150), Harding (1991), hooks (1984), esp., the Introduction. See Harding (2004) for a representative collection of different standpoint theories and critical dialogue among and about them. See Wylie (2004) for an especially clear distinction between a 'social location' and a 'standpoint'.

This form of feminist standpoint epistemology starts from the recognitions that subjects of knowledge or epistemic agents occupy perspectives, that particular perspectives are partial and socially situated, and that a self-reflective perspective that includes a recognition of its (and others') partiality *as perspectives* is better, epistemically, than one that lacks reflexive awareness and naively presupposes its own impartiality. Further, it recognizes that an ability to critically shift between and compare distinct perspectives, especially those situated differently in relation to one another, makes possible a less partial, richer, 'thicker', more objective view than does limitation to a single perspective.[27] These general points about perspective seem to me to be uncontentious, even platitudinous, and not necessarily feminist.

The contentious, and potentially feminist, insights emerge when we consider the implications of those general observations about perspectives for a hierarchical society. The result is a picture on which, first, the hierarchy of power and social privilege heavily influences patterns by which epistemic authority is granted and withheld. This creates social and material obstacles for some, and smooth sailing for others, as they (try to) participate in epistemic practices and pursue their epistemic (and other) goals. Second – and here is the most contentious point – the perspectives of individuals situated differently in the hierarchy are differently enabled and limited epistemically, in virtue of their different 'locations', in ways that create epistemic challenges, with potential epistemic advantages, for the socially marginalized, and 'blinders' for the socially privileged. Those advantages include relatively easy access to a vantage point from which to 'take in' the social order itself, with its supporting framework of culturally inherited commitments, and the extent to which the conditions of one's own life are its consequences.

To see how marginalization offers that epistemic advantage, consider '*S*', a member of a subordinated class who desires to participate in mainstream socio-economic, intellectual, or political activities. In order to do so, she must negotiate her passage through an environment structured by, and from the perspective(s) of, the dominant class(es) from which she is excluded. So she must become attuned to the way things (including members of her class) look from the point of view of the dominant elite; and she must conform, to some degree, to the expectations and demands that make sense from perspectives other than her own simply to get through the day. Getting through the day requires (or is vastly facilitated by) developing a sense of how the privilege experience their worlds. As a result, *S* is in a position to develop 'double consciousness', that is, an ability to see things from her own marginalized perspective while also seeing things from the perspective of the privileged. This position offers *S* the opportunity to develop a critical consciousness of the partiality of their perspectives and of her own uncritical perspective; of their differences; of how their differences are produced by their different 'locations' in the hierarchy; all of which contributes to a richer, broader, less partial, more discriminating picture of things.

---

[27] Compare to Harding's notion of 'strong objectivity'; see especially (1991, 1993).

Since the privileged need not negotiate an environment structured by the marginalized, they need not even notice (or care about) perspectives different from their own, nor, then, notice (or care about) the partiality of their own perspectives. So the perspectives of the privileged easily conceal their status as perspectives (and so, as particular partial, socially conditioned, contingent perspectives among others). They have the power and privilege to represent, and impose into public discourse, their partial, relativized view of things as impartial, absolute, and normative. That power creates the conditions with which the marginalized must contend, and with it, the epistemic advantages of their 'outsider within' status. (In short: the socially marginalized and privileged stand in assymetrical relations with respect to 'knowing one another's minds').

To illustrate: imagine an African American man interviewing for a job at the corner bank in a predominately white community (let us call him 'job applicant of color', or 'JAC'). JAC needs to be attuned to how 'the successful' candidate will appear and behave, which is determined by the perspective of the white male interviewer who represents the company. JAC also needs to recognize (whether tacitly or explicitly) how that interviewer might see his (JAC's) appearance, behavior and mannerisms as being different from what is expected of 'the successful candidate', simply because they are *his* and he is African American. Hence JAC needs to be attuned to (usually unarticulated, tacit) expectations and standards derived from life experiences different from his own, and from cultural values that he may well not embrace. He needs to be able to 'see' himself as he is seen from the perspective of the powerful; to see how the powerful experience themselves in relation to him; and to recognize, at some level, how that is different from his own experience of himself and of his relation to them. In short, he needs to see their encounter as politically/culturally contextualized.

In our scenario, the white male applicant ('JOE', or 'job applicant of the elite') shares a social location with the interviewer, insofar as both are white and male. Of course there may be significant differences between them, but insofar as the default interpretation of desirable 'qualifications' for the position arises from the points of intersection in their social identities, JOE doesn't face the challenges which JAC faces in securing the job, some of which are epistemic. In contrast to JAC, JOE can get through his day, smoothly sailing from point *A* to point *B,* with no awareness of, or concerns about, how that route, or his passage along it, looks from a perspective different from his own. Because of JOE's position of social privilege, he encounters neither the epistemic challenges *S* does, nor the opportunities they offer.

3.7 Feminist Standpoint Epistemology, Contextualized

Notice that the potential epistemic privilege of a particular 'location' is context-contingent: it is not because JAC is a human being of African descent that he is poised to develop a standpoint; rather, it is because he is an African American man looking for work in a white-run company in a predominantly white community in a white supremacist culture. The distinction between social locations that do, and do not, enable epistemic privilege depends on cultural context. Therefore, social/historical facts conjoined with the contextual empiricist picture of how background commitments shape inquiry support the central claims of this form of standpoint epistemology. Contextual empiricism explains the interdependence of social politics, on the one hand, and knowledge practices and products, on the other. Standpoint theory traces out the consequences, showing how that interdependence yields significantly different epistemic perspectives depending on one's 'situatedness' in a hierarchical social environment, and explaining why that interdependence is so easily deniable by those whom it privileges materially, socially, and politically, while simultaneously evident to those whom it disadvantages materially, socially, and politically.

On this picture, then, standpoint theory and contextual empiricism are mutually enriching. Contextualism offers a theoretical framework for explaining and supporting the claims of standpoint theory, as here understood. Standpoint theory explains why the critical engagement by a diverse community required by contextual empiricism is epistemically necessary for achieving more objective (less partial, less limited) and therefore epistemically better understanding of things than would vetting by a uniformly socially privileged epistemic community. But here, again, we need to draw on virtue epistemology.

3.8 Contextualized Standpoint Theory, with Virtue

Notice that as so far characterized, standpoint epistemology is not necessarily feminist-friendly. Its thesis is that membership in politically marginalized groups creates potential epistemic privilege. This is consistent with the first two commitments of a feminist epistemology introduced earlier in this paper: it is agent- and context-sensitive. And although it is frequently pressed into service by and for feminist theorizing, it could also be pressed into service by and for anti-feminist, racist, homophobic, right-wing extremists. Along with women, men of color, LGBT individuals, the disabled, and so on, polygamist cultists, survivalists, Neo-Nazis, and Klan members are politically marginalized: they must navigate structures erected by a mainstream culture whose quasi-liberal values they abhor. It might seem that the same reasoning by which standpoint theory implies that women, men of color, the disabled, LGBT individuals, and other socially disenfranchised groups are epistemically privileged also implies that extreme and overt sexists, racists, homophobes, and so on, are epistemically privileged. This is clearly not what we want from a feminist epistemology!

This worry is based on an unnecessarily thin picture of standpoint theory, which becomes evident when approached from the perspective of feminist virtue epistemology. Remember that the epistemic privilege enabled by marginalization is not 'automatic', but requires development through self-conscious critical work. Marginalization positions an epistemic agent to *develop* a self-reflective, critical awareness of multiple perspectives, the relations among them, and their relations to social structures. But it does not guarantee that individual agents will take advantage of its epistemic opportunities. Whether or not the agent develops the epistemic potential of marginalization depends on her motivation, values, perceptual acuity, balance of confidence and humility, courage, and so on. That is, it depends on her epistemic (and moral!) virtues, and especially for the purposes of inquiry, the epistemic self-reflection required for epistemic trustworthiness.

An epistemically trustworthy agent knows and accepts the value and responsibility of subjecting her views to critical scrutiny. Her goals include truth and understanding, which require being willing and able to suspend commitment to an ideology that obstructs or predetermines the outcome of (genuine) inquiry. In contrast, the Neo-Nazi explicitly embraces an ideology critically insulated by dogmatic adherence to the traditional social ontology that naturalizes socially constructed (or imagined) differences among 'categories' of people. It deliberately seeks to enforce its white supremacist patriarchal belief/value system and to invalidate any perspective other than its own. Doing so requires, among other epistemic and moral transgressions, dogmatically rejecting, or willfully ignoring, the cumulative results of three decades of research and reflection in the life and social sciences,[28] and deliberately, dogmatically dismissing the testimony of those who express views that challenge one's own. The neo-Nazi's perspective is thus dogmatically closed-minded in precisely the ways in which the perspective of a standpoint is critically open-minded: it insulates itself from the process of critical engagement with multiple perspectives required for developing a standpoint, and further, it would not *survive* if it were so engaged. This is in part because the Neo-Nazi's epistemic practices are *uncritically exclusive*: their standards are organized as tools to perpetuate dogma and deflect critical scrutiny; they require that epistemic agents be epistemically untrustworthy. In contrast, the practices associated with a standpoint are *critically inclusive*: their standards are organized as tools to discern, interrogate, and displace dogma. A feminist virtue epistemology therefore can support the thesis that some politically marginalized locations provide a potential for epistemic privilege that others do not provide.[29]

[28] See, for example, Bleier (1984), Gould (1996), Fausto-Sterling (1992), Lewontin et al. (1984).

[29] I've only recently discovered Laura Ruetsche's (2004) *Hypatia* article and am happy to find a convergence between our views that I take to support my approach. In Ruetsche's discussion of what an 'Aristotelean feminism' (that is, a feminist virtue theory) can add to a traditional conception of justification in the philosophy of science, she says: 'cast in terms of second natures [enduring dispositions that can be virtues], the standpoint theorist claims that gender-involved second natures do important epistemic work,' (95), and she locates that work in the insight into different perspectives made possible by contingencies of some (i.e., marginalized) individuals' social histories that, like all contingencies marking differences among individuals, are filtered out by a traditional approach to the epistemology of science. Reutsche goes on to suggest that her virtue-theoretic interpretation of standpoint theory differs 'only in emphasis' from Longino's 'sophisticated feminist empiricism' (96). There is also an interesting and promising connection here with Alison Wylie's suggestion (2004) that 'objectivity' be thought of as 'a loosely defined family of epistemic virtues.' (345).

It also provides a route from that mere potential to its actualization. The process of developing liberatory, oppositional epistemic agency by constructing a standpoint from a perspective of marginalization requires work, including the critical self-and-other-understandings necessary for developing epistemic trustworthiness: e.g., learning to spot the kinds of situation in which one tends toward overconfidence, wishful thinking, self-deceit, undue diffidence, gullibility, excessive humility, undue deference toward authority figures, and so on (all of which develop and evolve through epistemic engagement within a similarly aspiring community). In short: the marginalized epistemic agent develops a standpoint by developing epistemic trustworthiness. The profile of her epistemic psychology, so to speak, is that of the epistemic agent envisioned by feminist virtue epistemology.

3.9 Conclusion

Feminist epistemologies reject the idea, central to traditional epistemology, of an ahistorical, universalizable, disembodied subject of knowledge. They see that genders, in their multiple forms and expressions, matter to experiences, perspectives, epistemic questioning, practices, and products (which, in turn, matter to everything!). And they explore how genders matter to knowledge, with the ultimate aim of furthering (broadly understood) feminist social and political goals. In doing feminist epistemology we therefore face a complex challenge: to engage in, and integrate, different dimensions of epistemic analysis – agential, local-contextual, social/political, cultural – without assuming that any 'dimension' is monological, or is constitutively or functionally independent of the others.

My proposal in this paper is that such a challenge recommends pluralism, that is, differently focused epistemological/political projects that evolve collaboratively (and critically) around shared political goals, integrated by a shared conception of epistemic agencies as social, flexible, diversely enabled and limited, responsive, politically situated, and politically efficacious. A virtue epistemology centered on epistemic trustworthiness provides one such conception, flexible and 'thick' enough to support, enrich, and solidify connections among other forms of theorizing in feminist epistemology. It lends itself both to accounting for how epistemic agencies are typically contextually/culturally constructed so as to reinforce the traditional power structure, and of how, by developing critical self-awareness regarding how and to whom we speak, listen and respond (and why), we can develop and express feminist epistemic agencies that disrupt and reshape that structure.

I want to close with several brief acknowledgments: first, insofar as my discussion in this paper frequently focuses on the epistemic agency of *individuals*, it risks overstating the distinction between individual and community epistemic agency. But the reflections here are intended to explore epistemic virtues of individuals while continually affirming the causal and logical/constitutive interdependence of individual and community epistemic agency. Second, my discussion is theoretically artificial in its exclusive focus on *epistemic* agency, which is not really separable from other aspects of agency. The goal has been to tease out features of our lives and selves that matter for epistemology and for feminism, that otherwise may go unanalyzed precisely because they don't exist and function on their own. That artificiality carries some risks, especially the risk of reproducing the traditional distinction between 'the affective' and 'the epistemic'. The trick now is to put the pieces back together, wholistically. Finally, my discussion has continually returned to the core role that beliefs and assumptions about ourselves and others as epistemic agents play in testimonial exchange (and here is a key spot where I artificially separate belief from affective states). This suggests that knowledge of (or beliefs and affects concerning) self and others is implicated in knowledge of (beliefs/affects concerning) much else. The better we understand 'our own minds' and 'other minds' (and I take these projects to be inseparable) the better we understand other aspects of our epistemic lives.[30]

Acknowledgments Many thanks to Heidi Grasswick for her perceptive comments on the originally submitted version of this essay. Thanks also to audience members from the Society for Analytical Feminism conference in Lexington, KY (April, 2008) for thought-provoking discussion of an earlier version of this paper, and to Lisa McLeod, Jonathan Malino, and Vance Ricks for helpful discussion of a still earlier draft.

References

Anderson, Elizabeth. 2009. Feminist epistemology and philosophy of science. In *The Stanford encyclopedia of philosophy*. Winter 2009 edition: Edward N. Zalta ed. Forthcoming. http://plato.stanford.edu/archives/win2009/entries/feminism-epistemology/.

Biology and Gender Study Group. 1988. The importance of feminist critique for contemporary cell biology. *Hypatia* 3: 61–76.

Bleier, Ruth. 1984. *Science and gender: A critique of biology and its theories on women*. Elmsford: Pergamom Press.

Braatan, Jane. 1990. Towards a feminist reassessment of intellectual virtue. *Hypatia* 5(3): 1–14.

Code, Lorraine. 1987. *Epistemic responsibility*. Hanover: University Press of New England/Brown University Press.

Code, Lorraine. 1991. *What can she know? Feminist theory and the construction of knowledge*. Ithaca: Cornell University Press.

Code, Lorraine. 1994. Responsibility and rhetoric. *Hypatia* 9(1): 1–20.

Code, Lorraine. 1995. *Rhetorical spaces: Essays on gendered locations*. New York/London: Routledge.

Code, Lorraine. 1998. Epistemology. In *A companion to feminist philosophy*, ed. Alison Jaggar and Iris Marion Young, 173–184. Malden/London: Blackwell Publishers.

[30] In (1991), Code stresses the importance of knowing other people to epistemology in general. See especially Chaps. 2 and 4, where she argues that knowledge of other people would make a better paradigm case for epistemology than knowledge of inanimate objects, as is typical in mainstream ('malestream') epistemology. Code's conception of 'ecological epistemology' grows out of that proposal, as I understand it. See (1991) chapter 7, and (2006).

Code, Lorraine. 2006. *Ecological thinking: The politics of epistemic location*. Oxford: Oxford University Press.

Cohen, Stewart. 1987. Knowledge, context, and social standards. *Synthese* 73: 3–26.

Collins, Patricia Hill. 1986. Learning from the outsider within: The sociological significance of black feminist thought. *Social Problems* 33(6): 14–32.

Collins, Patricia Hill. 1991. *Black feminist thought*. New York/London: Routledge.

Daukas, Nancy. 2002. Skepticism, contextualism, and the epistemic "ordinary". *Philosophical Forum* 31(1): 63–79.

Daukas, Nancy. 2006. Epistemic trust and social location. *Episteme: A Journal of Social Epistemology* 3(1): 109–124.

DeRose, Keith. 1999. Contextualism: An explanation and defense. In *The Blackwell guide to epistemology*, ed. John Greco and Ernest Sosa, 187–205. Malden/Oxford: Blackwell.

Fausto-Sterling, Anne. 1992. *Myths of gender: Biological theories about women and men*, 2nd ed. New York: Basic Books.

Fausto-Sterling, Anne. 2000. *Sexing the body: Gender politics and the construction of sexuality*. New York: Basic Books.

Fricker, Miranda. 2003. Epistemic injustice and a role for virtue in the politics of knowing. In *Moral and epistemic virtues*, ed. Michael Brady and Duncan Pritchard, 139–158. Malden/Oxford: Blackwell.

Fricker, Miranda. 2007. *Epistemic injustice: Power & the ethics of knowing*. Oxford: Oxford University Press.

Frye, Marilyn. 1983. *The politics of reality*. Freedom: The Crossing Press.

Gould, Stephen Jay. 1996. *The mismeasure of Man*, 2nd ed. New York/London: W. W. Norton.

Grasswick, Heidi. 2008. Feminist social epistemology. In *The Stanford encyclopedia of philosophy*. Fall 2008 edition: Edward N. Zalta ed. http://plato.stanford.edu/archives/fall2008/entries/feminist-social-epistemology.

Greco, John, and Turri John. 2010. Virtue epistemology. In *The Stanford encyclopedia of philosophy*. Spring 2010 edition: Edward N. Zalta ed. http://plato.stanford.edu/archives/spr2010/entries/epistemology-virtue/.

Harding, Sandra. 1991. *Whose science? Whose knowledge?* Ithaca: Cornell University Press.

Harding, Sandra. 1993. Rethinking standpoint epistemology: "What is strong objectivity"? In *Feminist epistemologies*, ed. Linda Alcoff and Elizabeth Potter, 49–82. New York/London: Routledge.

Harding, Sandra (ed.). 2004. *The feminist standpoint theory reader*. New York/London: Routledge.

Hooks, Bell. 1984. *Feminist theory: From margin to center*. Boston: South End Press.

Hubbard, Ruth. 1990. *The politics of women's biology*. New Brunswick: Rutgers University Press.

Jones, Karen. 2002. The politics of credibility. In *A mind of one's own: Feminist essays on reason and objectivity*, 2nd ed, ed. Louise Antony and Charlotte Witt, 154–176. Boulder: Westview Press.

Keller, Evelyn Fox. 1984. *Reflections on gender and science*. New Haven: Yale University Press.

Lehrer, Keith. 2006. Testimony and trustworthiness. In *The epistemology of testimony*, ed. Jennifer Lackey and Ernest Sosa, 145–159. Oxford: Clarendon.

Lewis, David. 1979. Scorekeeping in a language game. *Journal of Philosophical Logic* 8: 339–359.

Lewontin, Richard C., Steven Rose, and Leon J. Kamin (eds.). 1984. *Not in our genes*. New York: Pantheon.

Lloyd, Genevieve. 1984. *The man of reason*. London: Methuen.

Longino, Helen. 1989. Can there be a feminist science? In *Feminism & science*, ed. Nancy Tuana, 45–57. Bloomington: Indiana University Press.

Longino, Helen. 1990. *Science as social knowledge*. Princeton: Princeton University Press.

Longino, Helen. 2002. *The fate of knowledge*. Princeton: Princeton University Press.

Longino, Helen, and Doell Ruth. 1983. Body, bias and behavior: A comparative analysis of reasoning in two areas of biological science. *Signs* 9: 206–227.

Mackenzie, Catriona, and Natalie Stoljar (eds.). 2000. *Relational autonomy: Feminist perspectives on autonomy, agency, and the social self*. New York/Oxford: Oxford University Press.

Martin, Emily. 1991. The egg and the sperm: How science has constructed a romance based on stereotypical male-female roles. *Signs* 16(3): 485–581.

Meyers, Diana T. (ed.). 1997. *Feminists rethink the self*. Boulder: Westview Press.

Montmarquet, James. 1993. *Epistemic virtue and doxastic responsibility*. Lanham: Rowman & Littlefield.

Moody-Adams, Michelle. 1998. Self/other. In *A companion to feminist philosophy*, ed. Alison Jaggar and Iris Marion Young, 255–262. Malden/London: Blackwell Publishers.

Nelson, Lynn. 1993. Epistemological communities. In *Feminist epistemologies*, ed. Linda Alcoff and Elizabeth Potter, 121–159. New York: Routledge.

Paludi, Michelle A., and Lisa A. Strayer. 1985. What's in an author's name? Differential evaluations of performance as a function of author's name. *Sex Roles* 12: 353–360.

Plumwood, Val. 1993. *Feminism and the mastery of nature*. London/New York: Routledge.

Potter, Elizabeth. 1993. Gender and epistemic negotiation. In *Feminist epistemologies*, ed. Linda Alcoff and Elizabeth Potter, 161–186. New York: Routledge.

Rooney, Phyllis. 1991. Gendered reason: Sex metaphor and conceptions of reason. *Hypatia* 6(2): 77–103.

Ruetsche, Laura. 2004. Virtue and contingent history: Possibilities for feminist epistemology. *Hypatia* 19(1): 73–101.

Sosa, Ernest. 1991. *Knowledge in perspective*. Cambridge: Cambridge University Press.

Spanier, Bonnie. 1995. *Im/partial science: Gender ideology in molecular biology*. Bloomington: Indiana University Press.

Williams, Michael. 1996. *Unnatural doubts*. Princeton: Princeton University Press.

Williams, Bernard. 2002. *Truth and truthfulness*. Princeton: Princeton University Press.

Wylie, Alison. 2004. Why standpoint matters. In *The feminist standpoint theory reader*, ed. Sandra Harding, 339–351. New York/London: Routledge.

Wylie, Alison, Janet R. Jakobsen, and Gisela Fosado. 2007. Women, work and the academy. The Barnard Center for Research on Women, *New Feminist Solutions* 2. www.barnard.edu/crow/newfeministsolutions/index.htm.

Zagzebski, Linda. 1996. *Virtues of the mind*. Cambridge: Cambridge University Press.

Chapter 4
The Implications of the New Materialisms for Feminist Epistemology

Samantha Frost

Abstract Feminists drawing on the physical and biological sciences increasingly repudiate the notion that biology and matter are passive or inert and instead recognize the agency of biology or matter in worldly phenomena and social and political behavior. Such 'new materialist' work challenges the linear models of causation that underlie constructivist analyses of the ways power shapes the subjects and objects of knowledge. It provokes feminist epistemologists to develop models of causation and explanation that can account for the complex interactions through which the social, the biological, and the physical emerge, persist, and transform.

Keywords Agency • Causation • Complexity • Materiality • New materialisms

In a recent argument detailing the ways the mineral content of bones is shaped by the interaction between gendered cultural practices and sexual endocrinology, Anne Fausto-Sterling invites feminists to 'accept the body as simultaneously composed of genes, hormones, cells, and organs – all of which shape health and behavior – and of culture and history' (Fausto-Sterling 2005, 1495). This invitation to include the biology of the body in cultural and political analysis might seem a bit bizarre considered at the phenomenological level of daily living – who, after all, could deny the effects of hormone swings, blood sugar, sleep deprivation, and aging as we live, work, think, and play? Yet, at the philosophical or theoretical level, the invitation is less bizarre and instead rather interesting. For feminist philosophers and theorists, the body as a living organism is a vexed object, so vexed, in fact, that in philosophical and theoretical work, it is often sidelined, bracketed, or ignored. In such a context, Fausto-Sterling's solicitation is a provocation and a challenge: in suggesting that feminists should consider intellectually the biology they cannot but

S. Frost (✉)
Department of Political Science, Gender and Women's Studies Program, Unit for Criticism and Interpretive Theory, University of Illinois, 361 Lincoln Hall, 702 S. Wright Street, Urbana, IL 61801, USA
e-mail: frost@illinois.edu

H.E. Grasswick (ed.), *Feminist Epistemology and Philosophy of Science: Power in Knowledge*, DOI 10.1007/978-1-4020-6835-5_4,

acknowledge in their daily lives, Fausto-Sterling also demands that feminists rethink some of the deeply held assumptions about causation that have structured feminist critique for a number of generations.

Of course, in many respects, a scholarly and theoretical focus on the body's materiality is nothing new: for several decades, feminists have denaturalized both embodiment and material objects, analyzing and specifying the manifold discursive practices through which bodies and matter are constituted as intelligible. The focus of such work has been on elucidating the processes through which norms and power relations are incorporated as forms of subjectivity or materialized in institutions, cultural practice, and facts. Recently, however, a group of scholars including Fausto-Sterling, Elizabeth Grosz, and Karen Barad have begun to try to include in such analyses the movements, forces, and processes peculiar to matter and biology. These 'new materialists' consider matter or the body not only as they are formed by the forces of language, culture, and politics but also as they are formative. That is, they conceive of matter or the body as having a peculiar and distinctive kind of agency, one that is neither a direct nor an incidental outgrowth of human intentionality but rather one with its own impetus and trajectory.

In seeking to re-introduce biological and material agency into feminist analysis, new materialists do not advocate that feminists renounce insights into the ways in which power infuses bodies and matter to make them into socially and politically intelligible subjects and objects. Quite to the contrary, they are alert to the awful political uses to which biological essentialism has been put historically. What they ask is that feminists leaven our analyses of the discursive constitution of embodiment and material objects with an acknowledgment of the forces, processes, capacities, and resiliencies with which bodies, organisms, and material objects act both independently of and in response to discursive provocations and constraints. For example, in her innovative re-reading of Darwinian evolution, Elizabeth Grosz suggests that it is because feminists are interested in the ways in which bodies are inscribed by culture that we must also ask 'what these bodies are such that inscription is possible, what it is in the *nature* of bodies, in biological evolution, that opens them up to cultural transcription, social immersion, and production, that is, to political, cultural, and conceptual evolution' (Grosz 2004, 2). Making a similar point, Karen Barad suggests that feminists consider 'how the body's materiality – for example, its anatomy and physiology – and other material forces actively matter to the processes of materialization' (Barad 2003, 809). If we do so, she claims, we will better apprehend how the body in 'its very materiality plays an *active* role in the workings of power' (Barad 2003, 809). These new materialists, then, explore how the forces of matter and the processes of organic life contribute to the play of power or provide elements or modes of resistance to it.

This is an exciting and provocative development in interdisciplinary feminist scholarship, for it represents an effort to supplement cultural or discursive analysis of social and political phenomena with scientific insights about biological, physical, or chemical processes. But of course, it is also a project that likely raises some alarm among feminists whose insightful analyses of gender, racial, and sexual politics have proceeded through the careful delineation of the processes through which normative imperatives have been naturalized to support arguments that social and political formations arise through the agency of nature or biology. In working against biological essentialism, feminists quite understandably have tended to deny matter or biology any agency at all in shaping social or political relations. And many are likely to be suspicious of any 'biologizing' move that might, advertently or inadvertently, dress up power relations and disciplining norms as a force of nature or a biological imperative.

However, the problem raised by new materialists is not the problem of essentialism: it is only when we think about causation in simple linear terms that essentialism can be seen as the inevitable outcome of an attempt to think about the agency of matter or biology. Indeed, the new materialist work exposes the explanatory narrowness of the models of causation that underwrite feminist efforts at denaturalizing power relations. New materialists aim to shift feminist critical analysis from a framework within which the agency of bodies and material objects is understood largely as an effect of power – a unidirectional account of agency – to a framework within which, for example, culture and biology have reciprocal agentive effects upon one another.[1] In calling for feminists to acknowledge that matter and biology are active in their own right, new materialists push feminists to relinquish the unidirectional model of causation in which *either* culture *or* biology is determinative and instead to adopt a model in which causation is conceived as complex, recursive, and multi-linear. To shift our understanding or model of causation in this way represents a huge challenge: feminists will have to retool their theories of explanation and political critique so that they encompass both an awareness of the ways in which power is discursively naturalized and an appreciation of the distinctive and effective agency of organisms, ecosystems, and matter. This in turn will demand that feminists rethink how to apportion responsibility for injustice and assess the possibilities for and paths toward social and political transformation.

To understand the stakes and the implications of the new materialisms, it is perhaps helpful to distinguish them from other approaches to thinking about matter, most notably the Cartesian account of matter as essentially inert and the historical materialist understanding of matter as transformed and given agency by humans' labor and cultural practices. In neither of these latter two cases does matter have a distinctive agency of its own. Rather, as Barad rightly observes, 'matter is figured as passive and immutable, or at best inherits a potential for change derivatively from language and culture' (Barad 2003, 801).

For René Descartes, matter is passive, unmoving in itself and subject to the mechanistic laws of physical cause and effect when compelled to move by an external force. This conception of matter is central to his notorious metaphysical dualism and his claim that the thinking self is an immaterial substance ontologically distinct from the embodied, material self.[2] According to this framework, thinking is a

[1]For a small sample of works not otherwise discussed in this essay, see Alcoff, *Visible Identities* (2006); Barad, *Meeting the Universe Halfway* (2007); Braidotti, *Metamorphoses* (2002); Capra, *The Web of Life* (1996); Haraway, *When Species Meet* (2008); Hayles, 'Computing the Human' (2005); Kirby, *Telling Flesh* (1997); Masters, 'Biology and Politics' (2001); Oyama, *The Ontogeny of Information* (2000a); Young, *On Female Embodied Experience* (2005).

[2]See René Descartes, *Philosophical Writings* (1985).

purely rational intellectual activity distinct from the passions and opposed to the provocations and arousals of the body-in-the-world. Indeed, Descartes's portrayal of the body as essentially unthinking underpins the modern understanding of the human self as a rational, free, and self-determining agent.[3]

As feminist scholars have pointed out, historically, this Cartesian understanding of the passivity of matter was figured in racialized, gendered, and class terms that in turn were used to justify racial, gender, and class inequities.[4] Women, the lower classes, and people of various cultural or national origins were construed as trapped in and by the body because they were perceived as lacking the wherewithal to distance themselves from the body's operations and to steer a rationally-defined course for their behavior and actions. That is to say, the 'others' of modernity were construed both as subject to the determinations of the biological or animal functions of the body and as vulnerable to a kind of a behavioral determinism, a vulnerability which derived from the inability of a weak intellect to protect the volitional faculty from the solicitations, seductions, and predations of the social and cultural milieu. Feminists have not only elucidated the historical and ideological basis of such figurations, tracing the power relations and institutions conditioned and sustained by the presumption that certain classes of humans are by nature irrational, bound by their bodies and emotions, or vulnerable to the pressures of social forces. They have also asserted the equality of all humans in their capacity to reason and know and revalued the passions, experience, and social wisdom as forms of insight and knowledge.[5] For new materialists, however, it is not enough to assert the rationality of modernity's others, to revalue the passions of the body or phenomenological experience. They seek also to challenge the very notion that matter is passive and unthinking, to undo the opposition between reason and passions, and to question the distinction between self and world that positions individuals as separate from yet in relation to the contexts of their actions (Wilson 1998; Brennan 2004).

For historical materialists, matter is less inert and more plastic than it is for Cartesian substance dualists. Marx suggests in *The German Ideology* that humans have a peculiarly intimate relationship with the material world: their thoughts and their experiences of themselves are formed and transformed by the activities through which they work upon and transform matter for their own or others' purposes (Marx 1978, 150). But this mutually formative relationship between humans and the matter upon which they work is not one in which humans are possessed of complete self mastery. For as Marx notes in his analyses of capital, commodities,

---

[3]For feminist critiques of Descartes's rationalism, see Bordo, *The Flight to Objectivity* (1987); Lloyd, *The Man of Reason* (1984). For new materialist critiques of the way a Cartesian account of matter underwrites modern accounts of the self, see Lloyd and Gatens, *Collective Imagining* (1999) and Frost, *Lessons from a Materialist Thinker* (2008).

[4]See, for example, Wiegman, *American Anatomies* (1995); Mills, *The Racial Contract* (1997); McClintock, *Imperial Leather* (1995); Pateman, *The Sexual Contract* (1988); Macpherson, *The Political Theory of Possessive Individualism* (1962). See also Schiebinger, *Nature's Body* (1993).

[5]Eg. Bryson, 'Mary Astell' (1998); Cornell, *At the Heart of Freedom* (1998); Dalmiya and Alcoff, 'Are "Old Wives' Tales" Justified?'(1993); Archer, *Being Human* (2001).

and the social and political relations that emerge through productive activities, the products of labor become constitutive elements of the economic and political structures that direct, constrain, and compel individuals' behavior. In other words, when it is worked upon and transformed by human labor, matter can be an agent by proxy, absorbing and translating the agency of individuals in ways that exceed each agent's deliberate intentions. The agency of matter, here, is an indirect extension and aggregate effect of the productive activities of the humans who work upon it. Invested with and animated by this agency, matter consolidates the social and political relations that are the historical condition of its productive transformation. Yet, material objects and institutions do not *necessarily* confront humans as alien, constraining, and determinative conditions for human action. As Georg Lukács points out in 'The Standpoint of the Proletariat,' the labor activities demanded by the capitalist production process generate experiences for the proletariat that contradict the governing ideological forms of self-understanding. Through these contradictions, individuals and classes of people can develop critical awareness of the ways in which the agency of matter is actually their agency absorbed and translated into concrete social, political, and economic structures, an awareness that forms the basis of a revolutionary class consciousness (Lukács 1971).

Feminist and critical race theorists found in historical materialism an epistemology that can generate critical standpoints from which to analyze the sexual and racial dimensions of the division of labor. In thus appropriating historical materialism, they have articulated forms of oppositional political subjectivity and challenged the entrenchment of gender, racial, and colonial power relations in the institutions and material practices that structure and organize our lives.[6] Other theorists have wrested the insights of historical materialism from their basis in a critique of political economy and used them to generate a broader constructionist understanding of the creative and constraining force of human activity with respect to matter. Within this broader constructionist view, matter is more completely saturated with power: institutions, objects, and bodies themselves quite literally materialize or incorporate the imperatives that drive power relations. The norms and cultural formations that arise through historical practice not only constrain but also invite us to discipline our behavior, shaping our desires, our physical posture and gestures, and our phenomenological experience of self.[7] In keeping with this shift, feminist epistemologists have not only analyzed the gendered and racialized assumptions implicit in the disavowal of the role or place of embodiment, emotions, and intersubjectivity in the production of knowledge.[8] They have also elucidated the embodied and

[6]Eg. Hartsock, 'The Feminist Standpoint' (1983); Combahee River Collective, 'A Black Feminist Statement' (1983); Collins, 'The Social Construction of Black Feminist Thought' (1989); Mohanty, 'Women Workers and Capitalist Scripts' (1997); Sandoval, *Methodology of the Oppressed* (2000); Harding, *The Feminist Standpoint Theory Reader* (2004).

[7]Eg. Butler, *Bodies That Matter* (1993); Stoler, *Race and the Education of Desire* (1995); Ahmed, *The Cultural Politics of Emotion* (2004); Alcoff, *Visible Identities* (2006).

[8]Eg. Code, *What Can She Know?* (1991); Grosz, *Volatile Bodies* (1994); Damasio, *Descartes' Error* (2000).

socially and historically situated character of epistemological subjects and the constituting powers of language, institutional arrangements, and interpersonal interactions.[9] Further, they have sought to develop and articulate modes of knowing or knowledge production mindful of their own occlusions and elisions.[10]

Constructivism has been tremendously useful for feminist epistemologists in their efforts to denaturalize and politicize knowledge claims that disavow the historicity of empirical facts – that refuse to acknowledge the mediation of perception by language and culture, the identification and demarcation of objects of knowledge through social practice, or the production of knowing subjects through the elaboration of norms and disciplinary procedures. Indeed, as a critical project, constructivism has prompted the exhaustive search for the mark and agency of the social in any knowledge claim, a quest not simply to identify the social, linguistic, or cultural dimensions of perception but also to specify the social and political relations, negotiations, and practices through which both subjects and objects of knowledge come to be constituted as such. From studies of the economic, imperial, and political forces that historically have shaped biological classifications of sex and race, to analyses of the ways in which political and cultural imperatives shape the movements of identification and desire, to explorations of the extent to which social and cultural practices transform bone and flesh, the insights and methods of constructivism have been crucial to feminist challenges to claims that import, encode, and at the same time deny power relations by presenting propositions as true or certain knowledge or as objective or natural fact. But importantly, such insights into the materialization and embodiment of power remain rooted in the historical materialist sense that the agency of matter is derivative of deliberate human activity.

New materialists aim to counter the figuration of matter as an agent only by virtue of its receptivity to human agency. They try to specify and trace the distinctive agency of matter and biology, elucidate the reciprocal imbrication of flesh, culture, and cognition, investigate the porosity of the body in relation to the environment in which it exists, and map the conditions and technologies that shape, constrain, and enhance the possibilities for knowledge and action.[11]

However, in bringing the processes, movements, and activities of biology and matter into their analyses, they must often confront the suspicion that they might be suffering from a political amnesia and intellectual myopia through which the essentialisms of old might reassert themselves. The concerns here are twofold and related.

The first concern is about the political obtuseness of generality. The feminist insight into the implicit normativity of metaphysical categories has resulted in an ever-more refined specification of the various power relations through which

---

[9]Eg. Fausto-Sterling, *Sexing the Body* (2000); Potter, *Gender and Boyle's Law of Gases* (2001).

[10]Eg. Harding, *Whose Science, Whose Knowledge?* (1991); Haraway, *Simians, Cyborgs, and Women* (1991a); Hubbard, *The Politics of Women's Biology* (1990); Longino, *The Fate of Knowledge* (2002); Nelson, *Who Knows* (1990).

[11]For a sampling of such work, see Coole and Frost, *New Materialisms* (2010).

particular forms of matter, materiality, or embodiment are rendered intelligible. Feminists have argued that there is no 'matter' in general, no 'human body' in general, nor even 'women's bodies' in general. Rather, there are particular bodies produced or constituted through a complex interplay of racial and sexual economies of power, language and ideology, historically and geographically contested cultural formations, and psychological identifications and resistances. From this perspective, to talk of matter, biology, or the body in the register of the singular or general is to occlude these manifold and historically specific constituents of objects and embodiment, to obscure or even perpetuate the power relations that both make possible and produce facts, things, and subjectivities.

The second concern is about the ineluctable mediation of perception and knowledge by language, culture, and power. The worry is that in their efforts to consider the peculiar agency of organic or inorganic matter, new materialists might, wittingly or unwittingly, read linguistic, cultural, or political facts and meanings into the material – that they might misrepresent *as* biological, physiological or natural what is actually social and historical. And of course, such misrepresentations would be problematic because they would naturalize social and political artifacts, which is to say that they would essentialize gender and race.[12]

These concerns about power and essentialism are extremely important. But as Susan Oyama points out, incredulity toward the real and anxiety about essentialism are part of the legacy of Cartesian dualism and they sometimes function as 'traps' that shut off paths of intellectual inquiry.[13] Indeed, Elizabeth Wilson claims that the 'compulsive antiessentialism' that underlies such criticisms demands that new materialists acknowledge the inevitable power and pervasiveness of culture or discourse in their very effort to consider what might condition culture or discourse (Wilson 1998, 1).[14] It is possible, however, to recast or re-examine these concerns from within a different framework. Feminist scientists and historians of science have done a marvelous job breaking down the modern binary of nature and culture by showing how the natural environment or aspects of biological processes and behavior are shaped by the social and cultural. Non-scientific feminists, however, have been wary of if not downright resistant to reconsidering biology or materiality as anything but discursive formations, as historically specific products of power relations, linguistic practices, and cultural beliefs. As Lynda Birke notes provocatively, even scholars who critically engage the sciences of genetics or reproduction

[12]For a fascinating on-going discussion about the dangers of racial and sexual essentialism in genetic sciences, see the contributions to the on-going Social Science Research Council forum 'Is Race "Real"?'.

[13]As Oyama observes critically, within such a framework, 'if one voices skepticism about some "biological" interpretation, then, one is assumed to be an environmental determinist, and vice versa. This assumption is a trap, and it is better to dismantle traps than to step into them (or, for that matter, to set them for others)' (Oyama, *Evolution's Eye*, 2000b: 154).

[14]In an ironic formulation, Wilson claims that 'compulsive antiessentialism' is a 'disciplining compulsion' that has 'been naturalized not simply as good critical practice, but as the sine qua non of criticism itself' (Wilson 1998, 1–2).

can forget 'the "meat"' that 'is busily reconstructing itself' while they elaborate their discursive analyses (Birke 2000, 145). To put the point differently, feminists have been more comfortable with denaturalizing nature than with what we might call 'deculturalizing culture' – or admitting that matter or biology might have a form of agency or force that shapes, enhances, conditions, or delimits the agency of culture. Yet, this wary reluctance, understandable as it is given historical precedent, is structured by an understanding of causation that binds feminists to the binaries they have otherwise been deconstructing.

First, as noted in the discussion above, feminist epistemologists in the West have generally aligned themselves with arguments that any social or political significance attributed to bodily differences is a social and political construct. They proceed with the sense, if not a consensus, that biological organisms in themselves can be objects of theoretical or epistemological indifference because biology has no political entailments. However, their evident sense of the danger involved in the effort to explore, identify, or specify how different aspects of biology might shape behavior reveals an implicit concern that sexual or racial differences, if specified, might in fact entail particular social policies or political relations. Fueling this concern is the assumption that causation can only be unilinear and unidirectional: either the one or the other, biology or culture, is the causal agent in social phenomena. Within such a causal framework, the argument about the social constructedness of race and gender can be maintained only if the social is granted complete immunity to the biological. Conversely, to acknowledge that biology might have some agency would entail giving up the claim to construction – which is political ground that simply cannot be ceded. It is as if implicit in the fear of essentialism is the worry that were we to release biology from the conceptual confines of the role of absorbing – and perhaps transmitting – cultural mandates, it would hijack the causal arrow, run rampant in social and political institutions and practices, and effectively steal from us our rational agency and our capacity for individual and collective self-determination. In other words, underwriting the concern about 'the risk of essentialism' is, paradoxically, the presumption that a material, biological agency would override and overwhelm the effects of culture and politics and would end up being the determinative force in our lives no matter what kinds of efforts we might undertake to make it otherwise (Oyama, *Evolution's Eye* 2000b, 164–165).

Second, and related, when the determinist dangers associated with claims about the possible agency of the biological propel feminists away from biology to focus on the cultural, the linguistic, or the discursive formation of embodiment, the subsequent focus on construction reinstates the modern terms of subject-formation as an exercise in self-creation. To be sure, the creation at stake here is conceived as social, cultural, and/or political in character. But each of these forms of construction or constitution recenters the human as the definitive agent of order, meaning, and action (Smith and Jenks 2005, 147). In turning to culture to evade the determinism implicitly associated with the biological body, feminists recapitulate the modern fantasy of freedom, autonomy, and self-determination that they have otherwise so carefully dismantled. That is, the concern about unwitting essentialism is bound by the terms of Cartesian dualism that put rationality, freedom, and agency on one side of an ontological divide and matter, passivity, and determinism on the other.

In their quest to unravel the ubiquitous threads of Cartesian dualism by reconceptualizing matter and embodiment, the scholars and theorists of the new materialisms concur with many of the insights about science, philosophy, and politics shared by feminist epistemologists of various ilks. However, they also present feminist epistemologists with some suggestive points of departure for rethinking their models of causation as they continue their critical and reconstructive work.

In their effort to denaturalize nature *and* deculturalize culture, new materialists push feminists to decenter human intentionality and design in the conceptualization of the relationship between nature and culture. In tracing the dynamic interactive processes that constitute objects and organisms as at once '100% nature and 100% nurture' (Fausto-Sterling 2005, 1510), they insist that we attend to both the agency of the human or cultural upon the biological or natural and the agency of the natural or biological upon the human or cultural. Indeed, to pose the issue in just such a way – as if there are two agencies that are distinct from one another and that interact in relation while maintaining their integrity as distinct entities – does not quite capture the reciprocally transformative nature of the relationship.[15] Susan Oyama contends that neither biology nor culture operates as a pure unfettered force. To the contrary, there is a 'stunning array of processes, entities, and environments – chemical and mechanical, micro- and macroscopic, social and geological' that shape and are shaped by biological constitution and social behavior. If we attend to the 'interdependence of organism and environment,' then we can elucidate the ways in which 'organisms and their environments define the relevant aspects of, and can affect, each other' (Oyama *Evolution's Eye*, 2000b, 3). Making a similar point in her reconceptualization of the interactions between biology and culture involved in evolution, Elizabeth Grosz argues that 'biology does not limit social, political, and personal life: it not only makes them possible, it ensures that they endlessly transform themselves and thus stimulate biology into further transformations. The natural world prefigures, contains, and opens up social and cultural existence to endless becoming; in turn, cultural transformation provides further impetus for biological becoming' (Grosz 2004, 1–2). As Grosz suggests, then, to admit into our analyses the ways in which biology prefigures culture is neither to delimit nor to predetermine cultural possibilities. To the contrary, she claims that the patterns of adaptation, innovation, and diversification that are the hallmark of evolution undermine the commonplace that nature is a constraint, a hindrance, or an obstacle to cultural creativity: 'Nature is open to any kind of culture, to any kind of "artificiality," for culture itself does not find pre-given biological resources, but makes them for its own needs, as does nature itself' (Grosz 2004, 72). The key insight in work by the likes of Fausto-Sterling, Oyama, and Grosz is that biology and culture, organisms and contexts, are co-emergent; they provoke, challenge, and consequently shape one another.

---

[15] Karen Barad suggests the term 'agential intra-action' to capture such a relational ontology. See Barad, 'Posthumanist Performativity' (2003), 814.

Clearly, to conceive of causation in singular, linear, and unidirectional terms is to elide the mutual and on-going transfigurations, the serendipitous, surprising, and sometimes anomalous developments that emerge through the kinds of interactions highlighted by these new materialists. The dynamic interactive processes that constitute organisms, objects, and environments require feminists to develop a theoretical vocabulary for talking about the complexity of causation. If we are to do justice to the ways in which objects, organisms, and cultural forms and practices emerge and transform through relationships that develop and reconfigure themselves over time, we must adjust the terms we use to capture or represent the multiplicity, the recursivity, and the varied temporality of causes and effects.

The complexity of causal processes brought to the fore by new materialists also brings into more emphatic focus the interdependencies that define the contexts in which both objects and knowers exist. In fact, the innumerable networks of interdependencies that constitute and shape the interactions between subjects and objects suggests that, methodologically, feminists must think ecologically not only about objects of knowledge but also about individual knowers and their epistemological communities. To think ecologically is not to simply note the broad context as a background against which objects exist or are known nor is it to delineate the forms of cultural and political embeddedness that shape and constrain what a subject may know. Rather, as Jane Bennett explains, 'to call something ecological is to draw attention to its necessary implication in a network of relations, to mark its persistent tendency to enter into a working system' that is 'more or less mobile, more or less transient, more or less conflictual' (Bennett 2004, 365). It is to emphasize that 'humans are always in composition with nonhumanity, never outside of a sticky web of connections' (Bennett 2004, 365). Or as Lorraine Code suggests, to think ecologically is to think in terms of 'diverse, complex, multiply interconnected *milieux*' when we conceive of epistemological subjects, to consider not simply 'individuals' or 'communities' but rather the dynamic interrelationships between subjects, objects, and habitats and the transformative effects of those relationships upon subjects, objects and the successively larger eco-systems within which they exist and interact (Code 2006, 27).[16]

And finally, the shift towards thinking in terms of complex causation and interdependencies brings into focus a form of ignorance or a limit to knowledge that challenges the aspiration towards cognitive and practical mastery over the world. As suggested above, for new materialists, objects always exist in dynamic 'assemblages' and connections that affect what they are and how they behave. Accordingly, it does not make sense to conceive of an object as a bounded and distinct thing – as if it existed in isolation from other objects and humans (Bennett 2004, 365).[17] Indeed, in their admission of the agencies and interdependencies of

[16]For similar efforts, see also Braidotti, 'Feminist Epistemology after Postmodernism' (2007); Grasswick, 'Individuals-in-communities' (2004).

[17]Bennett notes that in her interest in the agency of material objects, she considers not 'the thing as it stands alone, but rather the not-fully-humanized dimensions of a thing as it manifests itself amidst other entities and forces' (Bennett 2004, 366).

matter and organisms, new materialists find themselves confronted by an important kind of epistemological impossibility: the impossibility of complete and predictive knowledge of complex causal processes. Because the complex causal relations at issue here are multi-directional and recursive, the manifold interacting elements of an open system can '"spontaneously" develop collective properties or patterns... that do not seem implicit...within the individual components'(Urry 2005, 5). According to John Urry, '[s]uch emergent characteristics emerge from, but are not reducible to, the micro-dynamics of the phenomenon in question' (Urry 2005, 5). The irreducibility of such complex interactions demands, as Monica Greco suggests, that 'we acknowledge, and learn to value as the source of qualitatively new questions, the possibility of a form of ignorance that cannot simply be deferred to future knowledge' (Greco 2005, 24). That is to say, we must learn to incorporate the possibility of an impossibility of knowing into our epistemologies that is not indexed to the limits of perception or to the development of technology but rather intrinsic to the complexity of objects or processes themselves.

What is at issue in this impossibility is not the partiality of perspective that is so central to the various iterations of standpoint theory – although, as Donna Haraway has pointed out, the recognition of such partiality is both a useful prompt to political humility in the face of diversity and a goad to coalition building (Haraway 1991b). Nor is it the politically productive forms of ignorance that are implicated in the orders of knowledge through which disciplinary political powers are elaborated and made effective – although feminist philosophers and theorists are beginning to do fascinating work on the political work accomplished by regimes of ignorance.[18] Rather, what is at stake in thinking in terms of complexity, interdependence, and ecology broadly construed is epistemological and political humility in the face of the organic and inorganic world: an acknowledgment of the impossibility of full and definitive knowledge and a corollary surrender of the teleological assumption that we might possibly, at some future point, achieve full mastery over ourselves and the world around us. Of course, to acknowledge a zone of necessary ignorance in complexity is not tantamount to giving up on knowledge altogether: we do not need the promise of full knowledge as the backdrop for scientific investigations. As Karen Barad points out, in our investigations into how things work, we perform 'agential cuts' that effect a separation between objects and subjects existing in interdependence and that thereby constitute each as having distinct, determinate boundaries (Barad 2003, 815).[19] Any explanation or generalization entails a simplification of the causal field: some simplifications are pragmatically useful in that they are replicable and can be instrumentalized to further our purposes. But even tried and true and seemingly innocuous replications and instrumentalizations

---

[18]See Tuana and Sullivan, *Feminist Epistemologies of Ignorance* (2006). See also Sullivan and Tuana, *Race and Epistemologies of Ignorance* (2007).

[19]Diana Coole offers a similar account of the ways in which human agents carve a space of 'agency' out of a complex array of other organic and inorganic agents. See Coole, 'Rethinking Agency' (2005).

sometimes issue in surprises and unexpected consequences. The key, then, is to remember that we have produced rather than found distinct objects, that we have artificially reduced complexity and not mastered it. As Jane Bennett muses, perhaps recognition of the complexities of the objects and interactions in any given ecology will provoke 'a more cautious, intelligent approach to our interventions in that ecology' (Bennett 2004, 349).

If feminists can figure out theoretically how to acknowledge the manifold recursive interactions through which nature and culture develop and evolve, if they can learn to account for the dynamism, the temporal breadth, the spatial breadth – the complexity – of organic and inorganic materiality, in short, if they can rethink the terms of causation, they may find they have the conceptual tools to engage and criticize essentialism. To acknowledge complexity in causation requires a shift from thinking about essentialism in terms of misattribution ('you're describing the cause incorrectly') to thinking in terms of reductionism ('you're ascribing causes too narrowly'). Whereas the framework of misattribution makes us wary of misrepresenting actions or effects as caused by one kind of cause rather than another, the framework of reductionism makes us wary of over-simplification.[20] In both instances, we can tackle essentialism. But the first strategy conceives of essentialism as a malicious or ignorant misdesignation of a cause that demands the revelation of misrepresentation and (in historical practice) an accompanying disavowal of any relationship between biology and culture. In contrast, the second strategy conceives of essentialism as a reduction of many causes to a single linear one. The response demanded by latter formulation is the rejection of the simplification and the specification and elaboration of the complex, creative, and sometimes surprising interplay between biology and culture. This may be a slower and more difficult task, but it is also surely one to which the critical and political skills feminists have developed are particularly well-suited.

References

Ahmed, Sara. 2004. *The cultural politics of emotion*. New York: Routledge.

Alcoff, Linda. 2006. *Visible identities: Race, gender and the self*. New York: Oxford University Press.

Archer, Margaret. 2001. *Being human: The problem of agency*. Cambridge: Cambridge University Press.

[20] For instance, Elizabeth Grosz asserts that contrary to the seeming wishes of some western feminists, sexual difference is real and not simply a social construction. Yet, this assertion is not an essentialism in the way that feminists have habitually and suspiciously been wont to claim. As Grosz explains, 'There is an irreducible difference between the sexes, and this difference is not only irreducible to one of its terms, in the case of sociobiology, its reproductive cells; it is also irreducible to any other level, whether cellular, morphological, cultural, or historical. Sexual difference is irreducible difference, yet it is not a measurable, definable difference between given entities with their own characteristics but an incalculable difference that reveals itself only through its temporal elaboration' (Grosz 2004, 67).

Barad, Karen. 2003. Posthumanist performativity: Toward an understanding of how matter comes to matter. *SIGNS: Journal of Women in Culture and Society* 28(3): 801–831.

Barad, Karen. 2007. *Meeting the universe halfway: Quantum physics and the entanglement of matter and meaning*. Durham: Duke University Press.

Bennett, Jane. 2004. The force of things: Steps toward an ecology of matter. *Political Theory* 32(3): 347–372.

Birke, Lynda. 2000. *Feminism and the biological body*. New Brunswick: Rutgers University Press.

Bordo, Susan. 1987. *The flight to objectivity: Essays on cartesianism and culture*. Albany: SUNY.

Braidotti, Rosi. 2002. *Metamorphoses: Towards a materialist theory of becoming*. Cambridge: Polity.

Braidotti, Rosi. 2007. Feminist epistemology after postmodernism: Critiquing science, technology, and globalization. *Interdisciplinary Science Reviews* 32(1): 65–74.

Brennan, Teresa. 2004. *The transmission of affect*. Ithaca: Cornell University Press.

Bryson, Cynthia. 1998. Mary Astell: Defender of the "disembodied mind". *Hypatia* 13(4): 40–62.

Butler, Judith. 1993. *Bodies that matter: On the discursive limits of 'sex'*. New York: Routledge.

Capra, Fritjof. 1996. *The web of life: A new scientific understanding of living systems*. New York: Anchor Books.

Code, Lorraine. 1991. *What can she know? Feminist theory and the construction of knowledge*. Ithaca: Cornell University Press.

Code, Lorraine. 2006. *Ecological thinking: Thinking the politics of epistemic location*. Oxford: Oxford University Press.

Collins, Patricia Hill. 1989. The social construction of black feminist thought. *SIGNS: Journal of Women in Culture and Society* 14(4): 745–773.

Combahee River Collective. 1983. A black feminist statement. In *This bridge called my back: Writings by radical women of color*, ed. Cherrie Moraga and Gloria Anzaldua, 210–218. New York: Kitchen Table Press.

Coole, Diana. 2005. Rethinking agency: A phenomenological approach to embodiment and agentic capacities. *Political Studies* 53(1): 124–142.

Coole, Diana, and Samantha Frost (eds.). 2010. *New materialisms: Ontology, agency, and politics*. Durham: Duke University Press.

Cornell, Drucilla. 1998. *At the heart of freedom: Feminism, sex, and equality*. Princeton: Princeton University Press.

Dalmiya, Vrinda, and Linda Alcoff. 1993. Are "Old wives' tales" justified? In *Feminist epistemologies*, ed. Linda Alcoff and Elizabeth Potter, 217–244. New York: Routledge.

Damasio, Antonio. 2000. *Descartes' error: Emotion, reason, and the human brain*. New York: Quill.

Descartes René. 1985. *Philosophical Writings*, vols. 1–2, trans. and ed. John Cottingham, Robert Stoothoff, and Dugald Murdoch. New York: Cambridge University Press (1985 (vol. 1), 1984 (vol. 2)).

Fausto-Sterling, Anne. 2000. *Sexing the body: Gender politics and the construction of sexuality*. New York: Basic Books.

Fausto-Sterling, Anne. 2005. "The bare bones of sex": Part 1 – Sex and gender. *SIGNS* 30(2): 1491–1528.

Frost, Samantha. 2008. *Lessons from a materialist thinker: Hobbesian reflections on ethics and politics*. Palo Alto: Stanford.

Grasswick, Heidi. 2004. Individuals-in-communities: The search for a feminist model of epistemic subjects. *Hypatia* 19(3): 85–120.

Greco, Monica. 2005. On the vitality of vitalism. *Theory, Culture and Society* 22(1): 15–27.

Grosz, Elizabeth. 1994. *Volatile bodies: Toward a corporeal feminism*. Bloomington: Indiana University Press.

Grosz, Elizabeth. 2004. *The nick of time: Politics, evolution, and the untimely*. Durham: Duke University Press.

Haraway, Donna. 1991a. *Simians, cyborgs, and women: The reinvention of nature*. New York: Routledge.

Haraway, Donna. 1991b. Situated knowledges: The science question in feminism and the privilege of partial perspective. In *Simians, cyborgs, and women: The reinvention of nature*. New York: Routledge.

Haraway, Donna. 2008. *When species meet*. Minneapolis: University of Minnesota Press.

Harding, Sandra. 1991. *Whose science, whose knowledge? Thinking from women's lives*. Ithaca: Cornell University Press.

Harding, Sandra (ed.). 2004. *The feminist standpoint theory reader: Intellectual and political controversies*. New York: Routledge.

Hartsock, Nancy. 1983. The feminist standpoint: Developing the ground for a specifically feminist historical materialism. In *Discovering reality: Feminist perspectives on epistemology, metaphysics, methodology, and the philosophy of science*, ed. Sandra Harding and Merrill B. Hintikka. Dordrecht: Reidel.

Hayles, N.Katherine. 2005. Computing the human. *Theory, Culture and Society* 22(1): 131–151.

Hubbard, Ruth. 1990. *The politics of women's biology*. New Brunswick: Rutgers University Press.

Kirby, Vicki. 1997. *Telling flesh: The substance of the corporeal*. New York: Routledge.

Lloyd, Genevieve. 1984. *The man of reason: 'Male' and 'female' in Western philosophy*. London: Routledge.

Lloyd, Genevieve, and Moira Gatens. 1999. *Collective imagining: Spinoza, past and present*. New York: Routledge.

Longino, Helen. 2002. *The fate of knowledge*. Princeton: Princeton University Press.

Lukács, Georg. 1971. The standpoint of the proletariat. In *History and class consciousness: Studies in Marxist dialectics*. London: Merlin Press.

Macpherson, C.B. 1962. *The political theory of possessive individualism: Hobbes to Locke*. New York: Oxford University Press.

Marx, Karl. 1978. The German ideology: Part I. In *The Marx-Engels reader*, 2nd ed, ed. Robert C. Tucker. New York: Norton.

Masters, Roger. 2001. Biology and politics: Linking nature and nurture. *Annual Review of Political Science* 4: 345–369.

McClintock, Anne. 1995. *Imperial leather: Race, gender and sexuality in the colonial contest*. New York: Routledge.

Mills, Charles. 1997. *The racial contract*. Ithaca: Cornell University Press.

Mohanty, Chandra Talpade. 1997. Women workers and capitalist scripts: Ideologies of domination, common interests, and the politics of solidarity. In *Feminist genealogies, colonial legacies, democratic futures*, ed. M. Jacqui Alexander and Chandra Talpade Mohanty, 3–29. New York: Routledge.

Nelson, Lynn Hankinson. 1990. *Who knows: From Quine to a feminist empiricism*. Philadelphia: Temple University Press.

Oyama, Susan. 2000a. *The ontogeny of information: Developmental systems and biology*. Durham: Duke University Press.

Oyama, Susan. 2000b. *Evolution's eye: A system's view of the biology-culture divide*. Durham: Duke University Press.

Pateman, Carole. 1988. *The sexual contract*. Palo Alto: Stanford.

Potter, Elizabeth. 2001. *Gender and Boyle's law of gases*. Bloomington: Indiana University Press.

Sandoval, Chela. 2000. *Methodology of the oppressed*. Minneapolis: University of Minnesota Press.

Schiebinger, Londa. 1993. *Nature's body: Gender in the making of modern science*. Boston: Beacon.

Smith, John, and Chris Jenks. 2005. Complexity, ecology, and the materiality of information. *Theory, Culture and Society* 22(5): 141–163.

Social Science Research Council. Is race "real"? A web forum organized by the Social Science Research Council. http://raceandgenomics.ssrc.org.

Stoler, Ann. 1995. *Race and the education of desire: Foucault's history of sexuality and the colonial order of things*. Durham: Duke University Press.

Sullivan, Shannon, and Nancy Tuana (eds.). 2007. *Race and epistemologies of ignorance*. Albany: SUNY.

Tuana, Nancy and Shannon Sullivan, eds. 2006. *Feminist epistemologies of ignorance*. A special issue of *Hypatia: A Journal of Feminist Philosophy* 21(3): 1–236.

Urry, John. 2005. The complexity turn. *Theory, Culture and Society* 22(5): 1–14.

Wiegman, Robyn. 1995. *American anatomies: Theorizing race and gender*. Durham: Duke University Press.

Wilson, Elizabeth. 1998. *Neural geographies: Feminism and the microstructure of cognition*. New York: Routledge.

Young, Iris. 2005. *On female body experience: Throwing like a girl and other essays*. New York: Oxford University Press.

Chapter 5
Interrogating the Modernity vs. Tradition Contrast: Whose Science and Technology for Whose Social Progress?

Sandra Harding

Abstract Western feminist philosophies of science have pursued two critical directions in reevaluating the familiar contrast of modern with traditional societies. They have argued that women have made important contributions to pre-modern empirical knowledge of nature and social relations, and they have identified how modern ideals of scientific rationality, objectivity, and good method are shaped by a familiar stereotype of manliness. Yet there is a third strategy which is worth pursuing: interrogating the way gender stereotypes constitute the very project of the contrast between modernity and tradition.

In this third project, one can see that modern sciences and technologies become the 'motor' for transporting men from loyalties to their traditional social worlds into commitments to modernity and its projects, and for extracting economic and political systems from the (now labeled) private sphere of the household into the public sphere. Men and social institutions must be freed from women's worlds. Yet from the standpoint of women's experiences of modernity one can see that men's modern worlds are assigned little or no responsibility for the flourishing of women's supposedly traditional worlds. Thus modern sciences and technologies function as the one-way-only time-travel machines for externalizing men from women's household responsibilities and excluding women from full participation in the direction and management of economic and political projects. The discussion here concludes with one suggestion – admittedly, a provocation – for research projects which can produce resources for blocking modernity's commitments to male- and Western-supremacy. Such projects should start out conceptualizing their research designs – all of them – from the lives of women in households. Unreasonable as this may sound, it can be well-supported.

Keywords Gender • Science • Modernity • Philosophy of science • Postcoloniality

---

S. Harding (✉)
Graduate School of Education and Information Studies, University of California, Los Angeles, CA 90095, USA
e-mail: sharding@gseis.ucla.edu

H.E. Grasswick (ed.), *Feminist Epistemology and Philosophy of Science: Power in Knowledge*, DOI 10.1007/978-1-4020-6835-5_5,

Western feminist philosophies of science have pursued two critical directions in reevaluating the movement from tradition to modernity in which modern sciences and technologies have played such a central role. Alongside other feminist science studies, they have argued that women have made important contributions to what is thought of as pre-modern empirical knowledge of nature and social relations. Additionally, they have identified how modern ideals of scientific rationality and objectivity are shaped by a familiar stereotype of manliness. Yet there is a third strategy worth pursuing. This one centers women's experiences of Western modernity and interrogates the way gender stereotypes constitute the very project of the contrast between tradition and modernity.[1]

From the standpoint of women's experiences one can see that men's modern worlds have no responsibility for the flourishing of women's supposedly traditional worlds. The connections men feel with these women's worlds are regarded as obstacles to their personal achievement of autonomy and rationality and to their society's attainment of modernity. Modern sciences and technologies are routinely claimed to be the desirable motor – the one-way only 'time travel machine' – of this externalization of men from women's lives and from households, and of women from the direction and management of public sphere institutions. At the same time, these narratives obscure how men's supposedly individual achievements and those of modernity in general always remain dependent upon women's activities of maintaining responsibility for children, households, kin relations, and the flourishing of the communities and natural environments upon which such activities and social relations depend.

Of course much of this argument has long been made by feminist critics of the public vs. private spheres of Liberal democratic theory. What is different here is the exploration of how this kind of argument can and must be directed to the modernity vs. tradition contrast itself, and to the roles philosophies of modern sciences and technologies play in maintaining this contrast. Philosophies of modern sciences and technologies are much more pointedly implicated in male-supremacist and Western-supremacist theories and practices of modern social life in this argument. One must take note, however, that it has only recently become possible to make this kind of argument. This third kind of feminist critical study of tradition and modernity could not emerge until postcolonial critics showed that modernization is not identical to Westernization. That is, there are multiple forms of modernization, and each can occur only through the incorporation of local traditional cultural features. Thus it becomes possible to begin to see how modernity consistently appropriates and internalizes selective features of local traditions while simultaneously both disvaluing tradition and depending upon it for its own successes.

[1]This essay was written in 2006. It draws on themes in *Sciences From Below: Feminisms, Postcolonialities and Modernities*. (Durham: Duke University Press, 2008), especially in Chaps. 8 and 9.

For helpful comments on earlier drafts I thank Heidi Grasswick, Francoise Lionnet and Sharon Traweek, and audiences at the UCLA Center for the Study of Women (April 2007) and the Western Political Science Association (March 2007).

Thus it becomes clear that prevailing Western modern notions of social progress are even more deeply implicated in androcentric and Eurocentric projects than had been imagined.

Here I focus on the emergence of this third kind of critical perspective on modernity's ideals and practices, and on one strategy for methodological approaches to research which can enable moving forward against and past male-supremacist and West-supremacist notions of modernity, tradition, and social progress.

5.1 Three Feminist Critical Approaches to Traditional and Modern Scientific Knowledge

On the one hand, we Western feminists have pointed to women's active development and preservation of health, medical, and environmental knowledge in traditional societies. Women seem to play a more important role in creating and disseminating indigenous knowledge than they do in producing modern scientific and technological knowledge. It is often, or even usually, women who develop and preserve local pharmacological knowledge, health practices and medical treatments. (Appleton et al. 1995) Women developed effective gathering, agricultural and cooking practices, as well as weaving, pottery making and all of the associated technologies required for the manufacture of goods needed for daily life. This makes sense once one recollects that in many – perhaps most – societies it is women who are assigned responsibility for the daily provisioning of children, other household members and dependent kin, and for daily maintenance of their health. These kinds of traditional knowledge and traditional environmental knowledge are also important forms of empirical knowledge, we have argued. One early development of standpoint epistemology specifically grounded its arguments in the reliability of women's (culturally mediated) experience of our own bodies – our traditional or 'indigenous' knowledge of our bodily processes, one could say. This reliability (but not, of course, infallibility) contrasted with that of the dominant medical and health sciences which had been created by male professionals who lacked such experience and the knowledge it could generate. And, of course, it is not just the health of women's own bodies for which women are assigned the daily responsibility. Traces of this insight can be found in Hartsock's and Smith's accounts also (Hartsock 1983; Rose 1983; Smith 1987, cf. also Martin 1987; Yoon 1995). So, one could say, this strategy has sought to justify women's traditional knowledge as a kind of real science. In doing so, it contributes to the ongoing process of expanding what gets to count as the production of scientific knowledge (Beck 1997; Nowotny et al. 2001).

Yet this strategy has not been convincing to those who take modern sciences such as physics, chemistry, engineering and biology as different in kind from traditional knowledge, and who think that very difference is what certifies the promise of modern sciences to deliver social progress. Is it different in kind? Can such modern sciences deliver social progress to women and their dependents today? Does traditional knowledge make any important contributions to human flourishing today?

A second feminist critical strategy identified historic fingerprints on the purportedly trans-cultural standards of objectivity and rationality for which modern science was to provide the most perfect exemplar. Rationality and objectivity have persistently been associated with masculinity, even as standards for these concepts have themselves shifted over time (Bordo 1987; Jaggar 1989; Keller 1984; Lloyd 1984; MacKinnon 1982). This feminist strategy could also draw on the valuable critical resources of poststructuralism which delineated additional features of modernity's delusions of historical and cultural transcendance (Flax 1990; Haraway 1991). Moreover, a few Western feminists and many from the Third World have participated in postcolonial projects of identifying distinctively Eurocentric fingerprints on such standards (Braidotti et al. 1994; Haraway 1991; Harding 1998, 2006, 2008; Shiva 1989, 1993, 1994). So one could conclude that modern scientific work per se is not culturally neutral and could not in principle attain such neutrality as long as it retains distinctively androcentric and Eurocentric conceptions of rationality and objectivity. Yet that neutrality and the methodological practices designed to attain it were supposed to constitute the significant differences between modern and traditional knowledge projects.

This second strategy leaves many philosophers and most scientists cold, since they cannot see how such 'associations,' interpretations, or meanings (for example, of the masculinity of modern Western rationality), regretable as they might be, have effects on contemporary research methods or results. They cannot – or refuse to – recognize that such meanings shape scientific practices. Such associations affect how scientists conceptualize and interact with nature's order. To the philosophers and scientists scientific work can still retain its trans-cultural value-neutrality regardless of these cultural meanings or interpretations of it. What is scientific about modern sciences does not include whatever meanings or interpretations people in one cultural context or another assign to scientific methods, standards, or the facts scientific research produces. Culture is conventionally conceptualized as an obstacle over which scientific method, its standards and production of facts must triumph. It is precisely these methods, standards, facts, and the prediction and control of nature's regularities that they enable that certifies scientificity. The regularities of nature's order will have their effects on us regardless if one is a woman or man, Hindu or Muslim, they argue.

Such conventional critics ignore how some four decades of sociology, history, and ethnography of modern sciences have again and again identified precisely how cultural meanings of nature and scientific practice have shaped what and how Western sciences come to know (Biagioli 1999; Hackett et al. 2008). Yet most of these science studies have tended to a very limited conception of the social, avoiding the most controversial aspects of Western social relations in which modern sciences have been deeply implicated, such as gender, empire, colonialism, and the systematic abandonment of concern for whatever modernity defines as traditional worlds. Such issues have been left for the disvalued fringes of science studies, represented by feminist and postcolonial accounts, and by critical accounts of the alarmingly tight fit between the projects of modern sciences on the one hand, and of national security and capitalist expansion on the other.[2]

Yet there is a third strategy which feminist science studies and its philosophic projects have left undone. This is critical exposure of the gendered modernity of the modernity vs. tradition binary itself – not just the masculinity of standards of objectivity and rationality, but also the femininization of tradition. It is not just that women make important contributions to traditional knowledge and that modern rationality and objectivity are associated with models of masculinity. In this third case the focus is on how tradition and its knowledge projects vs. modernity and its knowledge projects are in modernity's narratives conceptualized from the start in terms of gender stereotypes. Moreover, this contrast is also 'orientalized' or 'colonialized' from the start, since modernity is associated exclusively with projects originating in Europe. Thus gender and colonial stereotypes constitute the content and moral desirability of modernity and the disvalue of tradition. Modernity is not a free-standing 'thing' which can be understood or explained by examining it alone. Rather it is always half of a relationship. Like masculinity's relation to femininity (Flax 1990), modernity always defines itself as not its other – 'not tradition.' Modernity is obsessively preoccupied with this contrast; the feminine and the primitive always appear in modernity's narratives as the negatives to modernity's positives. Any traces of tradition which remain in modern societies, such as discrimination against women, for example, are regarded as unfortunate residues which will wither away as modernity more thoroughly disseminates throughout social relations.

This kind of interrogation of tradition and its relation to modernity could not occur until postcolonial scholars had demonstrated that the West did not and will not ever have a corner on modernity. Modernization is not the same as Westernization. That is, an interrogation of tradition to match the feminist and poststructuralist interrogation of modernity becomes possible only with the recognition that many other societies around the world have developed their own forms of modernity.[3] They have done so self-consciously *within* cultural beliefs and practices that the modern West regards as traditional. Thus, for example, overtly Hindu and Islamic *modern* sciences have been developed (Prakash 1999; Sardar 1997). Moreover, such studies then also reveal the many ways in which modernity reproduces tradition and depends upon it for its own successes, contrary to modernity's claims that tradition is always only an obstacle to its successes, that modernity is completely incompatible with tradition.

---

[2]For the latter see, for example Hessen (1970) on the fit of Newtonian projects with the economic needs of the emerging modern European society, Forman (1987) on how U.S. physics was shaped by national security needs in the World War II era, and Mirowski (2005) on the fit between three generations of U.S. philosophy of science (Dewey, Reichenbach, and Kitcher) and dominant political/economic projects of the U.S. in global politics.

[3]For such primarily Western recognitions of this phenomenon see Eisenstadt (2000), Giddens (1994).

5.2 Obscured Paternal Narratives of Modernization

Modernization theory originates in the attempts of such Nineteenth Century social theorists as Weber, Marx, and Durkheim to understand the industrialization and urbanization which were then sweeping across Europe. It was reinvigorated by post-World War II sociologists, such as Alex Inkeles and W.W. Rostow, who were concerned to understand and justify the economic, political and social changes occurring as the West's economic resources were being redirected from war efforts to social transformations within Cold War dynamics. Its most powerful critics in the West have been Marxian- inspired world systems and 'dependencia' theorists, such as Andre Gunder Frank, Celso Furtado and Immanuel Wallerstein (Frank 1969; Furtado 1970; Wallerstein 1974). Though, as we will see, these critics could not identify the gender dynamics of modernization theory, nor could they avoid reproducing such dynamics in their own work.[4]

Four themes in modernization narratives show how they are grounded in gender stereotypes, and how the benefits of Western sciences and technologies deliver models of masculinity which are male-supremacist. Then the following section turns to identify some silences and gaps in logic exhibited in these representations of scientific rationality and technical expertise.

5.2.1 *Men Separate from the Past*

One theme is that if men would become active agents of their own histories and of social progress in general, they must escape the pull of women and everything associated with them. Modernity's normative power is always carried by the struggle to separate from the past – from childhood, nature, the premodern, tradition, the emotions which attach one to childhood, women, kin and the past, and whatever else is metaphorically conceptualized as part of women's worlds (Felski 1995, 13). That is, it is precisely this struggle which marks modernity as progressive (Jardine 1985, 33).

5.2.2 *Creation of Exclusively Masculine Public Sphere of Economics and Politics*

A second theme is that political and economic activities must be removed from the private realm of households and exercised solely in the public sphere, which is thereby defined as men's space (Felski 1995; Jardine 1985; Scott 1995). Indeed, this

[4]Their work predated all but the very earliest stirrings of feminist criticisms of modernization theory.

disarticulation of social institutions is always cited as a significant mark of modernity and its social process (Eisenstadt 2000). Yet for women, it is precisely their exclusion from these now public activities which insures their lesser social status and power (Kelly Gadol 1976). Hence the continuing importance of women's efforts to gain access to public institutions and, especially, their governing positions. And one can understand, too, why families then are the sites of greatest resistance to external economic, political, and social intrusions and invasions, whether these be slave families in the American south, or African families during European imperialism and colonialism (Davis 1971; Caulfield 1974). In pre-modern societies, it is kinship networks centered in households and tribes through which economic and political relations are organized. Women are central to the creation and maintenance of kinship relations. Moreover, when economic and political activities are located at the sites where women are, women have greater participation in them (Kelly Gadol 1976).

Thus women are much more excluded from participation in the management of economic and political relations in modern than in most traditional social relations. 'Women's worlds' of household and kin responsibilities from which the organization and management of social and political relations have been extracted are not traditional at all; they are an intentional creation of modernization projects.[5] This is not overtly themed in modernization's own accounts, but becomes visible when modernization theory and practice is examined from the standpoint of women's experiences of them.

As feminist critics have argued for decades, women lose status and power during the formation of states and with the institution of democratic procedures and practices. As historian Joan Kelly-Gadol famously pointed out in the early 1970s, women lost social status and power not only in eras historians claim as achievements of social progress 'for humanity,' but precisely because of the nature of those achievements themselves. It was whatever marked the era as progressive which was itself responsible for women's status and power regress (Kelly Gadol 1976). Of course, women are not the only group to lose status and power at such moments.

5.2.3 *The Need for Scientific Rationality and Technical Expertise*

A third theme in the modernization accounts is that it is only scientific rationality and technical expertise that can direct economic and political activities in the modern public sphere. Thus it is boys and men who receive scientific and technological training. Moreover science and technology are overtly 'mission directed' in modernization contexts. However devotedly scientists and engineers conceive the purity of their search for knowledge, it is clear that the economic and political justification

[5]Of course there have been and still are traditional societies where women have little power or status. But what is surprising to Westerners is that in both the pre-modern past of Western societies and in so many other societies today this is not the case.

for providing the huge public material and social resources required for modern sciences and technologies is to advance economic and political projects in the name of national or supposedly universal social progress (Forman 1987).

To be sure, scientific rationality and technical expertise have been deployed in the private sector. Male experts appropriated control of obstetrics and gynecology from midwives. Such rationality and expertise have even been deployed by the great male chefs and the male authors of books on child-raising, and even by 'how to' authors on the topic of sex. Moreover, in the early part of the last century some socialist feminists developed 'technocratic feminism' which tried to rationalize the design of homes and the nature of domestic activities.[6] Yet this spill-over of public sector rationality and expertise into the private sector did not either eliminate the gender-stereotypes which structured the modern division of social life into public vs. private spheres, nor did it shorten the hours women were expected to spend on traditional domestic labor (Cowan 1983).

5.2.4 *Evolutionary Progress*

Finally, it is an evolutionary model of social and political change which is to govern these transformations. Development was conceptualized by the modernization theorists as the struggle for maturity and thus the achievement of dominance of men and their modernity over nature and women. '…in using an evolutionary model, they portray development as the ever-widening ability of men to create and transform their environment' (Scott 1995, 24). This was to be accomplished through political and economic policies shaped by scientific and technological practices. These would 'leave women behind' in the supposedly private and natural world of the household. 'Women's continued subordination in fact defines male citizenship' (Scott 1995, 24).

Moreover, throughout this work, 'the comparison of the liberated and independent woman of the West with the tradition-bound woman of the Third World also informs many accounts of the psychosocial requisites of modernity' (Scott 1995, 25). Third World women are presented as 'uniformly oppressed by men and family structure' (Scott 1995, 25). 'Such contrasts not only serve to establish a Western sense of difference and superiority (and complacency about women's rights in the West); they also mark women, in Mohanty's terms, as 'third world (read: ignorant, poor, uneducated, tradition-bound, domestic, family-oriented, victimized, etc.)'. As the most 'backward' group in society, women serve as an implicit contrast between Western modernity and non-Western tradition' (Scott 1995, 26). This delusion, which Western women no less than modernization theorists have promoted,

[6] A confession: my Father worked briefly in the 1920s for one of the early founders of time-budget studies (Frank Gilbreth). I still have amusing memories from several decades later of my Mother's frustration upon returning home to discover that he had once again rearranged the kitchen appliances and furniture in order to reduce by a few seconds or so the time she spent getting from the refrigerator to the stove or the table to the sink.

has long been the target of such Third World feminist scholars as Chandra Mohanty, Uma Narayan, and Gayatri Spivak (Mohanty 2003; Narayan 1997; Spivak 1988).

5.3 Silences and Gaps in Logic: Others' Questions

Feminist and postcolonial readings identify revealing silences and gaps in the logic of these modernization narratives. While explicit references to women or femininity only rarely appear in these texts, when they do they tend to signify whatever 'primitive' or traditional features modernity finds unintelligible, illogical, irrational or ridiculous. Women and femininity are encountered as whatever is inconceivable for modernity to think. Let us look more closely at what's wrong with some of the assumptions and claims of modernization theory that lead to such conclusions.

5.3.1 Is Modernity Incompatible with Tradition?

Not at all. Modernity narratives are always about time and historical change. Modernity defines itself against the past. So the past must be invoked and represented so that the modern may be contrasted with it. In this way modernity obsessively recuperates feminized 'tradition' within its own projects in order to define its own different, manly and Western progressive features. Consequently, women and femininity are invoked to emphasize the borders or horizons of the modern (Jardine 1985, 34). Through such processes, modernity appropriates selectively traditional features that it encounters, including social hierarchies. It reshapes them to suit its goals, and then obsessively reproduces them. Postcolonial critics have pointed out how modernity narratives radically reshape so-called traditions and reposition them in ways that make modernity look desirable. For example, suttee, the Indian widow-burning practice, was restricted to only a small section of India prior to the arrival of the British. It spread in resistance to the British attempts to eliminate it as an example of the backwards and savage nature of Indian tradition (Spivak 1988; Mohanty 2003; Narayan 1997). In this way feminized tradition becomes conceptually internal to modernity. Thus when notions of modernity are built upon stereotypes of the masculine, modernity's 'others,' regardless of who they are and of the content of that contrast, are always feminized.

Moreover, modernity depends on the survival and flourishing of traditional activities in additional ways. The new must always be 'sutured' into prevailing cultural practices and made acceptable and desirable to groups potentially hostile to it. Murata (2003) shows the neccessity of traditional craft labor to create the infrastructure in Japan for the introduction of such Western modern technologies as railway systems. 'Practice theorists' in philosophy, such as Rouse (1987, 2002), have pointed to ways in which the careful labeling and organizing practices of the laboratory have escaped into our kitchens and medicine closets. We can then see that in such cases it is mostly women's labor which must create the recommended laboratory-conditions in the household, and mediate acceptable social relations between traditional pharmacological or cooking practices on the one hand, and the new scientific ones on the other hand. Modern conceptions and beliefs would never come into practical existence without the on-going traditional artisanal labors of women and other groups thought of as pre-modern.

Thus tradition is neither entirely in the past, nor is it fixed and static. Once modern tendencies appear in a society, tradition becomes internal to modernity and evolves along with other modern institutions. Tradition is not identical to the pre-modern.

5.3.2 *Who Is Responsible for the Flourishing of Households?*

Not modern men or their institutions, it seems. Both modernization theorists and their marxian critics focus on the public realm of production and, for the modernization theorists, Liberal democratic politics as the motor of social progress. As indicated, both economic production and democratic politics are to be designed and managed by scientific rationality and technical expertise. It is transformations in the public realm which will bring about social progress for humanity. But who is to be responsible for the flourishing of households? Care of children, family, and kin relations, cooking, shopping, and other household tasks, the emotional labor which keeps us sane, and the maintenance of community relations are prerequisites for life as wage laborers and as managers and administrators of social relations of the public sphere. Yet apparently the flourishing of this supposedly private sphere is not the responsibility of modern institutions or their managers. Liberal economic theory explicitly conceptualizes such necessities for life itself as 'externalities,' concern for which weakens the powers of the market. So women, unsupported by modern institutions or their immense resources, remain responsible for 'women's worlds' – worlds that are no longer the pre-modern 'traditional' ones, since they no longer possess the economic and political resources characteristic of those older households.

In both modernization theory and the marxian alternatives, the continued domination and exploitation of women in modern societies is persistently claimed to be an unfortunate residue of pre-modern beliefs and practices which will disappear when women can participate in wage labor and democratic politics in the public realm. An influential study of socialism in Czechoslovakia pointed out that while socialism delivered many benefits to women, it ultimately failed to liberate them and enable them to achieve equity. It thereby weakened its other projects because as long as women remained responsible for a double day of work and had to suffer oppression within the household, women could not participate equally in the public sphere (Scott 1974).

There has been a mysterious silence in the Marxian accounts about the realm of consumption. The success of production depends upon effective organization of the labor of consumption of the goods and services which are produced. Without effective and every-changing organization of consumption, production fails as gluts of products pile up unbought and unused (Jardine 1985; Felski 1995). Consumption includes not only shopping, but also such tasks as turning the raw fish, potatoes, and artichokes into edible and culturally desirable foods, and other products for household use into desirable ones. What would our understanding of economic relations look like if one started off thinking about it from the lives of those who do the labor of organizing and processing for consumption already produced goods and services instead of only from the lives of those who produce, as is characteristic of all mainstream forms of economic theory?

Today in the West, it is armies of largely immigrant and foreign 'visitor' or guest-worker women and men from the Third World who take care of the children, households, and gardens of professional and other working women and men, and who perform the service work in restaurants, hospitals, schools, and offices (Cf. Sassen 1998). These workers often are maintaining also their own families, households, kin-networks, and communities in the Philippines, Latin America, and elsewhere as they work for us in the industrialized societies. I return below to reflect further on the emergence in global restructuring ('globalization') of these globally-distributed households.

5.3.3 Are Modern Men Autonomous, and Themselves Responsible for Their Own Achievements?

Of course not, as a generation of feminist accounts have detailed. Their ability to engage in managerial, administrative, and professional labor requires the armies of workers tending their bodies, their families, and the local places where they work and live (See Sect. 5.3.1 above). Modern men are freed from having to tend their bodies by the body-work everyone else must do. These others are mostly but not entirely women (Cf. Smith 1987).

5.3.4 Why Do Gender and Race/Imperial Discourses Mutually Circulate in Modernization Theories?

In short, each needs the other for its own successes. As the critics point out, those horizons that modernity creates to cordon itself off from its now unintelligible past are always both patriarchal and racial/colonial/ imperial. Women are consistently represented as outside history and society. They are primitive, incomplete, immature forms of the human. They appear as an a-historical other, like the noble (or sometimes ignoble) savage. Women represent nature, tradition, the emotions, as well as the Mother's body. 'The redemptive maternal body constitutes the ahistorical other and the other of history against which modern identity is defined' (Felski 1995, 38). Consequently it should come as no surprise to discover that pre-modern peoples and practices are themselves coded feminine in modernity narratives (Stepan 1986). Women are like 'savages' and 'savages' are like women in such accounts. Hence attempts to transform only one of these two kinds of hierarchies are doomed to fail since each can continue to exercise its powers inside the other. Thus gender and science projects cannot succeed in eliminating gender hierarchy in the sciences as long as the hierarchical contrast between the West vs. other races and cultures thrives. A parallel claim can be supported about postcolonial science projects which do not address the gender stereotypes structuring their own narratives and consequent policies.

This recognition can be discouraging. On the other hand, it reveals the excellent positions that those feminist and postcolonialism/anti-racism projects which conjoin in coalitions already occupy for taking on social transformations that will have better chances for success. Social movements that can recognize how the 'crisis of the West' is simultaneously a 'crisis of masculinity' can enable incomparable opportunities for kinds of social transformation which are not visible to less-alert progressive moments.

5.3.5 Is the Modern Model of the Ideal Relation Between Speech and Authority Suitable for Human Social Progress?

No. The narrative of rationality and expertise recommended by and characteristic of modernity is a parochial monologue serving only the interests of the powerful (Harding 2006, Ch. 7). As legions of feminist, postcolonial, and other critics have argued, it lacks the critical resources to grasp its own irrationality and lack of expertise, and thus its own parochiality.[7] Whatever may have been the case in the past, today a society cannot reasonably be regarded as flourishing when the vast majority of it citizens and those of many millions of citizens of other societies must lose social status and power over their own lives in order that a small minority gain even greater amounts of such status and power. Prevailing philosophies of science that extol theories of rationality and expertise have duped us into accepting their false and self-serving accounts.

5.3.6 Why Does Modernity's Narrative Replicate the Freudian Narrative?

Finally there is one more point – no doubt controversial – which must be addressed. There is an eerie echo in both the modernization theories and their marxian alternatives of Freud's narrative about how a boy becomes a man. Feminist scholars point

[7]In addition to earlier feminist and postcolonial citations, see, for example, Beck (1997) and Nowotny et al. (2001).

to the startling fit between these narratives of modernization and the standard psychoanalytic-influenced narratives of child development (Scott 1995, 10–13). It is the 'mother world' against which children must struggle in order to separate and individuate into maturing children of one culture or another. They must learn to control their bodily processes (that is, nature). This requires a painful struggle against the mother and the women-organized world of kin and household. Achievement of rational adulthood separated from women's worlds, linked with the control of the natural world, are the goal of both Freud's young man and the modern man and his social institutions in modernization theories.[8]

To be sure, philosophy of science is probably just about as hostile an environment as one could find in which to raise such an issue. Of course, there are plenty of limitations and problems with such psychoanalytic accounts. Yet there are also good reasons to pay attention to them, I suggest. One such reason is that the psychoanalytic accounts direct attention to anxieties and fears that the modernization and marxian theorists evidently themselves find reasonable and, moreover, expect to be compelling to their audiences. The theorists do not much censor the gender-coding themes in their narratives. Such associations of modernity with manliness and tradition with femininity do not seem odd or inappropriate to the modernization theorists or their marxisn critics.[9] At one level they probably are not even aware of them since such gendered stereotypes are not invented by them but rather prevalent in their culture as well as probably part of their individual psychic structures. Feminist criticisms of gender stereotypes, though these already appeared among the Nineteenth Century women's movements (think of Sojourner Truth's challenges to bourgeois norms of femininity at the mid-Nineteenth Century beginnings of the U.S. Women's Movement), are not part of their cultures. So even if they were aware of them, they would probably not see anything wrong with them.

A second reason to think further about these echoes of narratives of child development in modernization theories is to be found in the suggestion of feminist psychoanalytic accounts that equitable social relations between the genders will remain an unsuccessful struggle for both women and men as long as it is to women that the primary responsibility for child care and to men that the direction of public life are assigned. The constitution of gendered public and private spheres, whatever its psychic origins, can support only gender inequities, as the feminist theorists consistently argue. So if the adult gender relations through which children's gender identities are formed do structure those children's adult fears and desires in ways

[8]For influential feminist accounts of the strengths and limitations of Freudian theories see Chodorow (1978), Dinnerstein (1976), Flax (1990).

[9]Of course we should recollect that the original modernization theorists were the great founders of sociology in the Nineteenth Century, before Freud's theories had appeared. Moreover, the post-World War II era when modernization theory was resurrected and re-energized was an era in which formal colonial rule was beginning to end and in which the second women's movement was about to gather steam in Europe and the U.S. The economic, political, social, and psychic pre-conditions for the rise of postcolonial and feminist criticisms of Western and male-supremacist ideals of social progress were already in place as women and soon-to-be ex-colonials began to imagine futures for themselves which had been virtually inconceivable in preceding decades.

which support social inequalities, universal social progress requires that the world of children's gender formation be addressed and probably changed.

Finally, a third reason to ponder the fit between Freudian and modernization theory is the possible light thrown on Freudian theory by this fit. Critics have always complained that Freudian child development theory did not hold for non-Western societies. Perhaps this relation to modernization in part explains such a limitation. Perhaps Freud, too, should be included in the pantheon of great, though flawed, Nineteenth Century modernization theorists.

What is to be done about the way gender stereotypes – past, present, and possibly future – give content, meaning, and moral energy to prevailing conceptions of modernity, tradition, and social progress?

5.4 Moving on: A Methodological Provocation

Rethinking modernity has itself been a challenging project for Westerners. Yet what has always remained unclear among the diverse Western critical perspectives brought to bear in such work is what the desirable alternatives to modernity could be.[10] Merely trying to add 'the excluded' to modernity's projects when the goals and practices of those projects are constituted in terms of their distance from 'the excluded' has turned out to be of limited value. Of course such additive projects absolutely must continue to be vigorously engaged, since 'the excluded' need the skills, knowledge, access and credentials that their entrance into dominant modern institutions provide. Moreover when they do gain such access to the ranks of administrators and managers of modern social institutions, their continuing roots in and links to the not-yet-included can make possible critical public dialogues which would not occur at all, or would be impoverished, without the benefits that even our less-than-progressive modernity can offer. Yet such minimal access and continuing dialogue without significant change in social institutions conceptual frameworks and practices has induced a disspiriting sense of despair among many progressive groups. So if adding the excluded to modern institutions and practices doesn't in itself transform modernity's ideals and practices, what is to be done? Let us consider two strategies that can make visible other possibilities: critically interrogating 'tradition,' and a distinctive kind of standpoint research methodology. The two strategies are related.

5.4.1 Interrogating 'Tradition'

If one focuses a critical eye on 'tradition,' its past, present, and future, one sees a significant reason for this unclarity and consequent despair. Tradition must also be interrogated to expose its male-supremacist and Eurocentric meanings for and

[10] See, for example, the way in which an influential science studies scholar who is critical of modernity stumbles around in such a project: Latour (1993).

practices in modernity. As we have seen, such interrogations reveal the ways that modernity's social dynamics depend upon continued appropriation and recuperation such of male-supremacist and Western-supremacist meanings of modernity's obsessive contrast with something it calls 'tradition.' To undertake such a task is to redefine ideals of social progress and the accompanying required practices which pathetically have remained stuck in endless repetitions of male supremacist and Western supremacist practices.

This 'crisis of the West' creates an exciting moment for such tasks, as numerous postcolonial thinkers have pointed out (Amin 1998; Third World Network 1988). For these thinkers it is not the West from which can emerge the cultural energies to generate valuable new directions in social progress. It is not up to us reading this text – 'the included,' whatever our social origins may be – to design universal sciences and their world systems which we should expect 'the excluded' to find satisfactory. They won't find them satisfactory and, therefore, for that reason alone, apart from their actual limitations, they won't be satisfactory. Yet there is another reason they won't be satisfactory: 'the included' are not in a social position to be able reliably to assess how nature and social relations work, let alone how they or anyone else could better interact with them. Their privileged social position systematically hides from them many realities of their own and others' lives.[11]

Yet 'the included' have important contributions to make to this promising new era. Dominant groups will not be passive recipients of tendencies disseminated 'from below' any more than the Third World was passive in its engagements with the dissemination of elements of Western cultures, or women have been passive when confronted with male-supremacy. As indicated above, feminisms in coalition with anti-racist postcolonialisms are in especially good positions to explore ways forward out of the impasses of Western modernity which became so deeply mired in gender and imperial/colonial hierarchies.

Here I offer only a methodological provocation pertinent to scientific research and its philosophy and epistemology which possibly could move us past the modernity vs. tradition binary through defining 'social progress' in more effectively pro-democratic ways. It begins in what may be an unexpected standpoint directive.

5.4.2 Women's Lives in Households as a Starting Point for All Research

Recollect the feminist standpoint directive to 'start off research (and politics) from women's lives,' rather than from the conceptual frameworks of the research disciplines, in order to create the kinds of knowledge that women need and want to empower themselves and their dependents – children, kin, households, and

[11] This is the starting point of standpoint theory. See below.

communities. Standpoint projects are designed to identify, explain, and transform the conceptual and material practices of power of the dominant social institutions, including our research disciplines, in ways that benefit those who are least advantaged by such institutions. They start off from the lives of the oppressed, but they do not end there as, for example, do traditional ethnographies. Their main task is to 'study up,' to identify and explain the material and conceptual practices of power which are for the most part undetectable by those who engage in them (Hartsock 1983; Smith 2005, see also Harding 2003).

The proposal here is to start off research on any and every topic not from women's lives in the many revealing ways this has been done, but specifically from how their lives are lived in their households. This is not to discredit all the other conditions and experiences of women's lives which have made such fruitful starting points for research, but only to cast a sharp light on a dimension of women's lives which is important for them and their entire societies but which, as we have seen, is at especially high risk of being neglected by social and science theorists. Here is the test which could usefully be applied to many – perhaps most or even all – social and natural science research projects: what can we learn about the research topic or the policies and practices likely to result from it by starting off thinking about such topics from the standpoint of women's lives in households? After all, research that will be useful for militaries and corporations virtually always will have consequences for households and women's lives in them. Think about the U.S. drone attacks in the mountains with their villages of Pakistan, or about agricultural practices which will produce foods dangerous especially to young children. Or think of automative industry practices that increase exposures to respiratory illnesses that will affect children who attend the schools that are located near freeways. As indicated above, this is not the only useful social location from which to start off progressive research. Rather, I am suggesting it is one which any research project seeking to advance the growth of knowledge and social justice should undertake at some point or other, but especially so for projects seeking to contribute to social theory. It is prerequisite to maximizing validity and objectivity, and also social justice.

5.4.3 *New Resources*

Of course such a methodological directive is controversial, a point to which I shortly turn. Yet I suggest that it will be valuable for several related reasons. First, it brings into a new kind of focus much of what the purported progressiveness of modernity and modernization theory has seen only as obstacles to its own successes. It starts off thinking about modernization, tradition, the global political economy, and science and technology projects within these phenomena precisely from those human activities that are the most disvalued in the modernity narratives and their models of social progress. Household life in every one of its global cultural settings and dimensions, including its ethics, responsibilities, priorities, and cultural meanings, has arguably the densest and psychically most compelling configurations of what modernity has defined itself against. Second and relatedly, the household is 'where patriarchy is at home,' as several feminist critics have put the point (Kelly-Gadol 1976; see also Jardine 1985; Felski 1995). Third, the household and its kin relations are where the most stubborn resistance is found to imperial and colonial projects as Caulfield (1974) and Davis (1971) argue. For these three reasons alone, it can seem a big mistake for progressive action groups to ignore women's lives in households as an origin of potential economic, political, and social insight and of progressive social transformation. By ignoring the conditions of women's lives in households and the effects of such conditions on public policies, and vice versa, such projects take sides, intentionally or not, with what is arguably the most resilient of patriarchal and imperial projects.

Fourth, such a project inserts the ethic of responsibility and care delineated by so many feminist political philosophers into the desired ontology of research and political projects. There are additional ways to do this, but this one has the virtue of locating such a directive at the level of research methods. Fifth, the temptation in 'starting off from women's lives' has often been to think of women as autonomous individuals who are only contingently the center of households, kin relations, community relations, and relations with certain parts of natural environments. This proposal reduces such residual individualism, which damages so much even progressive research and public policy.

Sixth, widespread progressive activist projects already do begin their research and activism from the standpoint of what happens in households. One such kind is the continuing violence against women projects around the globe which are focused on ending family violence – the child abuse which leaves such deadly residues in subsequent adult lives and the domestic battery and rape which so damages women and their children. Households and kin relations are not the only sites of such violence, of course; workplaces, the streets, and even churches are also not safe places for women or children. Yet households do remain one site of little-constrained opportunity for violence against women and children. Such violence has effects on how public institutions can function. Such issues have remained crucial to advocacy of women's rights within national and international organizations and agencies. A related focus is on how women and children fare in conflict zones. Here Western sciences and technologies long history of mutual dependency relations with militarism and nationalisms implicate them in women's daily lives in war zones. Of course it is not science and technology alone which bear such responsibility, and certainly not scientists and engineers alone. Yet, all are complicitous, intentionally or not, with the fate of the victims and survivors of social conflict.

5.4.4 *Objections*

I focus here on four objections to this methodological provocation.

'Havens in a heartless world'? We need not spend much time on the traditional male-supremacist view that homes are 'havens from the heartless world,' where private life can be protected from encroachments by the harsh politics and economics of the public sphere. Of course feminists have had a great deal to say about this delusion. The one familiar but relevant point here that I will mention, however, is that households are not only workplaces for producing things and services used by the household itself; they are also sites of production for exchange and for the organization of community political and social life more generally. As mentioned earlier, it was precisely the removal of such economic and political activities to the public sphere (as well as of education, care of the sick, and moral/religious education) that was a top priority for modernization policies. However, even within supposedly modern societies, production for sale of agricultural products, food, clothing, indigenous pharmaceuticals and other merchandise still occurs in households, as does domestic labor for other households, such as taking in washing and sewing, childcare, and healthcare such as midwifery. Such practices are not banished by modernization, for they remain widespread in the new 'serving classes' of modern, industrialized societies (Cf. Sassen 1998). Furthermore, the internet and cell phone have made possible many kinds of 'home work' at all economic levels in industrialized societies, from the artisanal work of dressmakers and caterers to craft manufacture for internet sale, home based financial and computer piecework and consultation, and the increasingly common office practice of working at home one or more days a week. These are just some of the contemporary practices which raise the interesting question of whether modern societies are becoming less modern as the purportedly required disaggregation of social institutions seems to be declining?[12] However, my point here is that households continue to play a much more important role in the economy than the modernization theorists recognize even within supposedly modern societies (Prugl 1999). Whatever they may be for men, they are not havens for women with respect to economic production, let alone in terms of bodily and emotional safety.

Collusion with conservatism? Feminists may object that this proposed methodological move colludes with conservative tendencies to see women only as mothers or housewives. It works against the huge effort feminists have made to get governments and agencies to see women is legitimate and valuable actors in public worlds also (Mies 1986). To be sure, the disembedding of women from motherhood and households has been an important counter to the persistence of the public/vs private spheres policies characteristic of dominant conceptions of social progress.

Moreover, women philosophers, like other professional women, should be expected to resist this proposal in light of their own experience. In order to gain access to our professions and to continue to thrive within them, those of us with partners and/or children have had to act as if our households do not contain more than ourselves and perhaps a completely self-sufficient partner, that is, one who presumedly does not require the daily 'provisioning' expected of our male colleagues

[12]This question arises with respect to the both porous and intrusive relations between scientific and other social institutions in, for example, the work of Beck (1992, 1997) and Nowotny et al. (2001).

from their female domestic partners. 'Whose children could those be crying outside the department meeting door?' Certainly no needy children, ailing parents, sick or even hungry partners or children, dirty floors or raggedy clothes ever would interfere with our devotion to a successful career in philosophy, we have had to imply. The struggle to get universities to provide sufficient childcare, let alone time off for pregnancies or elder care is one kind of evidence of the legitimacy of women professionals' probable hostility to my suggestion here. Another kind is provided by surveys and interviews of graduate students who find little or no support in their departments or universities for their struggles to figure out how to make professional careers compatible with family responsibilities (Watford 2007). Yet another kind is provided by the fact that even though wage discrimination against women in general in the U.S., for example, has decreased over the last few decades, such discrimination against mothers remains virulent (Correll et al. 2007; Crittenden 2001). Do we lose hard-won status and power by admitting that we do in fact have responsibilities to household members, and that these just might once in a while take priority over professional duties?

The nagging questions remain: who is to be responsible for the nourishment of children, kin relations, households and communities, and how are those responsible themselves to be provided with the resources so that they and their dependents can flourish? This is not an argument to put professional and other women working for wages 'back in households.' We are already there, wherever else in the public sphere we may be. The issue rather is how women can assert that public policy must take on responsibility for the flourishing of households without deteriorating women's hard-won achievements in the public sphere. Obviously one solution is to get men to take on such responsibilities, also. With wives and mothers working outside the household, many men already have. Yet their workplaces do not provide resources to them for taking on such responsibility any more than they do to women.[13] The proposal here is to address this question head-on by starting off any and all research from the conditions of women's lives in households, howsoever else one may start off research.

Women not only in households. Another objection is that women are in lots of places besides households, such as workplaces, community organizations, and national and international politics. So why center only their lives in households? This objection misstates my proposal. It is to center women's lives in households *also*, not *only*. Of course it is also important to start off research from women's lives in all the places women live them. One project here is to redefine what should count as social progress. It must not be valued in terms of the greater extent and depth of its level of neglect and exploitation of women, households, and loyalty to kin relations. Instead of conceptualizing women as individuals in the public sphere

[13] Heidi Hartmann (1981) decades ago demonstrated that the time women spend in household work increases by about 9 hours per week if the household includes an adult (i.e., over 14) male, and that this is so regardless of whether the woman works outside the household, there are children in the household, or the man contributes to taking on household responsibilities. I know of no recent data which reveals improvement in such conditions.

who also happen to have responsibility for households, as much feminist work has done, why not think of them as living in households and also doing important work in the public sphere. Public policy and practice, including science and technology policy and practice, handicap the flourishing of households.[14]

Not all women are in households. Yet another objection could be that not all women are in households or have such responsibilities at all anymore. Women in all economic classes now work outside households and often live alone. In the new domestic and immigrant 'serving classes' of support workers required by the new forms of global economic and political structures, many women certainly are not living in their own traditional households. Yet a closer look reveals that probably most women laboring far from their young children and kin are still the centers of new forms of 'distributed' households. In agriculture, manufacturing, health care, and service industries in the first world can be found armies of men and women laborers, often immigrants, who produce the food we consume, who manufacture our clothes, tend to us in doctor's offices and hospitals, and work in the service sector. They are the millions of housekeepers and childcare workers needed to perform the traditional work for households where wives and mothers work in the public sphere. Women have a high representation in these new classes of migratory international workers.[15] Women (and men) often leave their children and other dependents behind as they come to seek jobs in service industries at the global sites where technically elite work is done. Moreover, among technological elites in professional classes, such 'distributed' households are also not hard to find. Many of us working in professional jobs still have more than our brothers or husbands responsibilities for the needs of our adult children, parents, and kin networks. So even women not living in their households often have major responsibility for the flourishing of their children, parents, extended families, and the communities in which they reside.

5.4.5 New Households in Global Restructuring

Now we are at a point to be able to see that new forms of families and households have emerged in the course of modernization's latest stage: global restructuring.[16] It is families, households, and social communities that have also been globally restructured, not just the world of corporate profit-seeking and international relations.

[14]Nancy Folbre's (2001) recent study argues that contemporary data shows that whomsoever has such household responsibilities will be handicapped in the public economy.

[15]In the past, labor was represented as locationally fixed or stable and industries traveled to take advantage of it. Hence the 'runaway' industries and the phenomenon of out-sourcing manufacturing parts and services. This kind of labor relation certainly continues today. Yet it is also the case that labor now travels to where the work is (Afshar and Barrientos 1999; Peterson 2003; Prugl 1999; Sassen 1998; Sparr 1994; Visvanathan et al. 1997).

[16]I follow the practice of feminists critically examining the lot of women in new forms of the global political economy to prefer the term 'global restructuring' to the more euphimistic 'globalization.' See the preceding note.

What we have in both the immigrant and new technical classes are increasing numbers of distributed households, in which women from the South working in the North maintain far-flung households and kinship communities. They provide financial resources. They are actively involved in parenting through frequent telephone contact with their children and other kin. They engage in material and emotional care of parents and siblings, visiting as often as possible. Parallel kinds of household and kin relations are maintained by women in the technical classes as our children, parents, and siblings become distributed through regional and international political economy networks.[17]

My point here is that by 'the households' in which women 'live' I have in mind not only the conventional bourgeois model of the nuclear family, against the injustices of which feminists have long struggled, but all of the motley creative social arrangements women (and men) make to enable their dependents to survive and, better, flourish.[18] Why shouldn't the standards for social progress be measured also in terms of whether or not a society's households, for which women seem to be assigned responsibility around the globe, are flourishing ones?

5.5 Conclusion

My argument may seem to have strayed a long way from the concerns of prevailing feminist philosophies of science, not to mention from those of 'pre-feminist' philosophers. But that is part of my point. As long as the Western modern conceptual framework characteristic of philosophy (as well as of other research disciplines) assumes that what happens in households has nothing of interest to do with the advance of objective and reliable knowledge projects, or of the achievement of social justice, these philosophies are doomed to remain grounded in empirical falsehoods, conceptual irrelevancies, and politically regressive social projects. Women and modernity's other Others cannot achieve social justice and make social progress in their own terms as long as the realities of household life are conceptualized as obstacles to justice, progress, and modernity. What would modernities, including their scientific and technological projects, look like if they were designed to ensure the social and material well-being of households in which women and their dependents can flourish?

---

[17] I do not mean to suggest that the situations and resources available to immigrant low-paid workers are the same or equally desirable as are those available to professional women, but only that patterns of global restructuring are to be found in the organization of households and family relations no less than in economic, political, and public social relations.

[18] What about men? Good question. These global political economy processes creating today's 'new women' also creating 'new men.' There have been at least some attempts to identify and understand diverse forms of transformations in masculinities, at least some of which are highly resistant to conventional male supremacist ideals (Connell 1995; Connell et al. 2005). I cannot here pursue this topic beyond noting that the field of masculinity studies needs to be as fully as possible integrated into feminist studies.

References

Afshar, Haleh, and Stephanie Barrientos (eds.). 1999. *Women, globalization, and fragmentation in the developing world*. New York: St. Martin's Press.

Amin, Samir. 1998. *Capitalism in the age of globalization*. New York: Zed Books.

Appleton, Helen, et al. 1995. Claiming and using indigenous knowledge. In *Missing links*, ed. Gender Working Group, 55–82. Ottawa: International Development Research Centre.

Beck, Ulrich. 1992. *Risk society: Towards a new modernity*. London: Sage Publishing Co.

Beck, Ulrich. 1997. *The reinvention of politics: Rethinking modernity in the global social order*. Cambridge: Polity Press.

Biagioli, Mario (ed.). 1999. *The science studies reader*. New York: Routledge.

Bordo, Susan. 1987. *The flight to objectivity*. Albany: State University of New York Press.

Braidotti, Rosi, et al. 1994. *Women, the environment, and sustainable development*. Atlantic Highlands: Zed.

Caulfield, Mina Davis. 1974. Imperialism, the family, and cultures of resistance. *Socialist Review* 4(2): 67–86.

Chodorow, Nancy. 1978. *The reproduction of mothering*. Berkeley: University of California Press.

Connell, R.W. 1995. *Masculinities*. Cambridge: Polity Press.

Connell, Robert, Ingeborg Breines, and Ingrid Eide (eds.). 2005. *Male roles, masculinities, and violence*. Paris: UNESCO.

Correll, Shelly J., Stephen Benard, and In Paik. 2007. Getting a job: Is there a motherhood penalty? *The American Journal of Sociology* 112: 1297–1338.

Cowan, Ruth Schwartz. 1983. *More work for mother*. New York: Basic Books.

Crittenden, Ann. 2001. *The price of motherhood*. New York: Basic Books.

Davis, Angela. 1971. The black woman's role in the community of slaves. *Black Scholar* 2: 1–14.

Dinnerstein, Dorothy. 1976. *The mermaid and the minotaur*. New York: Harper and Row.

Eisenstadt, S.N. 2000. Multiple modernities. *Daedalus* 129(1): 1–30.

Felski, R. 1995. *The gender of modernity*. Cambridge: Harvard University Press.

Flax, Jane. 1990. *Thinking fragments: psychoanalysis, feminism and postmodernism in the contemporary west*. Berkeley: University of California Press.

Folbre, Nancy. 2001. *The invisible heart: Economics and family values*. New York: New Press.

Forman, Paul. 1987. Behind quantum electronics: National security as bases for physical research in the U.S., 1940–1960. *Historical Studies in Physical and Biological Sciences* 18: 149–229.

Frank, Andre Gunther. 1969. *Capitalism and underdevelopment in Latin America*. New York: Monthly Review Press.

Furtado, Celso. 1970. *Development of Latin America: a survey from colonial times to the Cuban revolution*. Cambridge: Cambridge University Press.

Giddens, Anthony. 1994. Living in a post-traditional society. In *Reflexive modernization*, ed. Ulrich Beck, Anthony Giddens, and Scott Lash. Cambridge: Polity Press.

Hackett Edward, J., et al. 2008. *The handbook of science and technology studies*, 3rd ed. Cambridge: MIT Press.

Haraway, Donna. 1991. Situated knowledges: The science question in feminism and the privilege of partial perspectives. In *Simians, Cyborgs, and Women*, ed. Donna Haraway, 183–202. New York: Routledge.

Harding, Sandra. 1998. *Is science multicultural? Postcolonialisms, feminisms, and epistemologies*. Bloomington: Indiana University Press.

Harding, Sandra (ed.). 2003. *The feminist standpoint theory reader: Intellectual and political controversies*. New York: Routledge.

Harding, Sandra. 2006. *Science and social inequality: Feminist and postcolonial perspectives*. Edwardsville: University of Illinois Press.

Harding, Sandra. 2008. *Sciences from below: Feminisms, postcolonialities, and modernities*. Durham: Duke University Press.

Hartmann, Heidi. 1981. The family as the locus of gender, class, and political struggle: The example of housework. *Signs: Journal of Women in Culture and Society* 8(4): 366–394.

Hartsock, Nancy. 2003 (1983). The Feminist Standpoint. In *Discovering reality*, 2nd ed, ed. Sandra Harding and Merrill Hintikka. Dordrecht: Reidel.

Hessen, Paul. 1970. *The economic roots of Newton's principia*. New York: Howard Fertig.

Jaggar, Alison. 1989. Love and knowledge: Emotion in feminist epistemology. In *Gender, body, knowledge: Feminist reconstructions of being and knowing*, ed. Alison Jaggar and Susan Bordo. New Brunswick: Rutgers University Press.

Jardine, Alice. 1985. *Gynesis: Configurations of woman and modernity*. Ithaca: Cornell University Press.

Keller, Evelyn Fox. 1984. *Reflections on gender and science*. New Haven: Yale University Press.

Kelly Gadol, Joan. 1976. The social relation of the sexes: Methodological implications of women's history. *Signs: Journal of Women in Culture and Society* 1(4): 810–823.

Latour, Bruno. 1993. *We have never been modern* (trans: Catherine Porter). Cambridge: Harvard University Press.

Lloyd, Genevieve. 1984. *The man of reason: 'Male' and 'Female' in western philosophy*. Minneapolis: University of Minnesota Press.

MacKinnon, Catherine. 1982. Feminism, marxism, method, and the state: An agenda for theory. *Signs: Journal of Women in Culture and Society* 7(3): 515–544.

Martin, Emily. 1987. *The woman in the body*. Boston: Beacon.

Mies, Maria. 1986. *Patriarchy and accumulation on a world scale: Women in the international division of labor*. Atlantic Highlands: Zed.

Mirowski, Philip. 2005. How positivism made a pact with the postwar social sciences in the United States. In *The politics of method in the human sciences: Positivism and its epistemological others*, ed. George Steinmetz, 142–172. Durham: Duke University Press.

Mohanty, Chandra Talpade. 2003. *Feminism without borders: Decolonizing theory, practicing solidarity*. Durham: Duke University Press.

Murata, Junichi. 2003. Creativity of technology and the modernization process of Japan. In *Science and other cultures: Issues in philosophies of science and technology*, ed. Robert Figueroa and Sandra Harding. New York: Routledge.

Narayan, Uma. 1997. *Dislocating cultures: Identities, traditions, and third world feminism*. New York: Routledge.

Nowotny, Helga, Peter Scott, and Michael Gibbons. 2001. *Re-thinking science: Knowledge and the public in an age of uncertainty*. (Cambridge: Polity Press)

Peterson, V.Spike. 2003. *A critical rewriting of global political economy: Integrating reproductive, productive, and virtual economies*. London: Routledge.

Prakash, Gyan. 1999. *Another reason: Science and the imagination of modern India*. Princeton: Princeton University Press.

Prugl, Elisabeth. 1999. *The global construction of gender: Home-based work in the political economy of the 20th century*. New York: Columbia University Press.

Rose, Hilary. 1983. Hand, brain and heart: A feminist epistemology for the natural sciences. *Signs: Journal of Women in Culture and Society* 9(1): 73–90.

Rouse, Joe. 1987. *Knowledge and power: Toward a political philosophy of science*. Ithaca: Cornell University Press.

Rouse, Joe. 2002. *How scientific practices matter: Reclaiming philosophical naturalism*. Chicago: University of Chicago Press.

Sardar, Ziauddin. 1997. Islamic science: The contemporary debate. In *Encyclopedia of the history of science, technology, and medicine in non-western cultures*, ed. Helaine Selin, 455–458. Dordrecht: Kluwer.

Sassen, Saskia. 1998. *Globalization and its discontents*. New York: The New Press.

Scott, Hilda. 1974. *Does socialism liberate women?* Boston: Beacon.

Scott, Catherine V. 1995. *Gender and development: Rethinking modernization and dependency theory*. Boulder: Lynne Rienner Publishers.

Shiva, Vandana. 1989. *Staying alive: Women, ecology and development*. London: Zed.
Shiva, Vandana. 1993. *Monocultures of the mind: Perspectives on biodiversity and biotechnology*. New York/Penang: Zed Books and Third World Network.
Shiva, Vandana. 1994. *Close to home: Women reconnect ecology, health, and development worldwide*. Philadelphia: New Society.
Smith, Dorothy. 1987. *The everyday world as problematic: A sociology for women*. Toronto: University of Toronto Press.
Smith, D. 2005. *Institutional ethnography: A sociology for people*. Lanham: Rowman and Littlefield.
Sparr, Pamela (ed.). 1994. *Mortgaging women's lives: Feminist critiques of structural adjustment*. London: Zed.
Spivak, Gayatri. 1988. Can the subaltern speak. In *Marxism and the interpretation of culture*, ed. Cary Nelson and Lawrence Grossberg, 271–313. Urbana: University of Illinois Press.
Stepan Nancy, Leys. 1986. Race and gender: The role of analogy in science. *Isis* 77: 261–277.
Third World Network. 1988. *Modern science in crisis: A third world response*. Penang: Third World Network.
Visvanathan, Nalini et al. (eds.). 1997. *The women, gender, and development reader*. London: Zed Books.
Wallerstein, Immanuel. 1974. *The modern world system*, 1st ed. New York: Academic.
Watford, Tara. 2007. *Looking Beyond Equal Representation: Perspectives of Gender Equity from the New Majority in Doctoral Education*, Ph.D. Thesis, University of California at Los Angeles.
Yoon, Soon-Young. 1995. Looking at health through women's eyes. In *Missing links*, ed. Gender Working Group, 129–158. Ottawa: International Development Research Centre.

Part II
Democracy and Diversity in Knowledge Practices

Chapter 6
Diversity and Dissent in Science: Does Democracy Always Serve Feminist Aims?

Kristen Intemann

Abstract Partly in response to feminist concerns about bias in science, there has been a recent trend towards viewing good science as democratic science (e.g., Kitcher 2001; Solomon 2001; Longino 2002). One similarity of these approaches is that democratic science is conceived of as what I will refer to as *Millian science*. Based on the epistemological views of John Stuart Mill (1859), ideal scientific communities are comprised of participants with diverse values who have equal authority in a 'free marketplace of ideas' to advocate for different research directions, theories, and interpretations of data. This model of science has been taken to be attractive from a feminist perspective insofar as it can help eliminate male-bias and explain the importance of feminist perspectives in science. Focusing on the work of Miriam Solomon (1994, 2001), and Helen Longino (1990, 2002), I argue that a Millian conception of democratic science cannot ultimately address several feminist concerns within philosophy of science. An alternative conception of democratic science is then offered.

Keywords Democracy • Diversity • Dissent • Feminist science • Science and values

6.1 Introduction

In light of feminist criticisms of science and philosophy of science, there has been a recent trend towards a normative conception of science as democratic science (e.g., Kitcher 2001; Solomon 2001, 2006; Longino 2002). Although there are different accounts of how science might be more democratic, there are some

---

K. Intemann (✉)
Department of History & Philosophy, Montana State University,
2-155 Wilson Hall, Bozeman, MT 59717, USA
e-mail: intemann@montana.edu

H.E. Grasswick (ed.), *Feminist Epistemology and Philosophy of Science: Power in Knowledge*, DOI 10.1007/978-1-4020-6835-5_6,

commonalities. In particular, there is a shared conception of democratic science as what I will refer to as *Millian science*. Reflecting the epistemological views of John Stuart Mill, ideal scientific communities will be those comprised of participants with diverse values and interests, who have equal authority to advocate for different research directions, theories, models, background assumptions, explanations, and interpretations of data. Mill held that because humans are fallible, allowing and encouraging dissenting views is epistemically important (Mill 1982 [1859], 143, 163). A dissenting view might be true (or at least partly true) and even if this is not the case, allowing dissent helps us to better understand the justification for our true beliefs (Mill 1982 [1859], 170). Mill argued, then, that scientific knowledge is best acquired by encouraging a 'free marketplace' of ideas in the sense of allowing a full range of hypotheses and exposing scientific reasoning to critical scrutiny by those with different values, perspectives, and interests.[1] Theories accepted by these democratic processes are less likely to be the product of any individual participant's values or interests, and more likely to be accurate. Thus, the Millian conception of democratic science has two key features. First, it places importance on having a *diverse* group of inquirers with different values and interests. Second, it takes dissent to be valuable in achieving scientific knowledge. These features of Mill's view show how science is social and value-laden, yet rational.

Focusing on Miriam Solomon's social empiricism (1994, 2001) and Helen Longino's contextual empiricism (1990, 2002), which I take to share key Millian features, I will evaluate the Millian conception of democratic science from a feminist perspective. Although neither of these authors claims that her version of empiricism is a *feminist* philosophy of science, each takes her view to address certain feminist criticisms of science and philosophy of science. In particular, these accounts are purported to address feminist concerns because they attempt to: (1) structure scientific practices in ways that are less likely to produce theories that are androcentric or male-biased, (2) account for why, from an epistemic perspective,[2] it is important to increase the participation of underrepresented groups in science, and (3) allow feminist perspectives to play an important role in science (through democratic processes). I argue, however, that such accounts are not ultimately successful in addressing feminist concerns in these three ways. Seeing why the Millian conception of democratic science is problematic from a feminist perspective, however, helps reveal how democracy might be alternatively conceived

[1] Mill (1982 [1895], 163) explicitly states that his general views in 'Of the Liberty of Thought and Discussion' apply to 'natural philosophy,' or the sciences.

[2] There are, of course, other non-epistemic reasons for thinking that the participation of underrepresented groups is important in science. For example, one might argue that it is important for achieving equal opportunity or social justice (for instance, by insuring that members of underrepresented groups will have role models and mentors within science). Democratic accounts of science, however, tend to make the further claim that the participation of underrepresented groups is also important for epistemological reasons.

so as to better promote the three feminist aims outlined above. I will begin by examining how Longino and Solomon's accounts seem to advocate a Millian conception of democratic science.

6.2 Millian Science in Longino's Contextual Empiricism

In *Science as Social Knowledge* (1990), and *The Fate of Knowledge* (2002), Longino defends *contextual empiricism*. Contextual empiricism maintains that it is not individual scientists, but scientific communities as a whole that are the locus of objectivity (Longino 2002, 51; 1990, 80). Longino also argues that there are no criteria for theory acceptance that are *guaranteed* to screen out the influence of values (Longino 1987, 55; 2002, 128). That is, there are no rules of theory choice that, even when followed correctly, necessarily prevent ethical and political values from influencing the reasoning of individual scientists. Scientists must rely on a host of background assumptions, many of which they cannot even be aware of, in testing theories. Moreover, ethical and political value judgments are often deeply held, and widely shared by those practicing science. As a result, it is very difficult for individual scientists to recognize when their work is being influenced by their own ethical and political values.

Yet, by structuring science so that scientific communities are diverse and criticism is encouraged, scientific communities as a whole can achieve a higher degree of objectivity. Scientific knowledge is objective, on Longino's view, insofar as the organization of a diverse scientific community satisfies the following four conditions:

1. there must be recognized avenues for the criticism of evidence, of methods, and of assumptions and reasoning;
2. there must exist shared standards that critics can evoke;
3. the community as a whole must be responsive to such criticism;
4. intellectual authority must be shared equally among qualified practitioners (Longino 1990, 76).

There must be recognized avenues for criticism so that members of the scientific community have the ability to point out problematic background assumptions or methods employed by other scientists. At the same time, there must be some shared standards of evidence and epistemic values that critics can appeal to so that their criticisms have force. Participants must also have equal intellectual authority in order for their criticisms and research to be taken seriously. Moreover, members of the scientific community must be responsive to criticism, particularly as it will appeal to the shared standards of evaluation that the scientific community as a whole accepts. When scientific communities are structured so as to meet these criteria (when research is adequately scrutinized by a diverse group of participants with equal intellectual authority), then any value-laden assumptions inappropriately influencing scientific reasoning are more likely to be caught (Longino 1990, 73–4, 80; 2002, 51). When values are different from one's own, it is easier to see when they are influencing scientific reasoning or description of data. Thus, a scientific community comprised of individuals with diverse ethical and political values will be able to identify or catch the ways that values influence the reasoning individual scientists.

Consider, for example, research on sex differences in visual-spatial abilities in the 1970s.[3] One experiment was the rod and frame test, where subjects in a dark room were presented with a lighted rod bisecting a frame tilted at an angle and had to instruct the experimenter where to reposition the rod so that it would be perpendicular to the floor. Differences in performance on this test led researchers to conclude that males had superior visual-spatial abilities in virtue of differences in their biology (Fausto-Sterling 1985). On Longino's view, one of the problems with these experiments was that the researchers were overwhelmingly males with homogeneous values and interests.[4] As a result, they failed to see ways in which the experiment design was biased so as to hinder the performance of females. As Fausto-Sterling (1985, 32) points out, female subjects were often tested in a dark room with male scientists, possibly making them uncomfortable and affecting performance. Scientists also neglected alternative hypotheses that would have accounted for the data equally well. For example, they never considered that because women are often socialized to be less assertive they might have been less comfortable in having the researcher make continual minute adjustments to the rod (Fausto-Sterling 1985, 32). A scientific community that had more scientists with feminist commitments might have been more likely to identify and challenge the problematic background assumptions operating in the experiment design and data interpretation. Feminist scientists might have offered alternative hypotheses (such as the hypothesis that differences in performance on the test were the result of differences in the way that males and females are socialized). Had the scientific community been comprised of a more diverse group of scientists with equal authority to publicly criticize assumptions made in visual-spatial research, biases and problematic background assumptions would have been more likely to be identified and corrected. Through the democratic processes of dissent within a diverse scientific community, more objective scientific conclusions could have been reached.

Longino's account of ideal science can be seen as a version of Millian democratic science. Like Mill, Longino places high value on diversity and dissent within scientific communities. On Longino's account, science is objective insofar as the scientific community is comprised of individuals with diverse values and interests who have the opportunity to propose, criticize, and advocate for different hypotheses, models, interpretations of data, and so on. Yet Longino takes the Millian conception even further by structuring the scientific 'marketplace of ideas' to ensure that participants

---

[3] Anne Fausto-Sterling (1985) provides a detailed analysis of this example.

[4] Dr. H. A. Witkin developed and popularized the rod and frame test with mostly male colleagues (Fausto-Sterling 1985, 31).

have equal intellectual authority so that their views and criticisms are taken seriously. Thus, minority views are given more genuine opportunity to compete. Through this process of transformative criticism by diverse participants, objective knowledge emerges.[5] In this way, Longino's version of democratic science provides a system of checks and balances so that the fallibility or biases of any individual epistemic agent will be less likely to lead to viable hypotheses being neglected or false hypotheses from being erroneously accepted. Those theories that survive the scrutiny of a community of diverse inquirers are most likely being accepted on the basis of explicitly shared epistemic values rather than on the basis of idiosyncratic individual values. As a result, the impact of individual biases is minimized.

6.3 Millian Science in Solomon's Social Empiricism

The key features of Millian democratic science can also be found in Solomon's *Social Empiricism* (despite the fact that there are also several significant differences between Solomon's and Longino's view). Like Longino, Solomon agrees that many different factors, including ethical and political commitments, can influence decisions about which theories ought to be accepted or rejected, but that science can still be rational. Solomon proposes the term 'decision vectors' to refer to all of the factors that influence theory acceptance (Solomon 2001, 53). This is intended to be an epistemically neutral term that does not reflect any assumptions about which decisions vectors are good or bad from an epistemic perspective. Decision vectors are further distinguished as either empirical or non-empirical. Empirical decision vectors, according to Solomon, influence the outcome of theory choice in ways that are conducive to empirical success (Solomon 2001, 56). For example, a preference for a theory with novel empirical successes is an empirical decision vector (Solomon 2006, 27). Non-empirical decision vectors are those factors influencing theory choice that do not track empirical adequacy (Solomon 2001, 56). Examples include a preference for simpler theories, holding onto a theory out of ideology or pride, or other factors unrelated to empirical success. Although Solomon concedes that non-empirical decision vectors cannot be eliminated from scientific reasoning, she argues that this need not hinder scientific rationality so long as empirically successful theories receive their fair share of the research effort (Solomon 2001, 76–77). A theory has received a fair share of the research effort relative to competitors when two conditions are met. First, the distribution of empirical decision vectors must be *equitable* relative to the theory's

[5]For Mill, it is 'truth' that emerges from this process, though Longino is not committed to this. Longino argues there are other aims to science such as explanation and extending knowledge in new ways (Longino 1990, 32–6). For both Mill and Longino, however, what comes out of appropriately structured democratic science is knowledge that best embodies the shared epistemic values that promote the aims of scientific inquiry.

empirical success (Solomon 2001, 77). The idea here is that theories should receive funding, attention, and resources in proportion to the theory's degree of empirical success. The more empirical success a theory has relative to its competitors, the more research effort it deserves. Second, any non-empirical decision vectors must be *equal* in number to those favoring competing theories (Solomon 2001, 77). If there are forces favoring a theory that have nothing to do to the theory's empirical success, then these must be balanced out by the same number of non-empirical decision vectors favoring its competitors. In this way, any additional attention, resources, or credibility that a theory may receive due to non-empirical decision vectors will be counter-acted by non-empirical decision vectors that favor competing theories.

This can again be illustrated by considering 1970s research on sex differences in visual-spatial ability. Much attention was given to the theory that biological sex differences cause differences in visual-spatial ability. Of course, an alternative hypothesis is that such differences are caused by differences in socialization. In the 1970s, both theories had empirical successes, but the biological hypothesis received far more research effort.[6] On Solomon's view, we can say that there were non-empirical decision vectors that caused the theory to receive more of the research effort than it deserved. Perhaps this was because scientists were influenced by sexist assumptions or implicit gender biases that led scientists to assume the biological hypothesis was more likely to be true. Or, scientists might have been more likely to believe the biological hypothesis because of the reductionist trend dominant at that time, which valued the explanatory power of low-level over high-level causal mechanisms. If any of these forces were at work, then non-empirical decision vectors (such as sexist attitudes or a psychological tendency towards reductionism) caused scientists to give the sociological hypothesis less attention than it deserved.

On Solomon's view, if there had been an equal number of other non-empirical decision vectors favoring the socialization hypothesis, then the research effort would have been appropriately distributed. Imagine, again, that there were more scientists with feminist political commitments working in the field at that time. Suppose that, because of their political values, these scientists were psychologically disposed to favor theories that were inconsistent with biological determinism. Because many feminists believe that biology ought not limit female potential, they may have been more inclined to discount purely biological hypotheses and examine alternative theories. In this case, the scientific community as a whole would have two non-empirical decision vectors (sexist attitudes and a tendency towards reductionism) that would favor the biological hypothesis and two non-empirical decision vectors (feminists' desire to change the status quo and a tendency against

[6] Between 1961 and 1976 the debate over gender differences in visual spatial ability was largely a debate over which of several different biological hypotheses was correct. During that time there were at least seven competing hypotheses posited to explain the exact biological mechanism responsible for gender differences, while research on social factors was largely absent from the literature (Fausto-Sterling 1985, 40–1).

reductionism) that would favor the socialization hypothesis regarding sex differences. So, the feminist and anti-feminist non-empirical decision vectors would 'balance each other out' so that the biological and sociological theories would each receive an appropriate amount of the research effort relative to their empirical success. Scientists with biases towards the biological hypothesis would still devote more time and resources towards pursuing it, but this would be counter-acted by scientists with biases favoring the sociological hypothesis. Each theory would be rigorously pursued by its advocates and the most empirically successful theory would eventually be the victor.

It is worth noting that there are several key differences between Solomon's social empiricism and Longino's contextual empiricism. First, unlike Longino, Solomon's empiricism is explicitly naturalistic. That is, Solomon takes herself to be identifying the factors that influenced historical successes and failures within science. Her account emerges from detailed case studies, both those that she uses as paradigm cases of a good distribution of research effort (e.g., evolutionary biology and the continental drift dispute) and improper distribution of research effort (e.g., genetics before the discovery of DNA and cancer virus research) (Solomon 2001, 68–95). So, Solomon takes her project to be distinct from Longino in that she denies she is offering an idealized conception of science. Yet, like Longino, her account is not purely descriptive. By identifying those factors that have led to good and bad distribution of research effort, Solomon is offering a normative prescription for how research effort ought to be distributed. That is, she is still giving an account of how scientific inquiry would be best structured.

Second, Solomon takes herself to have a different view of dissent than either Longino or Mill (Solomon 2001, 101). For Solomon, dissent is not always valuable. Dissent will be valuable in research contexts where there are multiple competing theories that all have empirical successes, but inappropriate when there is only one empirically successful theory (Solomon 2001, 101). Dissent is beneficial only when it helps to establish a fair distribution of research effort among empirically successful theories (Solomon 2001, 65–68, 100–101). Solomon interprets both Mill and Longino as maintaining that dissent is intrinsically valuable to inquiry. For Mill, consensus is dangerous because humans are fallible, and dissent helps us to be cognizant of why we accept the theories we do. For Longino, dissent is important to making sure that our grounds for theory acceptance is adequate (e.g., that it does not rest on any faulty background assumptions). But Longino also maintains that dissent can be an appropriate *result* of inquiry as there may be contexts in which there are multiple theoretical frameworks for explaining the same phenomenon. In other words, Longino endorses scientific pluralism (at least in some research contexts), which intrinsically values dissent as the only way to capture all of the facets of some phenomenon (Longino 2002, 178).[7]

[7] It is not clear that Solomon is correct in seeing Longino's view about dissent as so different from her own. Longino would agree that empirical success is an important constraint on dissent.

Finally, Solomon and Longino differ on the role that individuals play in scientific rationality, which is reflected in the normative claims each makes about what constitutes good science. Although both agree that the locus of objectivity is a scientific community, Longino takes objectivity to emerge from social practices of transformative criticism that help individual scientists reason better. Solomon has criticized Longino on this point, arguing that this conception of rationality is still too individualistic.[8] According to Solomon, 'social groups can work to attain and even recognize epistemic goals without individual rationality or individual cognizance of the overall epistemic situation' (Solomon 1994, 219). Thus, for Solomon, scientific rationality is determined by the distribution of decision vectors within scientific communities, of which individual scientists may not be aware. Longino, however, has argued that this overlooks the significance that individuals play in generating knowledge, as it is individuals who conduct experiments, make observations, and interpret data (Longino 1994, 143). Thus, Longino maintains that any account of scientific rationality must also account for the rationality of individual scientists (even if it is in relation to the rest of a scientific community).

Yet despite these differences, there are certain commonalities in virtue of which Solomon's social empiricism can, like Longino's contextual empiricism, be seen as a version of Millian democratic science. First, like Longino and Mill, diversity of the scientific community as a whole is epistemically important to achieving objectivity and rationality. An ideal scientific community is one that is comprised of scientists with diverse values and interests so that non-empirical decision vectors are balanced. With a diverse community, it is more likely that all empirically successful possibilities will be given their fair share of the research effort.

Second, although Solomon distinguishes her views about dissent from Mill, the reasons she has for valuing dissent in certain contexts are very much like Mill's. For Solomon, dissent plays a crucial role in contexts where there are multiple theories with empirical success. In those cases, dissent will be important to ensuring that all empirically successful programs receive their fair share of the research effort, so that we don't come to an inappropriate consensus on the basis of non-empirical decision vectors. Like Mill, Solomon sees dissent as a valuable tool in counteracting human fallibility. So, even though Solomon thinks that the 'freemarket place of ideas' should only include those theories that have empirical success, dissent will still be important to making sure that each viable theory is fully considered.

Thus, despite differences between social empiricism and contextual empiricism, both views maintain that scientific communities comprised of individuals with diverse values who have authority and resources to advocate for different hypotheses and background assumptions can reduce bias and produce the most empirically successful theories. I will now examine whether the Millian conception of democracy adequately addresses feminist concerns.

---

[8] Brad Wray (1999) has an excellent discussion of this debate between Longino and Solomon (where he ultimately defends Longino).

6.4 Evaluating Millian Science from a Feminist Perspective

In some respects, a Millian conception of democratic science might be seen as attractive from a feminist perspective. First, it attempts to structure scientific inquiry so as to reduce or eliminate sexist and androcentric bias in scientific reasoning. As employed by Longino and Solomon, the Millian framework puts in place mechanisms to protect and empower minority views. This helps address the epistemic fallibility of majority groups in science. By ensuring that those with diverse interests have the opportunity to participate and raise challenges to dominant assumptions that must be taken seriously, it is more likely that bias will be prevented or corrected. Second, such accounts purport to explain why it is important to have members of underrepresented groups participating in science. On the Millian view, science benefits from the inclusion of individuals who hold diverse values and interests. Insofar as members of underrepresented or marginalized groups are likely to have different interests than those in dominant groups, their participation will help increase the objectivity of scientific communities. It helps ensure that a full range of empirically successful theories will receive their fair share of the research effort (Solomon 2001) and that background assumptions will receive more rigorous scrutiny (Longino 1990; 2002). As a result, accepted scientific theories will be less likely to merely reflect the interests of dominant groups. Finally, Millian accounts appear to allow feminist ethical and political commitments to play a legitimate role in science by leading to new research directions, generating additional alternative hypotheses, and causing scientists to consider interpretations of data previously ignored. Because of their political commitments and interests, feminists may frame research questions in different ways and offer new models for collecting and analyzing data. And, in fact, there are several examples suggesting that the inclusion of feminist perspectives in science has provided these sorts of benefits (Wylie 2001; Anderson 2004; Wylie and Nelson 2007).

I will argue, however, that Millian accounts of democratic science (as recently conceived) cannot fully serve feminist aims. I will first explain what I take to be the central problem with Millian accounts, and then show how that problem has implications for the feminist aims outlined above.

While a Millian conception of democracy captures many feminist concerns regarding the importance of diversity and dissent within scientific communities, a central feature of the Millian framework is that it implicitly advocates a kind of value-neutrality. Millian science is 'value-neutral' in the sense that it does not endorse or privilege any particular ethical, political, or social values in virtue of their content. To be sure, Millian accounts endorse some broad values in structuring science (such as diversity, dissent, and democracy). Moreover, Millian accounts do not expect individual scientists to be neutral about all values (in fact both Longino and Solomon assume that individuals will be value-committed). In this sense, Millian accounts are not value 'free.' But, within a Millian framework, the scientific community as a whole is value-neutral in the sense that it does not privilege, endorse, or exclude the particular moral and political value judgments of any individual. Individual values, such as feminist values, anti-feminist values, liberalism, conservatism, and so forth, are merely instrumentally valuable towards contributing to a research community with diverse values and interests. That is, they serve as a means of distributing research effort (for Solomon) or achieving critical evaluation of background assumptions (for Longino). Having participants with diverse values and interests is thought to be important on Millian accounts, not because values that have influenced scientists in the past were *wrong* or unjustified, but because they caused scientists to ignore alternative areas of research, hypotheses, models, or potential problems with their background assumptions. Because it is the diversity of individual values, and not the content of those values that is important within a Millian framework, no particular set of values or interests should be endorsed or excluded within a Millian democracy. All values are equally important to generating potential challenges to scientific reasoning and practices. In this sense, Millian democracies are neutral with respect to the content of the ethical, political, and social values of individual scientists.

Neutrality about values (in this sense) is not a plausible feature for a model for feminist science for several reasons. First, it gives rise to a conception of diversity and dissent that is ineffective in addressing feminist concerns. The kind of diversity that Millian democracies prescribe is a diversity of interests and values of those practicing science. Because Millian accounts are value-neutral, there are no constraints on the sets of values and interests that should be represented in an ideally diverse scientific community. Sexist, racist, and heterosexist values will play an equally important role in contributing to a diverse scientific community. On Longino's view, it is the diversity of ethical and political values represented in the scientific community, and not the particular content of values that play a role in generating transformative criticism. Similarly, for Solomon, *all* ideological commitments, or 'social and political factors,' are non-empirical decision vectors that can help balance each other out so as to distribute the research effort (Solomon 2001, 58). Thus, sexist values would be just as important to good science, because they are needed to counteract the influence of feminist values in distributing research effort.

One worries this has the implication that research in astronomy should include and fund scientists that belong to the Flat Earth Society or the Ku Klux Klan in order to ensure that participants in the research hold a full spectrum of diverse values and interests. This seems counter-intuitive from a feminist perspective because it prescribes the representation of values that feminists presumably take to be unjustified as a part of the ideal scientific community. Moreover, it seems implausible to devote time and resources to racist, sexist, or creationist research programs given the limited resources that exist for conducting science.

In addition, on Longino's view, proponents of such values would have to be accorded equality of intellectual authority and have their views taken seriously. On the Millian view, representatives of all ethical and political commitments will have equal authority to raise challenges to research priorities, questions, assumptions, and interpretations of data. Scientists with sexist and racist values will have equal intellectual authority with those who hold feminist values, despite the fact that their value judgments are arguably less justified. Thus, challenges raised by creationists, racists, and sexists would need to be taken seriously by others within a scientific community.

To be fair, Solomon and Longino provide some constraints on diversity and dissent that might be understood to prevent these sorts of counter-intuitive consequences. For Solomon, funding diverse research projects and allowing dissent are only appropriate in those contexts when conflicting theories each have empirical success (Solomon 2006, 27). We need not, for example, fund research by members of the Flat Earth Society, even though they offer dissenting views, because the theories they advocate have arguably not had empirical success.

This, however, assumes that 'empirical success' is a neutral criterion that will be applied the same by those with diverse values and interests. Yet part of the debate between, for example, evolutionary theorists and creationists is that they disagree about what counts as empirical evidence for or against their theories. Their value judgments operate as background assumptions that allow them to disagree about what constitutes compelling empirical evidence for disputed claims. For example, because creationists value consistency with religious doctrine, they tend to argue that certain data should not be interpreted as evidence against creationism, but rather interpreted as evidence that God has made it look 'as though' evolution were true in order to test faith. They have also argued that methods such as carbon dating do not generate reliable empirical data. As a result, it is not clear that empirical success is a foundational criterion that can be used to distinguish 'reasonable' from 'unreasonable' dissent within scientific communities while remaining neutral about the value judgments that are operating as background assumptions.

Longino, however, has additional resources to respond to the worry that requiring diversity of values and interests will require representation of sexist, racist, or creationist values. One of Longino's four criteria for ideal scientific communities is that there must be shared standards for theory evaluation that critics can evoke (Longino 1990, 76). Creationists, for example, must share enough epistemic values with the rest of the scientific community in order to be legitimately included and granted equality of intellectual authority. It is possible that in disputing the reliability of carbon dating or reinterpreting what counts as empirically successful creationists are denying the very sorts of shared standards that Longino requires for objective scientific communities.

Yet this response is insufficient. Creationists may in fact share some empirical values, including empirical success, but the problem is that their *other* value judgments (e.g., that the literal text of *Bible* ought to be taken as a supreme authority, or that scientific theories should be consistent with religious doctrine) cause them to apply and interpret epistemic values in ways that are different than those who do not hold those religious values. In this case, we want to be able to say that creationist challenges to traditional theories are not reasonable because the value judgments they rely on to apply epistemic standards are themselves unjustified. That is, there are not good reasons for thinking that the *Bible* ought to be taken as an authority and operate as a constraint on how to interpret data. One might point to the fact that the *Bible* has been written, revised, and interpreted many times over several centuries, or that this value judgment is inconsistent insofar as creationists themselves do not generally take *everything* in the *Bible* literally. Thus, the ways in which creationists have historically interpreted empirical success are not justified because their value judgments are not justified. This, however, is inconsistent with the Millian commitment to value-neutrality. That is, excluding creationists from ideally diverse scientific communities requires us to endorse particular values and interests as those reasonable to represent within a scientific community, which the Millian framework does not seem to support.

Yet a further criterion that Longino has for increasing objectivity in a scientific community is that the community as a whole must be responsive to criticism. In her later work, she expands on this by arguing for a tempered equality of intellectual authority where members of scientific communities will only be granted intellectual authority insofar as they are willing to take challenges seriously and either reply or revise their views in light of those criticisms (Longino 2002, 131–134). Thus, Longino might maintain that creationists, racists, or sexists, can be legitimately excluded from scientific communities insofar as they have failed to engage in appropriate uptake of challenges to their own views. Sexists and racists, insofar as they do not recognize women and people of color as having equal intellectual authority may refuse to take seriously or respond to challenges raised by members of those groups. Creationists might also be guilty of dogmatism, or holding on to their theories regardless of any empirical data or criticisms that may arise.

But this is still a fairly weak constraint on equality of authority. Racists, sexists, and creationists have attempted to address challenges raised by feminists, anti-racists, and evolutionary theorists.[9] As mentioned earlier, when evolutionary theorists point out that we do not observe a uniform fossil layer (which we would expect to see if all species were created simultaneously), creationists have acknowledged and responded to that challenge by revising their background assumptions so as to revise their empirical predictions. They account for fossil observations by maintaining that God made it look as though evolutionary theory were true. Again, the problem is not that they are unwilling to take challenges seriously or make revisions in response to criticisms. The problem is that their responses often rely on problematic value judgments that lead to genuine disagreement about how to interpret data, what needs to be explained, or how best to frame research questions. Thus, one central problem with the Millian commitment to value-neutrality is that it appears to advocate the inclusion of representatives of even unjustified values that can lead to intractable disputes about which theories are empirically successful.

So far, I have argued that the Millian commitment to value-neutrality is unacceptable because it leads to conceptions of diversity and dissent that are problematic. This leads to a second, related, problem with Millian accounts. Part of the reason

---

[9] Though, one might question whether true equality of intellectual authority (in Longino's sense) has truly been achieved in scientific communities. If not, then it is not clear whether the right sort of debate has occurred between, for example, feminists and anti-feminists.

that Millians endorse neutrality about the ethical and political values of individual scientists is they assume that value judgments never play a legitimate *direct* role in scientific reasoning. For Solomon, values are instrumentally valuable to balancing non-empirical decision vectors and to pursuing the maximal range of empirically successful hypotheses. For Longino, values are instrumentally valuable to creating a diverse research community where there will be rigorous examination of background assumptions. Yet while values are viewed as instrumentally valuable in these ways, value judgments themselves can never provide good reasons in framing research questions, collecting and interpreting data, or judging a hypothesis to be more justified than its competitors.[10] This is why the *content* of the individual values of scientists is presumed to be irrelevant for evaluating the science that results.

The fact that Millian accounts assume that the content of value judgments plays no direct role in scientific reasoning is problematic insofar as there appear to be a growing number of case studies demonstrating that endorsing ethical, political, or social value judgments can be necessary and legitimate in generating or evaluating evidence for a theory. As I have already argued, there are cases where ethical and political value judgments seem to play a direct role in determining whether a theory can be considered 'empirically successful.' In addition, many have argued that some areas of research deal with normatively-laden concepts that require scientists to make value judgments in testing and evaluating hypotheses that contain those concepts. Elizabeth Anderson (2004), for example, has argued that concepts concerning what constitutes 'harmful effects' on children of divorce require making ethical judgments about what constitutes harms and benefits to children. John Dupré (2007) has argued that the concept of 'rape' employed by hypotheses in evolutionary psychology has an irreducibly normative component. I have argued elsewhere that clinical psychologists must make ethical and social value judgments in identifying and classifying mental disorders (Intemann 2001). These are just a few cases within a large body of work that criticizes the assumption that there is a sharp distinction between 'facts' (which are the proper subject of scientific hypotheses)

[10] It is important to note that in some of Longino's work, she seems to give value judgments a far more robust role in scientific reasoning. In two late 1990s papers, she argues that commitments to certain values can be directly beneficial for science, and can have implications for the criteria for theory choice. Specifically, if researchers are committed to 'revealing gender' in our explanations and models, then this will have implications for our epistemic criteria, including how we understand explanatory power, whether we prefer simpler theories and models to more complex ones, and whether we accord value to theories that are consistent with other existing confirmed scientific theories (Longino 1995, 1996). In these papers, Longino appears to offer a different account of ideal science than the kind of value-neutral Millian democracy that I have ascribed to her here and thus my criticism would not apply. Nonetheless, these arguments are entirely missing in Longino's book-length treatments of her view. In *Fate of Knowledge* (2002), she appears to return to the views of her earlier book, where social and political values are only instrumentally valuable to generating critical examination of background assumptions.

and 'values' (which are not the concern of science).[11] If there are cases where scientists must (and should) endorse ethical, political, or social values in scientific reasoning, then such values will play a direct, not merely indirect, role in science.

This problem is particularly significant from a feminist perspective, as the role that feminist perspectives currently play in Millian democratic science is fairly limited. If feminist ethical and political commitments are more justified than alternatives in decisions about how to measure 'harms' and 'benefits' to children of divorce or what counts as 'rape,' a Millian conception of science fails to provide any support for endorsing feminist perspective over alternatives. Within a Millian framework, sexist scientists should also be given resources to pursue research framed by their political commitments and have that work taken seriously.

This brings us to a final reason why a value-neutral conception of democracy is problematic. The reason that values are taken to only play an indirect role in scientific reasoning is that Millian accounts seem to treat values as non-rational psychological causes or desires. On Longino's view, social and political values are primarily desires that *motivate* scientists to pursue certain hypotheses, adopt certain background assumptions, or prefer interpretations of data (Longino 1990, 86). For Solomon, all 'ideological' commitments are 'non-empirical' decision vectors that potentially cause a theory to receive more than its fair share of the research effort and so must be balanced out. For both Solomon and Longino, such values can have good or bad causal effects. They can lead scientists to pursue hypotheses that have previously been ignored or they can cause scientists to disregard data that undermines their favored theory. Or, they can lead scientists to adopt one set of epistemic or cognitive values over another. But, as values, they are taken to be psychological forces that are distinct from scientific theories and auxiliary hypotheses that can be rationally justified. Rather, they are treated as psychological causal forces that influence which background assumptions are adopted or which theories are accepted. This is why Millian accounts focus on structuring science so as to minimize any negative effects of the values and interests of individual scientists.

This assumes, or at least reinforces, a certain metaethical view; namely, that values are *not* the sorts of things that can be rationally or empirically evaluated. This metaethical view is problematic from a feminist perspective as it denies feminists the resources to explain why their ethical and political commitments (for example, that sexism is wrong or ought to be abolished) are better, or more justified, than sexist and racist value judgments.

---

[11] See for example Dupré (2007), Putnam (2004), Campbell (1998). One might be concerned that the examples I have listed here are generally from social or 'soft' sciences. As Dupré argues, such value-laden concepts are more likely to occur in areas of science that 'matter to us' or that are linked to human interests. Nonetheless, they are still important as scientific concepts necessary to understanding the world, and these examples have normative implications for the role(s) that values can legitimately play, at least in some areas of science.

To summarize, recent Millian accounts of democracy in science are problematic for three main reasons. First, they endorse a kind of neutrality about values that gives rise to conceptions of diversity and dissent that put racist, sexist, and creationist values on par with feminist values, as all are equally instrumentally valuable within scientific communities. Thus, while Millian democracies can account for the importance of the participation of feminists and other underrepresented groups in science, they also require the participation of racists, sexists, and other groups with different interests and values. Second, Millian democracies assume that ethical and political values are only indirectly or instrumentally valuable to distributing research effort or scrutinizing scientific reasoning. This limits the kinds of criticisms that feminists can make about male-bias or androcentrism within science. According to Millian accounts, what makes the influence of sexist values objectionable has nothing to do with the fact that the values involved are sexist or androcentric or, more to the point, *unjustified* judgments. The problem is that they are *ethical and political values* (of any kind), which have been allowed to inappropriately influence research because the scientific community was not appropriately diverse (in Longino's terms) or because there were no countervailing non-empirical decision vectors that would have better distributed research effort (in Solomon's terms). Thus, neither view allows us to criticize the content of the values employed in sexist reasoning in explaining why such science is bad science. This is particularly troubling insofar as there are cases where ethical and political values are directly relevant to choice of methodology, testing hypotheses, or evaluating whether a hypothesis is empirically successful. Finally, Millian democracies treat values as distinct from the sorts of claims that can be rationally justified. This ultimately deprives feminists of the resources they need to argue that their value judgments are superior to anti-feminist ones.

6.5 Toward an Alternative Conception of Democracy in Science

Seeing how the Millian conception of democratic science fails to fully address feminist concerns about science reveals certain features that a more successful conception of democracy might possess. Clearly, the task of fully articulating and defending such an alternative conception is beyond the scope of what can be accomplished here. I will, however, attempt to offer several key features that a feminist account of democratic science might possess based the preceding analysis.

The initial motivation for the Millian conception of democratic science stems from the recognition of the epistemic fallibility of individuals. The central insight that justification is a social process that involves diverse individuals and allows for critical dissent is indeed valuable. The question is whether this insight can be accounted for in a way that avoids a commitment to value-neutrality and its counter-intuitive consequences.

6.5.1 Diversity

First, the kind of diversity that seems epistemically important to scientific communities is not a diversity of values and interests per se, but rather diversity of experiences that participants bring to the table. The reason why, for example, scientists may have failed to see potential problems with the rod and frame test that was supposed to measure visual-spatial ability is not that they were necessarily driven by sexist values or interests. It may have simply been that the male scientists had never had the sorts of experiences that would have made them aware of how uncomfortable it might be to be in a dark room with a strange man, or that one might feel bad making an experimenter continue to make minute adjustments to a rod. Scientists who *did* have such experiences would have been more likely to see the problems in the experimental design. Those who have had different experiences are more likely (at least in some cases) to think of different theories, models, or alternative explanations. In some cases (such as the rod and frame test) individuals with different experiences may have access to relevant evidence of which others may not be aware.

Historically, systems of oppression (such as racism, sexism, heterosexism, and classism) have influenced the material circumstances of individuals (such as their living conditions, their opportunities, and they way they are treated in a variety of social situations). As a result, it is likely that individuals from different social locations will have, to some extent, different experiences. Thus, by including individuals from relevantly diverse social positions, the resulting epistemic community will be comprised of scientists informed by diverse experiences that can increase the rigor of scrutiny brought to background assumptions, theories and models (Rolin 2006; Intemann 2009).

This is not to say that *all* experiences will be equally important in contributing to a research community that can rigorously scrutinize background assumptions. Some experiences will be more relevant in particular research contexts because of the content of the background assumptions relied upon in that research. People who are left-handed, for instance, may have different sorts of experiences than right-handed individuals, such as not being able to use easily the same scissors or knives as others, and this may even provide them with evidence for the belief that such tools are made for those who are right-handed. Yet these sorts of experiences are unlikely to be relevant to evaluating background assumptions related to research on climate change or racial differences in health disparities. Similarly, different experiences related to gender may not be relevant to work in theoretical physics. Which sorts of experiences are relevant depends partly on the content of the research.

In addition, certain sorts of experiences may be more valuable epistemically than others in particular research contexts. This is because there are certain experiences that have been historically underrepresented precisely because of the exclusion and marginalization of certain social groups. The experiences of women in developing countries have been historically underrepresented in research on the impacts of climate change and research on genetically engineered food (Harding

2008; Shiva 1999). As a result, they may be able to challenge assumptions identify limitations with existing models that do not capture ways in which women in developing countries are affected by these problems differently from men (Intemann 2009). In contrast, the experiences of whites living in a social context where whites are believed to be superior has not be excluded or underrepresented in research on racial differences. In fact, these sorts of experiences have significantly informed such research for some time. Thus, the inclusion of racist scientists is not likely to be as fruitful in terms of identifying or evaluating problematic background assumptions.

Moreover, empirical evidence may reveal that certain types of experiences produce unreliable beliefs. For example, the experience of those who have been raised in extreme isolation since children (such as those living in religious sub-communities having little contact with the outside world) may produce unreliable beliefs about the larger world.

Thus, diversity of experiences, merely for the sake of diversity, is not always valuable. The sorts of experiences that will be important to represent in a particular research context will depend upon the content and aims of the research, the sorts of experiences that have been historically underrepresented in that research, and whether those experiences are likely to yield epistemic benefits all things considered.

Consequently, this understanding of diversity does not require that we maximize the representation of different values, interests, or experiences. Rather, it requires promoting the sort of diversity that will be epistemically beneficial to a particular research context. There are some values and interests that are unreasonable or unjustified such that they do not deserve representation within scientific communities and there will be certain sorts of experiences that are irrelevant or unreliable with respect to certain areas of research. We need not seek to include representatives of, for example, the interests of cigarette companies in research on lung cancer as their values, interests, and experiences, have been established to be both unjustified and unreliable in producing accurate beliefs about smoking-related health problems.

Attempts to incorporate this kind of diversity within particular research groups is already promoted by organizations such as the NSF that evaluate grant proposals according to whether they involve the participation or training of underrepresented social groups, though, this aim has not been fully realized.

6.5.2 Mechanisms for Identifying and Endorsing Values with Multi-Expertise Feedback

I have argued there are cases where ethical and social value judgments will be directly relevant to collecting and interpreting data, particularly measuring empirical 'success'. This will be most likely to occur in cases where the aims of the research are tightly connected with certain social aims, such as biomedical research, or other areas of science that relate to public policy or the development of new technology.

In these cases it will be important for research groups to identify and agree about the social and ethical values that will guide research. Consider, for example, research on whether an oral contraceptive drug is ‘safe’ and ‘effective’ for women. Researchers will need to endorse ethical and social aims of the research. Is it to produce a drug that will be effective for women in developing countries, or only for those who can afford to buy it? What kinds of health risks will be serious enough to outweigh increased reproductive freedom (such that they ought to be examined)? Is it important to assess the long-term effects of drugs, and if so, what duration of study is necessary? In addressing these sorts of questions, certain ethical and social values will need to be adopted to guide the research.

Once a set of values has been endorsed, dissent about those values in the context of conducting the research would not be beneficial or productive. At the same time, the process of identifying and endorsing the social and epistemic values that will guide research is one that is also fallible and should be open to rigorous scrutiny. Moreover, it is not clear that sufficient scrutiny can be achieved by members of a scientific community or research group alone. At least at this point, scientists are not adequately trained to identify and evaluate many of the ethical and social value judgments relevant to their research. They are not encouraged to think about the social aims of their research (or what implications those aims might have for their methodological decisions). Nor are they trained to recognize and analyze normative concepts that may be central to their research (such as ‘harm’).

Thus, it is important for research groups to collaborate with a variety of experts from field that might be relevant to their area of investigation. Ideal research communities will be those not just with a diversity of life experiences, but also with diversity of skill sets or disciplinary expertise. In the oral contraceptives case, researchers might work with public health scientists, sociologists, and ethicists in developing methodologies that will best promote the social aims of the research and receiving feedback on identifying and evaluating the value judgments at stake. It might be possible for this sort of feedback to occur at the level of Institutional Review Boards (IRBs), though current IRBs tend to ignore the complex ways in which value judgments are present and relevant to decisions that are made throughout the research process.

Again, this does not imply that *all* skill sets or disciplinary expertise will be relevant to every research context. The sorts of expertise that will be important to include will depend on the content of the research and whether there are any other reasons for thinking that the collaboration would have other problematic consequences. For instance, an expert in therapeutic touch, who uses hand movements to address health problems by manipulating purported energy fields around the patient has very different ‘expertise’ than biomedical researchers. Yet there is much evidence that therapeutic touch is ineffective, even if it were in some sense relevant to background assumptions about, say, cancer research.

6.5.3 Dissent

Although it will also be important to allow room for dissent within scientific communities (again because of the epistemic fallibility of individuals), we need a conception of *reasonable dissent.* That is, there should be equality of intellectual authority among those within a scientific community to propose alternative hypotheses, methodologies, and interpretations of data, as well as raise challenges to the work of others within the scientific community, so long as that dissent relies on reasonable, plausible, or empirically viable claims (including value judgments). Of course, this immediately raises the question: what are *reasonable* claims or value judgments?

Which claims (including value judgments) are reasonable will depend upon the available evidence, widely shared considered judgments, as well as features of the particular research context. First, reasonable dissent should not depend on a claim or value judgment that has already been given careful consideration and rejected, provided there is no new evidence that would make it reasonable to reconsider it. This is why several of the challenges mounted by racists and creationists should not, at this time, be taken seriously. At this point, resources and attention have already been given to claims made by such groups and we have very good evidence that their theories and value judgments are unjustified. Dissent that revisits such disputes is not reasonable or productive given the fact that we have limited resources. Of course, this is not to say that the value judgments of racists, sexists, or creationists could never be considered or debated outside of the context of practicing scientific communities. If, at some point, new evidence suggested that those value judgments *were* plausible, this could change the status of them to among those reasonable to advocate within science.

Second, reasonable dissent will also depend on the accepted aims of the research that have been developed and endorsed by those with diverse experience and multi-disciplinary expertise. If a research program is designed aiming to develop an oral contraceptive drug for *all* women who are agreed to be deserving of full moral consideration, then only certain challenges will be reasonable within the context of conducting and interpreting that research. Within that context, for example, it would not be reasonable to question whether risks to women are acceptable that we would not deem acceptable for other moral beings. It might, however, be reasonable to question how various risks should be balanced or whether clinical trial methods are accurately testing the drug under conditions that apply to *all* women. Reasonable dissent must be consistent with the agreed aims of the research.

Third, dissent must be relevant to the epistemic and social aims of the research. Criticisms of research that point out alternative explanations or models, may be reasonable in relation to shared epistemic goals. However, dissent that is unrelated to such goals may be rejected. If scientists funded by cigarette companies rejected experimental results that found nicotine addictive simply because they did not like the results of the research, or because it would be consistent with their continued funding, this would be irrelevant to the widely shared epistemic and social aims of such research.

Whether particular challenges and criticisms are reasonable, of course, may change over time. Mistakes may also be made in making this assessment. For example, there may have been values and assumptions regarding race and sex that were plausible to assume in the 1800s, given the available evidence at the time, but for which we now have overwhelming evidence to reject. My claim, however, is that value judgments can be more or less rationally held[12] and that dissent within scientific communities should only include those values and interests that are well supported by the best available evidence.

The features I have identified capture the central insight of Millian science: that individual epistemic agents are fallible and that knowledge is best acquired through social processes that encourage diversity and dissent. But, they avoid any commitment to value-neutrality that has characterized recent Millian accounts.

This alternative account better advances feminist aims. First, it provides an explanation of why the inclusion of marginalized groups is particularly important to science. At the same time, it does not necessitate the representation of *all* sets of values and interests. Creationists should not have representation or be granted equal intellectual authority within research programs in astronomy, because the value judgments they rely on in order to mount challenges and measure empirical success are unjustified. Individuals who hold value judgments for which there is little rational support should not be given equal representation or intellectual authority within scientific communities. Second, this account still allows us to combat male-bias or androcentrism in science. Ideal scientific communities will include those with experiences that are relevant to recognizing and challenging androcentric assumptions. There will also be opportunities for those with expertise in a variety of related fields to help shape the aims and methods of research programs. Finally, this alternative account allows for the possibility that feminist perspectives will be more reliable or beneficial to science than anti-feminist ones. Feminist value judgments will be more beneficial in certain contexts insofar as they are more *justified*. In other words, my view rejects the kind of value-neutrality that has characterized recent Millian accounts. A feminist conception of democratic science should involve a conception of democracy where value judgments can be rationally evaluated, revised, or endorsed.

6.6 Conclusion

So, does democracy in science always serve feminist aims? I have argued that the answer to this question depends partly on the conception of democracy employed. A Millian conception of democracy that is neutral with respect to values, and

[12] Elizabeth Anderson (2004) also argues that value judgments are the sorts of claims that can be supported by evidence and, because of this, there is nothing that would logically prevent them from playing a role in scientific reasoning.

requires diversity and dissent for the sake of diversity and dissent does not always serve, and may even hinder, feminist aims. That is, it appears to put anti-feminist values and interests on par with feminist values and interests, which leads to several problematic consequences. Yet the central insight of Mill, that individual epistemic agents are fallible, is important and has implications for how epistemic communities might be best conceived. I take my alternative account to be consistent with that insight and, in that sense, 'Millian.' Yet unlike Mill, Longino, or Solomon, I take myself to be offering more rigorous constraints on how to understand diversity and reasonable dissent. These additional constraints, I believe, are ultimately necessary for addressing feminist concerns as well as generating normative prescriptions for good science.

Acknowledgments Special thanks to Heidi Grasswick, Carla Fehr, and Edrie Sobstyl for helpful comments and suggestions on an earlier draft of this paper, as well as Elizabeth Potter, Alison Wylie, and other audience members at the 2007 FEMMSS2 Conference. Any remaining problems are my own.

References

Anderson, Elizabeth. 2004. Uses of value judgments in science: A general argument, with lessons from a case study of feminist research on divorce. *Hypatia* 19(1): 1–24.

Campbell, Richmond. 1998. Illusions of Paradox: A feminist epistemology naturalized. Lanham: Rowman and Littlefield Publishers

Dupré, John. 2007. Fact and value. In *Value-free science? Ideals and illusions*, ed. Harold Kincaid, John Dupré, and Alison Wylie, 27–42. Oxford: Oxford University Press.

Fausto-Sterling, Anne. 1985. *Myths of gender*. New York: Basic Books.

Harding, Sandra. 2008. *Sciences from below: Feminisms, postcolonialities, and modernities*. Raleigh: Duke University Press.

Intemann, Kristen. 2009. Why diversity matters: Understanding and applying the diversity component of the NSF's broader impacts criterion. *Social Epistemology* 23(3–4): 249–266.

Intemann, Kristen. 2001. Science and values; Are moral judgments always irrelevant to the justification of scientific claims? *Philosophy of science,* [Proceedings of the Biennial Meeting of the Philosophy of Science Association, supplement to vol. 68, no. 3]: S506–S518.

Kitcher, Philip. 2001. *Science, truth, and democracy*. New York: Oxford University Press.

Longino, Helen. 1987. Can there be a feminist science? *Hypatia* 2(3): 51–64.

Longino, Helen. 1990. *Science as social knowledge*. Princeton: Princeton University Press.

Longino, Helen. 1994. The fate of knowledge in social theories of science. In *Socializing epistemology: The social dimensions of knowledge*, ed. Frederick Schmitt, 135–157. Lanham: Rowman and Littlefield.

Longino, Helen. 1995. Gender, politics, and the theoretical virtues. *Synthese* 104: 383–397.

Longino, Helen. 1996. Cognitive and non-cognitive values in science: Rethinking the dichotomy. In *Feminism, science, and the philosophy of science*, ed. Lynn Hankinson Nelson and Jack Nelson, 39–58. Dordrecht: Kluwer.

Longino, Helen. 2002. *The fate of knowledge*. Princeton: Princeton University Press.

Mill, John Stuart. [1859] 1982. *On liberty*. New York: Prometheus Books.

Putnam, Hilary. 2004. *The collapse of the fact/value dichotomy and other essays*. Boston: Harvard University Press.

Rolin, Kristina. 2006. The bias paradox in feminist standpoint epistemology. *Episteme* 1(2): 125–136.

Shiva, Vandana. 1999. *Stolen harvest: The hijacking of the global food supply*. Cambridge: South End Press).
Solomon, Miriam. 1994. A more social epistemology. In *Socializing epistemology: The social dimensions of knowledge*, ed. Frederick Schmitt, 217–233. Lanham: Rowman & Littlefield.
Solomon, Miriam. 2001. *Social empiricism*. Cambridge: MIT Press.
Solomon, Miriam. 2006. Norms of epistemic diversity. *Episteme: A Journal of Social Epistemology* 3(1): 23–36.
Wray, K. Brad. 1999. A defense of Longino's social epistemology. *Philosophy of science*, vol. 66. Supplemental Proceedings of the 1998 Biennial Meetings of the Philosophy of Science Association, Part I: Contributed papers, S538–S552.
Wylie, Alison. 2001. Doing social science as a feminist: The engendering of archaeology. In *Feminism in twentieth century science, technology, and medicine*, ed. Angela N.H. Creager, Elizabeth Lunbeck, and Londa Schiebinger, 23–45. Chicago: University of Chicago Press.
Wylie, Alison, and Lynn Hankinson Nelson. 2007. Coming to terms with the values of science: Insights from feminist science studies scholarship. In *Value-free science? Ideals and illusions*, ed. Harold Kincaid, John Dupré, and Alison Wylie, 58–86. Oxford: Oxford University Press.

Chapter 7
What Is in It for Me? The Benefits of Diversity in Scientific Communities

Carla Fehr

Abstract I investigate the reciprocal relationship between social accounts of knowledge production and efforts to increase the representation of women and some minorities in the academy. In particular, I consider the extent to which feminist social epistemologies such as Helen Longino's critical contextual empiricism can be employed to argue that it is in researchers' epistemic interest to take active steps to increase gender diversity. As it stands, critical contextual empiricism does not provide enough resources to succeed at this task. However, considering this view through an employment equity lens highlights areas where such theories need to be further developed. I argue that views such as Longino's ought to attend to nuances of community structure and cultural features that inhibit critical social interactions, if we are to maximize the epistemic as well as the ethical improvements associated with a social approach to knowing. These developments advance these epistemic theories for their own sake. They also help develop these theories into a tool that can be used by those calling for increased diversity in the academy.

Keywords Feminist philosophy of science • Social epistemology • Implicit bias • Employment equity • Workplace environment issues

7.1 What Is in It for Me?

A while ago I gave a lecture to science faculty members and university administrators regarding the underrepresentation of women and minorities in science, technology, engineering and mathematics (STEM) fields.[1] After my talk, an administrator, with

[1] In this paper I am primarily focusing on gender diversity. By doing so it is not my intention to minimize the epistemological and ethical concerns relating to the underrepresentation of members of other marginalized groups.

C. Fehr (✉)
Department of Philosophy, Iowa State University, Ames, IA 50011, USA
e-mail: cfehr@iastate.edu

H.E. Grasswick (ed.), *Feminist Epistemology and Philosophy of Science: Power in Knowledge*, DOI 10.1007/978-1-4020-6835-5_7,

demonstrated good will, gave me a 'tip.' He told me that if I 'wanted to get traction' when I was addressing the problem of the underrepresentation of women in STEM, I needed to 'answer the question that was in most of "these guys" minds – what is in it for me?' I was taken aback because the statistical data I presented painted a blunt and grim picture. While women have made up at least half of the undergraduate student body since the 1980s, these gains have not translated into corresponding increases in the proportion of women in the professoriate. Since the 1970s the number of women earning doctorates has tripled while the number of women who are full-time faculty has only increased by 1.5 times (West and Curtis 2006). At every stage of professional development a higher proportion of women faculty than men faculty leave the academy (NAS 2007; West and Curtis 2006). This is often referred to as the leaky pipeline.

Women faculty members tend to be concentrated in less prestigious institutions, at lower ranks and in less secure positions. According the National Academy of Sciences, at top research institutions, only 15.4% of the full professors in the social and behavioral sciences and 14.8% in the life sciences are women. The authors go on to write, 'these are the only fields in science and engineering where the proportion of women reaches into the double digits. Women from minority racial and ethnic backgrounds are virtually absent from the nation's leading science and engineering departments' (NAS 2007, S-2). Currently, 30% of women faculty members are in non-tenure track jobs, while only 18% of men faculty members hold these positions (West and Curtis 2006). At doctoral granting institutions, full time women faculty members are only half as likely to be tenured as full time men faculty members (West and Curtis 2006).[2] The demographic data show that there is indeed a problem with the underrepresentation of women and minorities. But according this administrator, the data alone, even though striking, were insufficient to motivate discussions of institutional change. When arguing for employment equity, it would be useful if, in addition to justice or ethics based arguments, we could also marshal arguments regarding the epistemic benefits that both women and men can garner from increasing the proportion of women and minority STEM faculty.

In this paper I investigate the reciprocal relationship between social accounts of knowledge production and efforts to improve the representation of women and some minorities in the academy. In particular, I consider the extent to which feminist social epistemologies such as Helen Longino's critical contextual empiricism, can be employed to argue that it is in researchers' epistemic interest to take active steps to increase the diversity in their communities. As it stands, critical contextual empiricism does not provide enough resources to answer the administrator's question. However, considering epistemologies such as Longino's through an employment equity lens highlights areas where such theories can be further developed. I argue

[2]I refer the reader to the following excellent sources of data on the representation of women and minorities in STEM and the academy more generally, and for summaries of social science research that shed light on the causes of these inequities: West and Curtis (2006), The National Academy of Sciences (2007), Wylie et al. (2007) and Xie and Shauman (2003).

that in order to answer the 'What is in it for me?' question, accounts such as Longino's require a more robust notion of community structure and more careful attention to the culture of our knowledge producing communities. The arguments provided here both advance the epistemic theory for its own sake and help develop the theory into a tool that can be used by those calling for increased diversity in the academy.[3]

7.2 An Example of 'Diversity Promotes Excellence' Theories

One possible response to the 'what is in it for me?' question arises out of work in feminist epistemology that points to communities rather than individuals as being the locus of knowledge production and argues that diverse communities have epistemic benefits that homogeneous communities lack (Longino 1990, 2002; Nelson 1990, 1993). Much of this feminist work can be represented by the catch phrase 'diversity promotes excellence.' The most developed diversity promotes excellence theory is Helen Longino's critical contextual empiricism. According to Longino, objectivity (1990) and justification (2002) require effective critical discourse within a diverse community. When a community is homogeneous with regard to the background assumptions, prejudices, and theoretical perspectives of its members, those assumptions, prejudices and perspectives can go unnoted and unchallenged. But when a community is diverse, the assumptions are more likely to be brought to light and subjected to explicit evaluation. Longino (2002, 129) writes that 'Effective critical interactions transform the subjective into the objective, not by canonizing one subjectivity over others, but by assuring that what is ratified as knowledge has survived criticism from multiple points of view.'

Longino has carefully described the characteristics of ideal epistemic communities that allow them to maximize the effectiveness of these critical interactions. According to Longino, an ideal community (1) has public venues for critical interactions, (2) has public standards for evaluating theories, hypotheses and data, (3) gives dissent uptake and (4) treats its members with tempered equality of intellectual authority. In later sections of the paper I will argue that uptake and equality of intellectual authority prove to be particularly challenging criteria to meet. *Uptake* points to the notion that successful communication requires not only that the speaker clearly state her views, but also that the listener is willing and able to pay attention to those views and engage those views with openness to the possibility of being convinced. Longino makes this point as follows, 'The community must not merely tolerate dissent, but its beliefs and theories must change over time in response to the critical discourse taking place within it' (2002, 129).

A community must also grant its members *tempered equality of intellectual authority*, meaning that assent is not forced by economic or social power and that

[3] See Wylie (this volume) for a similar strategic use and development of standpoint theory.

'every member of the community be regarded as capable of contributing to its constructive and critical dialogue' (2002, 132). 'The social position or economic power of an individual or group in a community ought not determine who or what perspectives are taken seriously in that community' (2002, 131).[4] The epistemological goal of tempered equality is to expose hypotheses to the widest range of critical scrutiny. Longino presents epistemic communities with a challenge:

> Thus a community must not only treat its acknowledged members as equally capable of providing persuasive and decisive reasons and must do more than be open to the expression of multiple points of view; it must also take active steps to ensure that alternative points of view are developed enough to be a source of criticism and new perspectives. Not only must potentially dissenting voices not be discounted; they must be cultivated (2002, 132).

There are many examples where the addition of women, with varying degrees of implicit or explicit feminist perspectives, have had a positive impact on our understanding of science, on the practice of science and on the products of scientific work. This can be seen in the critiques offered by feminist science studies scholars who are also working scientists. These critiques often involve revealing and questioning the role of gendered assumptions in the development of research questions, application of theories, choice of research methods and experimental design. The Biology and Gender Study Group (1989) describes feminist work as a control for gender influences. Examples include assumptions of female passivity and male activity that have structured investigations of prenatal development of sex differences (Birke 1986), the study of the mechanisms of fertilization (Martin 1991), the roles of hormones in the development of behavioral sex differences (Birke 1986; Longino and Doell 1983; Longino 1990) and the contributions of males and females to human evolution (Hubbard 1982; Hrdy 1986). Patricia Gowaty writes of her own research in evolutionary ecology, mating systems and sexual conflict,

> Feminism made the experimental designs better. Being self-conscious about my politics has made my experiments better than they might otherwise be, because I institute a variety of controls that others might also use, and would no doubt use, if they were more aware of their own biases. (2003, 917)

Donna Haraway's book *Primate Visions* (1989) documents the impact the incorporation of feminist women in research communities had on the study of primate behavior and animal behavior more generally. For example, primatologist Jeanne Altmann, instigated a quiet but powerful methodological revolution. In one of the most cited papers in the study of animal behavior, 'Observational study of behavior: Sampling methods' (1974), Altmann evaluated a range of sampling methods and in doing so developed a method, focal-animal sampling, that both undermined much previous research generating sexist accounts of leadership and control, and enabled research on topics such as mothering. Although Altmann reports that her location as a woman, a feminist and a mother were influential in her

---

[4]Longino is also sensitive to the fact that people have differences in training and ability that may grant them a cognitively privileged position in communities, but that does not impact the respect that ought to be shared among community members.

work, she did not write in an explicitly political way about these topics (Haraway 1989). The 'Sampling Methods' paper did not refer to gender at all, but rather to an analysis of the kinds of sampling that allow for good science. Focal animal sampling provides an effective method for studying the social behavior of female primates. This method has become an important approach in research about non-gendered as well as gendered phenomena. Altmann's methodological study had an impact on the practice of science that went beyond research on topics directly related to gender.

In Evelyn Fox Keller's (1983) biography of Barbara McClintock and in her subsequent book, *Reflections on Gender and Science* (1985), she reveals that McClintock was not conducting research on a gendered topic and did not identify as a feminist. However, Keller argues that the social experiences that came along with being a girl and woman affected her psychology in a way that made it possible for her to develop the close relationship with her study organisms that facilitated her Nobel Prize winning work on transposition. The addition of women, with varying degrees of feminist commitment, to scientific communities can uncover gendered assumptions, provide new or alternative methodologies and engage alternative perspectives that have bearing on research that relates to sex and gender and even research that does not.

Although there is a long list of cases where the addition of women to research communities allowed those communities to produce different and better science, the kinds of generalizations that one can draw from this list are not clear. It would be nice to be able use these examples of women making a difference in the practice of science, in conjunction with diversity promotes excellence theories, to argue that academic STEM departments ought to hire a diversity of candidates. It would be nice to be able to argue that departments ought to embrace hiring practices that increase diversity because the research produced by the department and its members will be improved. It will be more objective or better justified. Members of that community will have a better chance of spotting their assumptions, will have access to a wider range of methods, and will have access to those with a broader range of attitudes towards their work, if they are a member of a department where they can interact with people who are different from themselves. One cannot know beforehand what kinds of differences will be salient and so it is a good idea to nurture diversity in academic departments.

However, significant work needs to be done before we can make these nice arguments. First, members of academic departments can gain at least some of the benefits that arise out of diversity without hiring more women or minorities because they can 'free ride' off diversity that is present in other communities both inside and outside of the academy. In order to sort out the epistemic benefits of diversity to departments, I will develop an account of the kinds of communities of which scholars can be members, the interactions among those communities and the effects of status differences among community members. Second, in order for departments to gain maximal epistemic benefits from increasing diversity, they need cultures that enable women and minorities to effectively develop and express dissenting views. In order to assess the importance of these cultural changes, I will explore the relationship between social position and theoretical perspective. While Longino exhorts members of communities to grant each other's dissent uptake and to grant each other equality of intellectual authority, I will argue that it will take substantial cultural changes in order to meet these desiderata with regard to women scientists.

7.3 Epistemic Communities, Diversity Free Riders and Diversity Development

In order to assess the benefits that can arise from being a member of a diverse epistemic community we need a more detailed account of community structures and relations than the one Longino offers. Longino discusses communities in terms of groups of people who engage in critical interactions regarding their scholarship: a community is constructed in terms of who interacts with whom. When she discusses ideal communities she describes them as having public venues for critical interactions and some shared evaluative standards (Longino 1990, 2002). Lynn Hankinson Nelson offers a similar, yet thicker, definition of an epistemic community as a group that 'constructs and shares knowledge and standards of evidence' (1993, 124). Nelson goes on to write,

> [E]pistemological communities are multiple, historically contingent, and dynamic: they have fuzzy, often overlapping boundaries; they evolve, dissolve, and recombine; and they have a variety of purposes and projects which may include (as in the case of science communities) but frequently do not include (as a priority) the production of knowledge. (125)

That communities are multiple and dynamic and that they have overlapping boundaries has implications for the distribution of the benefits of epistemic diversity. Longino exhorts communities to 'cultivate potentially dissenting voices,' but communities are multiple and it is reasonable to consider which communities need to do this work in order for researchers to reap the benefits of diversity. I will argue that it is possible for a particular community to reap the epistemic benefits of diversity that Longino illustrates without cultivating dissenting voices.[5] It becomes important to spell out what cultivating dissenting voices means. If one holds an *inclusive sense of community* as simply being those with whom one interacts, then developing dissenting voices in a community can simply mean engaging in social interactions with people who have different social locations or theoretical perspectives than one's own. However, developing dissenting voices could also mean nurturing those

[5]Solomon and Richardson (2005) and Solomon (2006) also argue that Longino's conception of ideal epistemic communities is problematic. Solomon and Richardson argue that we lack historical and contemporary cases of scientific practice that meets these ideals; as a result we lack evidence that following them will lead to better science. Solomon (2006) argues that group deliberative processes can be influenced by biasing mechanisms associated with groupthink that are not transparent to members of groups and that her aggregative procedures lead to better epistemic outcomes than rational deliberative procedures such as Longino's. However see Wylie (2006) for arguments that Solomon's aggregative procedures as well as Longino's deliberative procedures are subject to implicit cognitive errors associated with gender schemas. I argue that views such as Longino's ought to attend to nuances of community dynamics and cultural features that inhibit critical social interactions, if we are to maximize the epistemic as well as the ethical improvements associated with a social approach to knowing.

with dissenting voices and working to increase the representation of those with dissenting voices in various specific communities within the academy. One can accomplish the former without working toward the later by free riding off existing diversity. This makes it difficult to use diversity promotes excellence theories, such as Longino's, to argue for increasing the representation of women and some minorities in particular communities, and is suggestive of ways that theories such as Longino's can be further developed.

7.3.1 Formal and Informal Communities

Faculty members are associated with numerous, overlapping, formal and informal epistemic communities. This distinction between formal and informal communities will often be a matter of degree and will be dependent on context. In this paper I am focusing on scientific or academic institutions. In this context a *formal community* is one that is institutionally recognized and conducts the kinds of activities acknowledged as contributions to the professional advancement of a faculty member. Membership in this sort of formal community is likely something that one could list on their vita. Formal communities can include academic departments (where members can be students, post doctoral researchers or faculty), committees or professional organizations. Departments are formal communities and are important because they are the primary place where scholars are paid for their epistemic labor. It is primarily departments that hire scholars to do scholarly work. While there is lots of scholarly work that happens outside of formal communities, workers could likely do more and better work if they were compensated for that labor by a formal community.

Informal communities can include professional networks of scholars at one's own institution or at other institutions. They can include networks of people outside of one's narrowly defined field but whose perspectives nonetheless influence one's research. They can also include networks of people who are not academics at all. Informal communities can be important sources of alternative perspectives and scholarly as well as personal support. One can gain the epistemic benefits that Longino describes by developing diversity in an informal community and without having to increase diversity in their formal communities, for example their departments, in order to gain these benefits.

7.3.2 Status – Marginal and Central Positions Within a Community

One can be a relatively marginal or central member of a formal or informal community or hold a perspective that is relatively marginal or central in terms of being valued by one's community. Longino's criteria of equality of intellectual authority is designed to ensure that the dissent arising from those who have little power or status in a community, a community's marginal members, is given uptake. However, as I will press in Sect. 7.5, it is likely that women's dissent is not given the same degree of uptake as men's because of unconscious cognitive errors or implicit biases that women are as likely to hold as men (Valian 1999). As a result of the unconscious nature of these biases, it is possible for people of good will to genuinely believe that they are giving women's dissent uptake and fairly rewarding women for epistemic diversity work (or for any of their professional accomplishments), when in fact they are not.

Donna Haraway's (1989) description of the revolutionary work of primatologist Jeanne Altmann exemplifies the multiplicity of communities of which researchers can be members as well as differences in the status of community members. At the time when Altmann published her 'Sampling methods' paper she had neither a doctorate nor any formal training in evolutionary or behavioral biology; she was a Research Associate in the Biology Department at the University of Chicago where her husband was a professor. This position was marginal in many senses, not the least of which was that she could not be invited to speak at conferences without her husband. On the other hand, she was a central figure at the long term Baboon study site in Amboseli National Park in Kenya. At Amboseli she collected many hours of field observations and this work was published in full collaboration with her husband. She was also deeply involved with many informal communities. Haraway writes that,

> Jeanne Altmann would get letters from students in the field with little training asking for methodological help; she gave it. Progressively, she became a node in a network or "invisible college" of field workers, including a growing network of young women. Jeanne Altmann became simultaneously a senior mentor and a peer contributor to reformulations of what could count as female for scientists and for their research subjects, animal and human. (1989, 308)

Altmann was a member of many communities that overlapped. Some of these communities gained epistemic benefits from her work while she held an official position that was marginal, was low-ranked and that positioned her as the wife of a professor rather than a professor herself. Although she went on to become an eminent scientist, even if she hadn't, the disciplines of primatology and animal behavior would still have benefitted from her work. Our sample of women or feminist scientists is biased because it is usually those who persist in their careers and become eminent whose stories are told. But, even among the group of women who persist in academic careers, and even in the case of an elite woman scientist such as Altmann, both formal and informal communities can benefit from their presence while these women are only marginal members of any formal community, or only members of an informal community.

7.3.3 Diversity Free Riding

Because epistemic communities are overlapping, one can glean the benefits of being a member of a diverse epistemic community by cultivating an informal community while being a member of an academic department, a formal community, that has no diversity at all. Although from an ethical perspective one should, one need not, confer rewards on the members of the informal community one cultivates. Further because of implicit gender biases it is likely the case that women accrue fewer of these rewards than men for the same kinds of work. I call this diversity free riding. A *diversity free rider* is either an individual or community who makes use of existing diversity without increasing the diversity of any formal community or without increasing the total representation of diverse voices. The notions of 'making use of diversity' and 'increasing diversity' need to be filled out; in doing so attention to the relatively marginal or central positions of community members and epistemic perspectives is important.

7.3.4 Making Use of Diversity

Making use of existing diversity involves using people with theoretical perspectives or social locations different from one's own as a means of doing better science by using them to increase the diversity of views in one's communities. This can be motivated by salutary reasons. One can imagine a researcher who suspects that his theoretical background, research design, decisions regarding the saliency of different sorts of data or interpretation of data may be based on gendered assumptions that he does not see. As a result he may seek out people with expertise with regard to gender or feminism (that results from their theoretical perspective or social location), develop an informal community that includes them, and use their perspective to uncover gendered assumptions in his research. This sort of reflexivity, of consciously using social interactions to uncover one's own assumptions, can result in creating maximally accurate knowledge of a gendered topic, which is a good thing and perhaps is not as common as we might like. Developing this informal community may also lead to the professional advancement of the scholar with expertise regarding gender, although this need not be the case. One can achieve these locally beneficial goals without contributing to the education or professional standing of the person who is being used as a source of diversity. The person being used as a source of diversity is doing what I call *epistemic diversity work*. Such work can be manifested in a number of ways, including talking to members of a community about their research and commenting on or reviewing grant applications or papers. These kinds of activities are often part of the everyday research and service activities of academics. However, epistemic diversity work is often performed in addition to the activities in which these scholars engage as part of their own research programs.

The impact of free riding on a diversity worker will depend on the social location of the worker and the kind of diversity that a worker adds to a community. For example, an esteemed developmental biologist may be called on to provide a different perspective on the research of a community of population geneticists, and the community of population geneticists may not reward the developmental biologist or do anything to increase the representation of developmental biologists in the academy. However, a diversity worker with high status and a valued research program will likely incur relatively small costs associated with being the object of free riding or with declining offers to do diversity work.

On the other hand, the situation can be very different for a diversity worker who is a marginal member of a formal or informal community or who is doing work that is not highly valued in those communities. As I will point out later, this is more likely to be the situation of a diversity worker who is a woman or feminist researcher, or who has a theoretical perspective that is related to gender. Free riding off such workers can have serious consequences for their professional development.

Insofar as there are not mechanisms in the academy that provide compensation for epistemic diversity work, even though diversity free riding may be motivated by salutary goals, it is problematic. It is problematic because this is time-consuming labor that need not contribute to the professional advancement of diversity workers and takes away energy that could be used by the diversity worker to advance their own career. If free riding is a common or continuing strategy, even if it is conducted for salutary reasons, it can have the effect of decreasing the pool of diversity workers since it can have negative effects on their career trajectories. In other words, free riding can lead to a lose-lose game, because it can decrease the overall pool of diversity workers.[6]

Of course epistemic diversity workers can choose to withhold their services and withdraw from communities, formal or informal, who are using their experiences, time and talents without offering reward. But choices about whether or not to provide epistemic diversity work are constrained in various ways and these constraints can differ depending on the status of the diversity worker and the kind of diversity that a worker is providing. First, even if this work does not lead to professional advancement it can be personally fulfilling or the diversity worker can feel a moral obligation to do this work. A developmental biologist may wish that communities of population geneticists conducted research that was more sensitive to developmental constraints and so may decide to provide diversity work for a community of population geneticists despite free riding. However, a woman or feminist scholar who is in a position to detect and possibly decrease the degree of gender bias in scientific research may be motivated to do this work because of the joint effects of producing better and less sexist science. In these cases, moral obligations can swamp considerations of professional advancement. Second, this kind of work may have the potential to build a network of contacts, a relatively stable informal community, which can be personally and professionally supportive. However, there is research showing that access to, and benefits of, these informal networks are not equitably distributed between men and women (Rosser 2004 and below).

For diversity workers who are marginal members of communities, both accepting and refusing to do diversity work can be risky. Refusal is risky because it

[6]This pattern can be especially prevalent with regard to institutional service work performed by women faculty (Bird et al. 2004) and faculty of color (Monture-Okanee 1995; Baez 2000).

involves withholding services from a person with greater power and authority than the diversity worker or from a community in which one is a marginal member. Since these kinds of social interactions are often thought to be included as a part of professional practice or good citizenship, refusing to do this work can be seen as refusing to do one's job or as being a bad citizen. Acceptance is also risky when the worker is in a marginal position. I recall making a brief comment on a senior colleague's work, saying that gender might play an interesting role. When he asked me to elaborate, my mind flashed to my upcoming tenure review and I realized that my response required not only philosophical acumen but a degree of diplomacy that I might not be able to muster. Epistemic diversity work can involve telling people things that they might not be inclined to hear. The power differences among members of formal and informal epistemic communities have professional and epistemic consequences.

If we employ an inclusive definition of a community as consisting of those with whom one interacts, an individual or community can cultivate diversity by seeking out and interacting with diversity workers. An individual or community can engage such workers and gain epistemic benefits from these engagements without the diversity worker benefiting from these interactions. If this is a consistent pattern or if the diversity work is onerous, this can retard the diversity worker's career advancement. As a result, instances of free riding that exploit diversity workers can lead to decreases in the pool of diversity workers.

While it is true that free riding is possible, if it is uncommon the situation would be less grave. Given the amount of care and effort that many senior faculty members spend mentoring students and junior colleagues, it may initially seem as if free riding off people with marginal social positions is unlikely. However, there is evidence that these important and well intentioned efforts are not equitably distributed between men and women recipients (see Trix and Psenka 2003 on letters of recommendation). A lack or ineffectiveness of formal and informal mentoring is one of the frequently cited barriers to the advancement of women in the academy (Rosser 2004). Further, both men and women tend to unconsciously undervalue the professional contributions and accomplishments of women academics (Valian 1999). We are often unaware when we are free riding off women, or when we are under valuing a woman's relative to a man's contribution to a professional community. A simple example that women often report is making a contribution during a meeting, having her contribution taken up by a male colleague and the meeting proceed as though the woman's colleague came up with the idea. While a single instance of this kind of usurping may be annoying, a persistent pattern can add up to significant devaluing of a woman's contributions to a community.

7.3.5 Some Feminist Concerns

Diversity promotes excellence theories that employ an inclusive sense of 'community' allow people to make use of diversity in ways that are consistent with the leaky pipeline and also with Harvard past-president Lawrence Summers's rehearsal of economic arguments that discrimination is not a factor in the underrepresentation of women in science at elite institutions. The leaky pipeline is troublingly consistent with 'diversity promotes excellence' theories that use an inclusive notion of community. Currently, there is a steady supply of women with STEM doctorates. Before women leak out of the pipeline they can be useful diversity workers. And the woman who is the well-trained and under-employed spouse of a scientist can be well situated to do diversity work, while being a marginal member of any formal community, or only a member of an informal community. While the supply of junior women academics can provide a source of diversity workers, steps taken to retain those women would make this pool larger, which could result in more dissenting voices and in diversity work being spread over a greater number of workers and thus having less negative impact on an individual diversity worker's career. It could also result in the mainstreaming of diversity work, which could lead to diversity work becoming part of a valued research program.

Theories that do not attend to differences between formal and informal communities have a troubling consistency with economic arguments such as those rehearsed by Harvard past-president Lawrence Summers (2005), to the effect that the underrepresentation of women in STEM is not due to discrimination, but rather to differences in the proportion of men and women with the talent and drive that it takes to succeed in science. The argument is based on the idea that if there were a pool of talented and under-employed women scientists, a university that saw this and hired them would have an advantage over institutions that did not hire women because of discrimination. However, an institution can gain epistemic benefits from diversity by free riding off diversity provided by women scholars in formal communities in other, less prestigious institutions and off of under- or unemployed women scholars. While it is true that such an institution will not garner prestige from employing these talented women, its members can benefit by free riding off of women scholars whom it does not support.[7]

These considerations of the details of kinds of communities and the positions of people within those communities raise important questions about the feminist nature of various social epistemology projects. A 'diversity promotes excellence' theory that uses an inclusive sense of community can be consistent with exploiting scholars who are marginal members of a formal community or only members of an informal community who do epistemic diversity work. It can also be consistent with inequitable employment patterns of women in the academy, with the leaky pipeline, and with women being employed at lower ranks and at lower ranking institutions than men. Furthermore, free riding has the potential to harm scholars in marginal social positions more than scholars in central social positions, epistemic diversity work can be riskier for marginal scholars, and marginal scholars face stronger constraints on their decisions regarding whether or not to perform epistemic diversity work. If we hold that a feminist theory ought to protect and lead to the advancement of those in

[7]Thanks to Heidi Grasswick for making this point.

marginal social positions, in particular women, then a feminist epistemology that does not attend to the nuances of kinds of communities and the social position of members of communities at least requires further development.

7.3.6 Increasing Diversity – Diversity Development Work

Recall that a diversity free rider is one who makes use of existing diversity without increasing the diversity of any formal community or the total representation of diverse voices. In previous sections I discussed the idea of making use of diversity. Here I discuss the notion of increasing diversity. Of course, it is possible to make use of diversity without exploiting diversity workers and this is most likely what feminist epistemologists such as Longino intend. One can avoid free riding by developing diversity in particular ways. Spelling out what developing diversity means may help clarify Longino's call for communities to take 'active steps to ensure that alternative points of view are developed enough to be a source of criticism and new perspectives' (2002, 132).

Communities and members of communities who cultivate those with dissenting perspectives by taking steps to train, hire, or retain those with underrepresented theoretical perspectives, social locations, or to reward them for their epistemic diversity work, or to nurture cultures that are conducive to the development of dissenting views are performing *diversity development work.*[8] Any one of these options can result in epistemic benefits to academic communities as well as improvements in the situation of women and minority scholars. Furthermore, all three of these options are interrelated. For example, rewarding someone for diversity work can help them build their vita and get a job. Increasing the representation of diversity workers within a department can support a culture where such workers can more effectively develop and articulate dissent. Creating a culture that is conducive to the development of dissenting views can have a positive impact on retention rates.

The relationships between kinds of diversity development work can also be negative. In the next section of the paper I will argue that the chilly climate for women and minorities in STEM fields and in the academy more generally is not conducive to the development and articulation of dissenting views. Although one can gain epistemic benefits from interacting with women and minority diversity workers in a chilly climate, this is far from an optimal ethical or epistemic situation.

The discussion above suggests that the answer to the 'what is in it for me if I embrace hiring practices that promote diversity?' question is more complicated than just saying that it is beneficial to be a member of a diverse epistemic community. Although one can benefit from interactions with diversity workers in one's own department, one can also obtain those benefits by engaging and possibly free riding off members of other communities, formal or informal, with which one is

[8]Thanks to Sandy Gahn for suggesting the name 'diversity development'.

connected. Hiring to promote diversity may be primarily a service to one's profession, because one is increasing the pool of potential diversity workers, not only for one's own use, but also for the use of members of other communities. One benefits from being a member of a profession in which other departments also perform this service thereby increasing the pool of diversity workers with whom one could engage. Considering the administrator's question makes clear the need for diversity promotes excellence theories to attend to the details of community structure.

In the next section of this paper I will explore the ways that the chilly climate impacts the epistemic benefits that a community or member of a community can accrue from diversity. Maximizing the epistemic benefits that can arise from diversity involves developing cultures that support diversity workers with alternative social locations and alternative theoretical perspectives.

7.4 Situational and Epistemic Diversity

In the previous section of this paper I referred to diversity workers as providing diversity in terms of their social locations or their theoretical perspectives, and I referred to diversity development work as a wide range of activities which include changes in hiring practices and cultures. In this section and the next one I will briefly unpack these different senses of diversity and diversity development work because each has bearing on the ways that communities, formal and informal, can cultivate and make best use of diversity. The epistemic benefits that can accrue from creating diverse communities arise not out of the inclusion of more women and minorities per se, but because of the different background assumptions and theoretical perspectives that these people may bring to critical discussions in virtue of their social location. This can be clarified by distinguishing between situational diversity and epistemic diversity. A community is *situationally diverse* when its membership consists of individuals with different social and material locations (gender, race, class, sexuality, etc.).[9] The failure of scientific communities to be situationally diverse is most often couched as an ethical problem. In terms of employment equity, these ethical issues come down to a matter of fairness. The relative lack of women and minorities in the academy is not the result of their lack of ability, commitment or drive, but because of institutional, social and psychological factors that function to exclude them (Wylie et al. 2007).

A community is *epistemically diverse* when it includes members who hold a range of different background assumptions, and theoretical and methodological perspectives. The failure of scientific communities to be epistemologically diverse is most often couched as a cognitive problem. It is a cognitive problem because all of the background assumptions that researchers use to determine the connection between theory and evidence do not announce themselves. Those assumptions can be brought to light through critical interactions with people who are aware of those

[9] I address situational and epistemic diversity in Fehr (2007).

assumptions or who hold different assumptions. The value of epistemic diversity can be realized in ways other than just making evidential relations explicit. Alternative perspectives can be fruitful in terms of providing alternative questions to ask, theories to test and methods with which to generate data. The relationship between situational and epistemic diversity reveals, what has been pointed out by feminist epistemologists, that our cognitive problems and ethical problems are often intertwined (Code 1991). The examples of women, with varying degrees of feminist engagement, such as Jeanne Altmann, Barbara McClintock, Ruth Hubbard, Sarah Hrdy and Ruth Bleier, demonstrate that situational diversity can have a significant impact on epistemic diversity, and that the epistemic diversity generated out of situational diversity with respect to gender can extend beyond investigations related to gender. One woman scientist interviewed by Sue Rosser reports that her social situation positively impacts her science:

> In the computer science discipline in which I work, respect is conferred upon those who possess knowledge obtained primarily through countless hours investigating the nuances of hardware and operating systems. To many in my peer group, this is a relaxing hobby and way of life. Though I learn these nuances as I need them for my research, outside of my work I read literature, am deeply interested in social issues and am committed to being involved in my child's life. I see this alternate experience base as an asset to my field. As Rob Pike of C language fame recently said, "Narrowness of experience leads to narrowness of imagination". (2004, xxiii)

Even though there are many cases where situational diversity has led to epistemic diversity, this does not mean that situational diversity necessarily results in epistemic diversity, nor that it always should do so. For example, one can imagine a woman, thoroughly professionalized in a traditional discipline, who uses traditional methods and a widely accepted theoretical approach to study a topic that may or may not be related to gender. A woman may not and need not bring any epistemic diversity, gendered or otherwise, to a community. Whether or not she brings epistemic diversity to a community can be influenced by a wide range of factors. Women as well as men can be curious about a wide range of topics and engaged by a wide range of approaches. Women as well as men are subjected to long apprenticeship-like training in central methods and approaches in their disciplines.

In summary, increasing situational diversity can and has led to increases in epistemic diversity, but it is not necessary that it do so. In the next section I will look at cultural factors that can block epistemic diversity from arising out of situational diversity with respect to gender. If communities are to glean maximal epistemic benefits from increasing situational diversity (e.g., departments hiring more women) it is important to discover and remove these cultural constraints.

7.5 From Ineffective to Effective Epistemic Diversity

When considering constraints that can block epistemic diversity from arising out of situational diversity, it is useful to point out a continuum between effective and ineffective epistemic diversity. *Effective epistemic diversity* describes a community that contains members who hold different background beliefs *and* makes use of that diversity to generate and evaluate theories, hypotheses and data from the widest possible range of perspectives. In this kind of community, members feel free to develop and offer dissenting views, their dissent is given uptake and they are granted equality of intellectual authority with those who hold more common or more central perspectives. In order for a community to reap the benefits described by diversity promotes excellence theories such as Longino's, they do not just require epistemic diversity, they require effective epistemic diversity. *Ineffective epistemic diversity* obtains when a community includes members who hold different background beliefs but *does not* use this diversity to generate and evaluate theories, hypotheses and data. Communities can be blocked from making use of the epistemic diversity that could be offered by their members. As will be discussed below, current research on women STEM faculty and on university cultures indicates that (1) there are likely factors acting that block women's dissent from getting uptake, (2) cultural patterns inhibit the kinds of social interactions required for women to offer dissent and (3) there are forces that inhibit women from developing dissenting views.

7.5.1 Women Can Face Challenges Gaining Uptake

Research in the social sciences reveals that women's professional accomplishments are undervalued relative to men's (Valian 1999). We can see this pattern in several studies that demonstrate gender bias in the evaluation of curriculum vitas that differ only in terms of the gender of the scientist being evaluated (Fidell 1970; Steinpreis et al. 1999). Vitas with a male name at the top of the page were evaluated as belonging to candidates who were more hirable and hirable at a higher rank than vitas with a woman's name at the top of the page. In one study where the vitas were returned to the experimenters, women's vitas had four times as many cautionary notes in written in the margins than identical men's vitas (Steinpreis et al. 1999). Academic vitas are the most objective descriptions of a faculty member's accomplishments. The data is presented in a stark, systematic and highly ritualized manner. What is evident is that this most basic form of evidence is not successful in providing objective data. A man's accomplishments as listed on his vita are more likely to get uptake than a woman's accomplishments.

Of course, there is a big difference between looking at a vita, and the kinds of face to face and written interactions that take place when members of a community engage in critical discourse involving dissent. However, similar patterns show themselves in a study of letters written for successful applicants for positions at a medical school (Trix and Psenka 2003). In general, the relationship between a candidate and a letter writer is closer than the relationship between a candidate and someone reviewing their vita. Trix and Psenka found that letters written for women tended to be shorter, were twice as likely to have a doubt raiser in the text, and one and a half times as likely to contain 'grindstone adjectives' as letters written for men. Women were less likely to be described as successful and their letters were much less likely to contain the words 'accomplishment' or 'achievement.' Letters written for women were much more likely to refer to the candidate's teaching and training, and much less likely to refer to her research and skills and abilities than letters written for men. In these letters we see different kinds of assessments of the professional accomplishments of women and men. This study suggests that the accomplishments that are most valued in the academy are often not granted uptake by letters writers, presumably some of whom are the applicant's mentors, the people in these women's scientific communities who know their work well and who have an interest in their continued professional success.

In 2005 the Statistical Research Center of the American Institute of Physics conducted a survey of 1,350 women physicists from more than 70 countries (Ivie and Guo 2006). Most of the women found their careers to be rewarding and 86% of respondents said that they would choose physics again. However, 43% of women respondents report being discouraged about physics because of the climate for women, 65% being discouraged by discrimination and 80% of women respondents report that attitudes about women in physics need improvement. Several of the respondent's comments speak directly to issues of uptake. One woman reported,

> It is difficult when you are, as I have been, the only woman in a conference. Or when people interrupt, or do not listen or even laugh at what you are saying, even if it is important. Or when advisors or mentors could not believe that I'd done the work myself. (Ivie and Guo 2006, 11)

If women's professional accomplishments don't make it into assessments of their professional ability by those assessing their vitas and by those who know their work well enough to write a recommendation letter, or if their contributions are mocked and ignored, the epistemic diversity, whether or not it relates to gender, that they may be able to offer their communities is wasted. Epistemic diversity may be present in the sense that there are members of a community who hold dissenting views or different background assumptions, but that diversity is ineffective if it does not get uptake and is not integrated into a community's critical interactions.

One might argue *pace* Kitcher (1993) that selfish motivations could overcome lack of professional respect and result in uptake of dissent that could improve one's research. However, research on the under valuing of women's relative to men's professional accomplishments shows that these biases are frequently implicit (Butler and Geis 1990; MIT 1999; Valian 1999). Not only is it common for subjects not to notice their biases but they often genuinely believe that they are not biased. Women and men are equally likely to undervalue women's contributions. This means that members of communities may not be aware that they are not granting dissent from women uptake. Further, as I will argue below, social arrangements inside departments and other academic units may limit the opportunities for interactions in which dissent can be raised, and women may be pushed to use traditional methods and approaches which may limit their ability to develop dissenting views. The fact that women face challenges getting uptake does not provide evidence that diversity promotes excellence theories such as Longino's ought to be abandoned. Rather, it shows that we need to address issues of culture and implicit gender bias in the evaluation of women's work and further develop Longino's views. There are both ethical as well as epistemological reasons to work toward ameliorating these issues.

7.5.2 *Lack of Social Interactions Required for Women to Offer Dissent – Isolation*

One well documented barrier to the retention and advancement of women STEM faculty is isolation and exclusion from networking opportunities. Much of the research on this topic focuses on women lacking knowledge of norms and practices required for tenure and promotion because they are excluded from professional networks. However, if we consider critical social interactions among members of a community as an epistemic desideratum, the isolation and exclusion of women becomes an epistemic issue. A formal community may be situationally diverse and epistemically diverse, but if women are systematically excluded from social interactions within that community, the functioning epistemic community may be an informal community from which women tend to be excluded. It is difficult to measure the impact of factors such as isolation because the vast majority of the data on this topic come from women who have persisted in academic careers. But even among those who persist and win national level competitive grants, isolation is still a factor. In Sue Rosser's study of women who received NSF POWRE grants, she found that in 2000, 30.5% of respondents cite problems with low numbers of women, isolation and lack of camaraderie/mentoring, and 21.9% report challenges gaining credibility/respectability from peers and administrators (Rosser 2004, Table 6, p. 36).[10] One of Rosser's respondents wrote, 'The biggest challenge that women face in planning a career in science is not being taken seriously. Often women have to go farther, work harder and accomplish more in order to be recognized' (Rosser 2004, 40). Similarly, one respondent to the American Institute of Physics survey wrote, 'The main reason [I've felt discouraged] is so often you are just made to feel like you shouldn't be there. You have to work twice as hard, do twice as much just to be considered half as qualified' (Ivie and Guo 2006, 11).

One way that isolation functions is that some women report trouble establishing collaborations with men. In Sonnert and Holton's (1996) study of women and men who won prestigious postdoctoral awards they found that when collaborating with men women were more often treated as subordinates rather than equal or senior research partners. Along similar lines, a woman from the American Institution of Physics study wrote, 'Interaction with colleagues has been the most difficult. I have often felt that I am ignored or discounted when I attempt to initiate collaborations with men' (Ivie and Guo 2006, 11). This comment speaks to uptake as well as intellectual isolation.

[10] Table 6 shows data ranging from 1997 to 2000. Although there is variation among these years, in all cases low numbers of women, isolation and lack of credibility and respect are identified as significant challenges facing women scientists.

Women can also be isolated in terms of their choice of research areas. Of the highly promising scientists that Sonnert and Holton studied, 40% of women and only 15.7% of men reported that their gender influences their choice of research topics. Sonnert and Holton report that women tend to adopt a niche approach; they tend to create their own pockets of research expertise. Several women report adopting this strategy to avoid taking part in a highly competitive culture in which researchers are racing with one another to solve a particular problem. Whether this choice is the result of women adaptively avoiding a hostile and aggressive work environment or it is simply a benign difference of research styles, this result, in combination with challenges that women report doing collaborative work, paints a picture of women being excluded from social interactions relevant to research as well as social interactions relevant to gaining knowledge regarding professional advancement. Reports of being an outsider or not feeling like a full member of professional communities are common. Mary Frank Fox writes that this has a wide range of impacts on women's careers:

> Within the same type of setting, women scientists can have fewer and different collaborative arrangements, claims to enabling administrative favors, *collegial opportunities for testing and developing ideas*, and entrees into the informal culture of science and scholarship (Fox 1991, 204 in Rosser 2004, 47 italics added).

Research on isolation and exclusion demonstrates that the effectiveness of diversity promotes excellence theories requires structural and cultural changes in the academy.

7.5.3 *Forces That Inhibit Women from Developing Dissenting Views*

Women are often solo or minority members of scientific communities and are relative newcomers to many contemporary professionalized academic disciplines. Women also tend to be in marginal positions within the academy. The increases of women scientists in the academy can be seen primarily in low-ranking institutions and at low academic ranks, and women are more likely than men to hold non tenure track positions (West and Curtis 2006; NAS 2007). These low numbers and marginality can impact the way women conduct their research.

Sonnert and Holton's survey data show that 34.8% of women and 9.9% of men thought their gender plays a role in the methods they use (1996). That one third of elite women scientists report that their gender influences their methodology might be initially suggestive of epistemic diversity. However, in the interviews respondents rarely reported that they used 'feminine methods,' or even methods different from those used by men. Interviewees rather report differences in the application of traditional methods in terms of using a greater degree of caution, carefulness, attention to detail and perfectionism. Sonnert and Holton write that,

> Rather than being iconoclasts, women tended to uphold to a particularly high degree the traditional methodological standards of science, such as carefulness, replicability and connection

> to fundamentals. As a group, women, as relative newcomers to science, adopted – or were taught to adhere to – an extra-high measure of conformity to the formal norms of conducting research. (8–9)

Sonnert and Holton postulate that this conservative research style can 'arise from a collegial environment particularly hostile to women who deviate from accepted standards' (9). One woman reported 'there's always somebody watching for me to make a mistake' and another said that women often find themselves 'under the magnifying glass' (9).

Being in a culture that many women describe as chilly or hostile, where they are marked as outsiders and where their low numbers result in surveillance for error or lack of conformity can push them to take more mainstream approaches to their research. This may have an effect of dampening the epistemic diversity that they are able or feel free to develop and to offer in critical discussions with members of their communities. There may be especially strong reasons to avoid developing approaches and offering dissent with respect to their gendered experiences, but it can also have an effect of limiting the dissent offered on any topic. If we take a social approach to knowledge construction, then it behooves us to look at the actual social arrangements within epistemic communities.

7.6 Conclusions

In this paper I identify several ways that diversity promotes excellence theories such as Longino's critical contextual empiricism, as they currently stand, fail to support arguments for increasing gender diversity in the academy. While increasing employment equity may not be the primary aim of these epistemological approaches, I argue that diversity promotes excellence theories can be further developed in ways that provide resources for epistemologists and activists alike.

One of the challenges I point out concerns the free rider problem. Developing an account of formal and informal communities and of power differences among members of communities reveals that one can nominally follow Longino's advice to cultivate diversity simply by engaging in social interactions with a person who holds a different epistemic position from one's own and without increasing the overall diversity in the academy. Focusing on this issue using employment inequities as a lens shows that a diversity promotes excellence theory can be consistent with the exploitation of members of marginalized groups and with inequitable employment patterns in the academy. This can provide a misleading answer to the 'what is in it for me?' question, because it seems to show that one can reap the benefits of epistemic diversity without employing women, or members of other underrepresented groups, and in fact without even rewarding them for their epistemic diversity work. This is not to say that one cannot benefit from including members of underrepresented groups within one's own department, just that one can find those epistemic benefits elsewhere. But, free riding off marginal members of communities is consistent with and contributes to a culture in which those who are in marginal social positions are undervalued. As a result, they are not in a position to contribute as much effective epistemic diversity to a community as they might otherwise be. Free riding might be common but it is not an optimal strategy. Although people can gain some epistemic benefits from free riding, in the long run they can likely gain greater epistemic benefits from doing diversity development work.

Longino's theory needs to be developed in a way that blocks this nominal interpretation of her work. Doing so is consistent with a richer sense of developing diversity and taking seriously her calls for members of communities to give dissent uptake and to treat each other with equality of intellectual authority. It is not easy work because, among other things, it involves addressing cultural issues. In many contemporary communities, women's voices are not given uptake and women are not treated with equality of intellectual authority. A richer sense of cultivating dissenting voices includes developing cultures that nurture epistemic diversity workers, both in their ability to explore and develop dissenting perspectives and in the social relations they share with other members of their communities.

I am interested in answering the 'what is in it for me?' question for two reasons. First, the underrepresentation of women and some minorities among STEM faculty, and in the academy more generally, is highly problematic, especially in light of social science research showing that this underrepresentation is not due to lack of the ability or drive to succeed in academic careers. It is apparent that those motivated to improve this situation need access to a wide range of arguments. Although diversity promotes excellence theories, such as Longino's critical contextual empiricism, are not designed to specifically address these employment inequities, they do provide an interesting avenue for addressing these problems. Second, focusing on the 'what is in it for me?' question provides an opportunity to explore ways that diversity promotes excellence theories can be developed both for their own sake and also to guide the activities of scientists seeking to improve their craft.

Acknowledgments I am grateful to Heidi Grasswick for generous help with various drafts of this paper and for unflagging professional and personal support. Warm thanks also to Sharon Bird, Sandra Gahn, Margaret Holmgren, Katie Plaisance, Anastasia Prokos, Diane Price-Herndl, Dave Saldana, Kate Padgett Walsh, Brad Wray, Alison Wylie and audiences in the philosophy departments at Duke University, Iowa State University and the University of Wisconsin-Madison. This material is based upon work supported by the National Science Foundation under Grant No. 0450821. Any opinions, findings, and conclusions or recommendations expressed in this material are those of the author and do not necessarily reflect the views of the National Science Foundation.

References

Altmann, Jeanne. 1974. Observational study of behavior: Sampling methods. *Behavior* 69: 227–263.

Baez, Benjamin. 2000. Race-related service and faculty of color: Conceptualizing agency in the academy. *Higher Education* 39(3): 363–391.

Bird, Sharon, Jackie Litt, and Yong Wang. 2004. Creating status of women reports: Institutional housekeeping as "Women's Work". *National Women's Studies Association Journal* 16(1): 194–206.

Birke, Lynda. 1986. *Women, feminism and biology: The feminist challenge*. Brighton: Wheatsheaf.
Butler, Dore, and Florence Geis. 1990. Nonverbal affect responses to male and female leaders: Implications for leadership evaluations. *Journal of Personality and Social Psychology* 58(1): 48–59.
Code, Lorraine. 1991. *What can she know? Feminist theory and the construction of knowledge*. Ithaca: Cornell University Press.
Fehr, Carla. 2007. Are smart men smarter than smart women? The epistemology of ignorance, women and the production of knowledge. In *The 'woman question' and higher education: Perspectives on gender and knowledge production in America*, ed. Ann Mari May, 102–116. Northampton/Cheltenham: Edward Elgar.
Fidell Linda, S. 1970. Empirical verification of sex discrimination in hiring practices in psychology. *American Psychologist* 25: 1094–1098.
Fox, Mary Frank. 1991. Gender, environmental milieu, and productivity in science. In *The outer circle: Women in the scientific community*, ed. John T. Brue, Jonathan R. Cole, and Harriet Zuckerman, 188–204. New York: W. W. Norton.
Gowaty, Patricia Adair. 2003. Sexual natures: How feminism changed evolutionary biology. *Signs* 28(3): 901–921.
Haraway, Donna. 1989. *Primate visions: Gender, race and nature in the world of modern science*. New York: Routledge.
Hrdy, Sarah. 1986. Empathy, polyandry, and the myth of the coy female. In *Feminist approaches to science*, ed. Ruth Bleier, 119–146. New York: Teachers College Press.
Hubbard, Ruth. 1982. Have only men evolved? In *Biological woman – the convenient myth: A collection of feminist essays and a comprehensive bibliography*, ed. Ruth Hubbard, M.S. Henifin, and B. Fried, 17–46. Cambridge: Schenkman.
Ivie, Rachel and Stacy, Guo. 2006. Women physicists speak again. College Park, MA: Statistical Research Center of the American Institute of Physics.
Keller, Evelyn Fox. 1983. *A feeling for the organism*. New York: W. H. Freeman and Company.
Keller, Evelyn Fox. 1985. *Reflections on gender and science*. New Haven: Yale University Press.
Kitcher, Philip. 1993. *The advancement of science*. New York: Oxford University Press.
Longino, Helen. 1990. *Science as social knowledge*. Princeton: Princeton University Press.
Longino, Helen. 2002. *The fate of knowledge*. Princeton: Princeton University Press.
Longino, Helen, and Ruth Doell. 1983. Body, bias and behavior: A comparative analysis of reasoning in two areas of biological science. *Signs* 9(2): 206–227.
Martin, Emily. 1991. The egg and the sperm: How science has constructed a romance based on stereotypical male-female roles. *Signs* 16(3): 485–501.
Massachusetts Institute of Technology. 1999. A study on the status of women faculty in science at MIT. *The MIT Faculty Newsletter* 11(4): 1–17.
Monture-Okanee, Patricia. 1995. Surviving the contradictions: Personal notes on academia. In *Breaking anonymity: The chilly climate for women faculty*, ed. Chilly Collective, 11–28. Waterloo: Wilfrid Laurier University Press.
National Association of Sciences. 2007. *Beyond bias and barriers: Fulfilling the potential of women in academic science and engineering*. Washington, DC: The National Academies Press.
Nelson, Lynn Hankinson. 1990. *Who knows: From Quine to a feminist empiricism*. Philadelphia: Temple University Press.
Nelson, Lynn Hankinson. 1993. Epistemological communities. In *Feminist epistemologies*, ed. Linda Alcoff and Elizabeth Potter, 121–159. New York: Routledge.
Rosser, Sue. 2004. *The science glass ceiling: Academic women scientists and the struggle to succeed*. New York: Routledge.
Solomon, Miriam. 2006. Groupthink versus the wisdom of the crowds: The social epistemology of deliberation and dissent. *Southern Journal of Philosophy* 44: 28–42.
Solomon, Miriam, and Alan Richardson. 2005. A critical context for Longino's critical contextual empiricism. *Studies in History and Philosophy of Science* 36: 211–222.

Sonnert, Gerhard, and Gerald Holton. 1996. Career patterns of men and women in the sciences. *American Scientist* 84: 63–71.

Steinpreis, Rhea, Katie Anders and Dawn Ritze. 1999. The impact of gender on the review of the curricula vitae of job applicants and tenure candidates: A national empirical study. *Sex Roles* 41(7/8): 509–528.

Summers, Lawrence. January 14, 2005. Remarks at NBER conference on diversifying the science & engineering workforce. In *National Board of Economic Research (NBER) conference diversifying the science & engineering workforce: Women, underrepresented minorities, and their S&E careers*. Cambridge: Harvard University, The Office of the President. http://www.president.harvard.edu/speeches/summers_2005/nber.php.

The Biology and Gender Study Group. 1989. The importance of feminist critique for contemporary cell biology. In *Feminism and science*, ed. Nancy Tuana, 172–187. Bloomington: Indiana University Press.

Trix, Frances, and Carolyn Psenka. 2003. Exploring the color of glass: Letters of recommendation for female and male medical faculty. *Discourse and Society* 14(2): 191–220.

Valian, Virginia. 1999. *Why so slow? The advancement of women*. (Cambridge: MIT Press).

West, Martha, and John Curtis. 2006. *AAUP faculty gender equity indicators 2006*. Washington, DC: American Association of University Professors.

Wylie, Alison. 2006. Socially naturalized norms of epistemic rationality: Aggregation and deliberation. *Southern Journal of Philosophy* 44: 43–48.

Wylie, Alison, Janet Jakobsen, and Gisela Fasado. 2007. *Women, work and the academy: Strategies for responding to 'Post-Civil Rights Era' gender discrimination*. NY: Barnard Center for Research on Women.

Xie, Yu, and Kimberlee Shauman. 2003. *Women in science: Career progress and outcomes*. Cambridge: Harvard University Press.

Chapter 8
What Knowers Know Well: Women, Work and the Academy

Alison Wylie

Abstract Research on the status and experience of women in academia in the last 30 years has challenged conventional explanations of persistent gender inequality, bringing into sharp focus the cumulative impact of small scale, often unintentional differences in recognition and response: the patterns of 'post-civil rights era' discrimination made famous by the 1999 report on the status of women in the MIT School of Science. I argue that feminist standpoint theory is a useful resource for understanding how this sea change in understanding gender inequity was realized. At the same time, close attention to activist research on workplace environment issues suggests ways in which our understanding of standpoint theory can fruitfully be refined. I focus on the implications of two sets of distinctions: between types of epistemic injustice (and correlative advantage) that may affect marginalized knowers; and between the resources of situated knowledge and those of a critical standpoint on knowledge production.

Keywords Employment equity • Epistemic injustice • Feminist philosophy of science • Standpoint theory • Workplace environment issues

When the MIT report, 'Women in the School of Science,' appeared in 1999, the terms of public debate about the status of women in science, and in academia generally, were fundamentally reframed. What the authors of this report declared, with electrifying effect, was that discrimination in the 'post-civil rights era' is subtle but no less effective for all that. They reported inequities in resources and support and, crucially, in outcomes for women that persist even in the absence of intentional discrimination. Discrimination in the 1990s, they argued, takes the form of innumerable small differences in uptake and response: 'a pattern of powerful but unrecognized attitudes and assumptions that work systematically against women

A. Wylie (✉)
Department of Philosophy, University of Washington, Box 353350, Seattle, WA 98104, USA
e-mail: aw26@uw.edu

H.E. Grasswick (ed.), *Feminist Epistemology and Philosophy of Science: Power in Knowledge*, DOI 10.1007/978-1-4020-6835-5_8,

despite good will' (MIT 1999, 11). Although individual incidents may seem to be trivial, cumulatively they translate into patterns of 'exclusion and invisibility' that can have a substantial impact on the quality of women's work life, their effectiveness in the workplace, and their career trajectories compared to those of similarly well trained and accomplished men (1999, 8).

The authors of the MIT report contrast these contemporary patterns of marginalization with the forms of explicit sex discrimination that had been addressed, in the United States, by executive orders (for federal contractors) and landmark equal opportunity legislation instituted in the late 1960s and early 1970s.[1] In the background is a conventional framework for explaining persistent gender inequality in academia that was articulated in particularly clear and influential terms by Jonathan Cole in *Fair Science* (1979), and reiterated nearly 30 years later by Lawrence Summers (then President of Harvard), in his infamous remarks about women's lack of capacity for careers in science (2005). On Cole's account, absent evidence of intentional discrimination, gender imbalances in the representation of women must reflect differences in the choices they make and in their accomplishments. Women must be self-selecting out of the sciences at higher rates than men and, when they do persist, he argued that gender differences in outcome (progress through the ranks, recognition, compensation) can all be explained by lower levels of productivity among women, compared to men, that cannot be accounted for by marital or parental status. Summers filled the explanatory lacunae in this account with the conventional wisdom that these patterns persist because women typically lack the necessary intellectual talent and drive to succeed in science (2005). Although Cole and Summers focus on women in science, these presuppositions surface, in field-specific terms, across academic disciplines and the professions.

Although Cole's account has been canonical in many contexts, he did face sharp criticism at the time. Margaret Rossiter published a prescient review of *Fair Science* in 1981, objecting that Cole 'seemed unwilling to face his own evidence' (101).[2] She reads his analysis against the grain, reinterpreting his statistical results – his distributional data – as evidence that women in science might be facing a persistent pattern of underestimation and marginalization such that 'the rate of exchange' (Cole's terms) by which they built research careers and reputations was different than for men; they received less recognition, compensation, and support for the same kinds of training, institutional affiliation, and track record of accomplishments, with

[1]For example, the Equal Employment Opportunity Act that was passed in 1972 struck down exemptions from the equal employment opportunity laws that had been granted to educational institutions under Title VII) (Rossiter 1995, 376), while Title IX extended the Equal Pay Act of 1963 to higher education and banned sex discrimination in any institution receiving federal funding (Rossiter 1995, 382). With this legislation in place there was tremendous optimism, for the next decade, that gains in the academic training pipeline would translate into steady improvement in the representation of women in the ranks of faculty.

[2]See also Harrison White who objected that, as a 'patriotic citizen of science,' Cole had begged the question of the fairness of science, lacking the data and the controls necessary to establish his favored conclusions (1982, 951).

ramifying consequences. She cites, in this connection, a growing body of research on evaluation bias that Cole had not considered,[3] and urged consideration of a richer set of explanatory possibilities for the patterns of underrepresentation he reported. It should be a priority, she argued, to 'try to understand the attitudes and behavior patterns that *lie behind* the distributional data' (1981, 103, emphasis added).

In the next 15 years an enormous body of grass roots activist research took shape that was animated by the suspicion, articulated by Rossiter, that the sciences, and academia generally, were not fair, not quite the level playing field that Cole maintained. Women reported innumerable ways in which institutional and disciplinary environments put them at a disadvantage, deflecting them from academic careers or marginalizing them within academia, even as they entered and succeeded in graduate programs at unprecedented rates. The MIT report is a recent and especially high profile outcome of a process that had unfolded over and over again in diverse academic and disciplinary settings since the early 1980s. A growing awareness of dissonance between their experience as women in academia and their expectations that academia is a meritocracy – that intellectual talent and contributions would be recognized and rewarded regardless of gender or race or other markers of social difference – focused their attention, often with great reluctance, on characteristic features of what came to be known as the 'chilly climate' for women in academia (Hall and Sandler 1982, 1984; Sandler 1986). Working groups and ad hoc committees undertook finegrained studies of local dynamics of interaction that might account for the persistent disparities in women's rates of appointment, promotion, compensation, and in other measures of academic outcome that were being documented at an institutional, discipline-wide, and national level. The results were typically reported in internal institutional self-studies, the reports of committees on the status of women, and pamphlets circulated by feminist research institutes. Often these reports provoked sharply hostile responses that reiterated, or presupposed, Cole's explanatory framework (e.g., Michell and Backhouse 1995, 138–142): if intention to discriminate could not be demonstrated, there were no grounds for attributing unfairness to the institutions of science or to the academic communities in which women continued to find themselves on the 'outer circle' (Zuckerman et al. 1991), 'outsiders in the sacred grove' (Aisenberg and Harrington 1988).

I argue that feminist standpoint theory is a useful resource for understanding the transformative shift in thinking about 'women, work and the academy'[4] marked by the MIT report, as well as resistance to the central insights of 'chilly climate' studies that continues even as these are vindicated by the results of mainstream

---

[3]As Rossiter points out, a number of controlled studies were available to Cole that documented systematic differences in the ways Curriculum Vitae are evaluated if they are attributed to women as opposed to men. One especially influential example was Lewin and Duchan's 1971 article in *Science*; Tosi and Einbender provide an overview of work along these lines that had appeared in the late 1960s and early 1970s (1985).

[4]The substantive details of this shift are outlined below, and summarized in report on the current state of research on equity issues for women in the academy, *Women, Work, and the Academy* (Wylie et al. 2007).

(professional) research in cognitive psychology and sociology. Reformulated in non-essentialist, pragmatic terms, feminist standpoint theory provides a framework for understanding why it has been so difficult to identify and delineate patterns of epistemic injustice in academic institutions and how, in some cases, these very patterns of marginalization have conferred epistemic advantage on those who are disadvantaged by them, putting them in a position to recognize and to document phenomena that are rendered inscrutable by the normative ideals of academic meritocracy.[5] At the same time, chilly climate research puts productive pressure on some key assumptions of standpoint theory, drawing attention to various kinds of epistemic advantage that may accrue to those who are marginalized in different ways (socially and epstemically), and sharpening an implicit distinction between the resources of situated knowledge and those of a critical standpoint on knowledge production. I turn first to a characterization of standpoint theory, then expand on the sketch I have given of how gender inequity has been reconceptualized in the last 25–30 years. I conclude with an analysis of the epistemic implications of this example of a hard-won shift in collective understanding that was mobilized by insights from the margins.

8.1 Standpoint Theory and Epistemic Injustice

I find it useful to think of standpoint theory as one instance of a broader genre: a form of social epistemology that focuses attention on the social conditions – the composition and dynamics of epistemic communities – by which knowledge production and authorization can be systematically skewed. It is, then, a theory of epistemic injustice in the sense usefully elaborated by Miranda Fricker (2006, 2007), that focuses attention on ways in which epistemic practice can be improved, given a robust appreciation of the epistemic advantages that may accrue to those who are otherwise marginalized.

Epistemic injustice is a form of systematic epistemic misrecognition; it arises, Fricker argues, when norms of credibility 'imitate structures of social power' (Fricker 1998, 170, 172), so that our socially inflected 'working indicators' of rational authority pick out the powerful and not necessarily the knowledgeable or the truthful. These patterns of misrecognition generate two kinds of epistemic injustice that are relevant for current purposes. The first takes root when members of socially recognized categories or communities – defined, for example, by gender, race, ethnic or religious affiliation, sexual identity, age, class – find that their competence is always in question, no matter what their epistemic credentials or track record. This is what Fricker describes as testimonial injustice (2007, 1, 9–29).

[5] In complementary analyses, Fehr (Chap. 7, this volume) considers the epistemic advantages that may accrue to situational diversity as well as the impediments to its uptake in a research community, and Rooney (Chap. 1, this volume) takes up these issues reflexively, with respect to epistemology.

In this case it is epistemic agents who are misrecognized (*qua* members of social kinds); they are not accorded the rational authority they deserve given their identification with subdominant or disvalued 'social kinds,' even if the epistemic claims they make take a form or have content that is conventionally recognized and valued as knowledge.[6] Sometimes such injustice is deliberate; it is a matter of intentional imposture as credible, or of a cynical refusal to attribute epistemic authority to those who are socially marginal, whatever evidence or arguments they may bring to bear.[7] Often such misrecognition is inadvertent; in cases of 'credibility overspill' attributions of competence overreach the limits of the expertise marked by working indicators, and the reverse in cases of credibility deficit, without anyone intending or even noticing (Fricker 1998, 169).

A second type of misrecognition, which Fricker refers to as 'hermeneutical injustice,' is a function of systematic gaps in the interpretive resources available to epistemic agents that put those who are marginal socially and materially at an epistemic disadvantage, not just testimonially but also conceptually and communicatively. Hermeneutical injustice (2007, 147–161) becomes entrenched when dominant norms of credibility and 'interpretive habits' render unintelligible any distinctive forms of experience or understanding that those in marked social categories may develop as a consequence of their social location. As Fricker describes this, 'relations of unequal power can skew shared hermeneutical resources so that the powerful tend to have appropriate understandings of their experience ready to drawn on as they make sense of their social experiences, whereas the powerless are more likely to find themselves... with at best ill-fitting meanings to draw in the effort to render [their experience] intelligible' (2007, 148). While conceptually distinct, these two types of epistemic injustice reinforce one another in obvious ways. For example, when there is pressure on norms of credibility to track power, the ability of those on the margins to advocate their knowledge and trustworthiness is diminished, especially in areas where what they know does not conform to dominant cultural norms (Fricker 1998, 169). By extension, when dominant groups are in a position to 'project their experience as representative of everyone in society... often [as] an unconscious act' (McConkey 2004, 202), they also project, well beyond the contexts where they originate, working indicators that sanction not just familiar kinds of knowers, but also the forms of knowledge and norms of plausibility associated with them. As McConkey's observation suggests, working indicators of competence and of plausibility are most effective, and most invidious – most impervious to change – when they take root in the attributional heuristics on which we depend to navigate the social world. In this case, 'habits of epistemically charged social perception' (Fricker 2007, 5) become socially charged habits of epistemic judgment

[6] I have in mind, here, a conception of social kinds as contingently constituted by looping effects of the sort characterized by Hacking (1999, 34, 103–104). See also Moya (2000).

[7] Derrick Bell's 'Rules of Racial Standing' is a particularly stark and compelling account of how epistemic injustice of this kind operates (1992, 109–126).

that are content-laden, tuned to interpretive resources that reflect the interests and experience of the powerful.

Misrecognition of both kinds – of subdominant knowers and of subdominant forms of knowledge – is properly described as epistemic injustice when, or to the extent that, socially defined categories of people and their distinctive forms of knowledge are systematically excluded from participation in an epistemic practice – from the 'rhetorical spaces' in which their claims could be heard and systematically adjudicated.[8]

The central tenets of feminist standpoint theory converge on this account of epistemic injustice at a number of key points.[9] First, and most important for the analysis that follows, both presuppose a *situated knowledge* thesis, where the situatedness of epistemic agents is construed in structural terms rather than as a matter of individual perspective or idiosyncratic skills and talents. In short: what individuals experience and understand is (contingently) shaped by systems of social differentiation that structure and are, in turn, structured by the material conditions of their lives, the relations of production and reproduction that condition their social interactions, and the cultural and conceptual resources available to them for representing and interpreting these relations.[10]

Second, standpoint theorists are typically concerned not only with the epistemic effects of positionality or situatedness (*social location*), but with our differential capacity to develop the kind of *standpoint on* knowledge production that is a 'project' (Weeks 1996, 101): a critical consciousness of the conditions under which knowledge is produced and authorized, and of the difference that our situatedness makes to epistemic agency.[11] Standpoint theory is itself such a project, animated by a commitment to understand how power relations inflect knowledge – what epistemic limitations or advantages accrue to epistemic agents as a function of their location in and negotiation of structured systems of social relations – especially where there is a mismatch between the epistemic resources of socially marginal or subdominant agents and the credibility ascribed to them on the basis of conventional norms of credibility.

---

[8] The term 'rhetorical spaces' comes from Code (1995). See Fricker (2007, 157–158) for a discussion of this requirement of systematicity in relation to hermeneutical injustice.

[9] This analysis and reformulation of feminist standpoint theory is developed in more detail in Wylie (2003).

[10] Fricker usefully distinguishes between three senses of 'structured' that figure in Hartsock's formulation of the central claims of standpoint theory (1983): an agent's resources may be structured in a material, an ontological, and an epistemic sense (2007, 147). I mean to indicate here the interdependence of structuring forces in these three senses.

[11] For example, Nancy Hartsock makes the point that 'a standpoint is not simply an interested position (interpreted as bias) but is interested in the sense of being engaged' (1983, 285); it is a matter of developing an 'oppositional consciousness... which takes nothing of the dominant culture as self-evidently true' (1997, 96–97). In a similar spirit, Fricker observes that a standpoint is 'the epistemic counterpart of a particular form of "engagement" with the world': it is 'not itself a social positioning... [but] something that is made available from the relevant social positioning' (1999, 194).

Finally, and most controversially, standpoint theory is characterized by an *inversion thesis*. A central tenet of standpoint theory is that those who are marginalized (socially, politically, economically), and who suffer epistemic injustice as a consequence may, in fact, be epistemically advantaged in key respects. This is a matter of shifting the emphasis from analysis of the epistemically disabling effects of systemic inequality to a consideration of epistemic resources (evidence and experience) that are ignored or discounted as a consequence of testimonial injustice,[12] and to distinctive insights and hermeneutic resources that may arise from non-mainstream experience and the struggle to understand and communicate it. In short, standpoint theory focuses attention on ways in which the experience of those on the margins may put them in a position to know different things, or to know some things better, than those who are comparatively privileged and whose status secures for them more automatic and more comprehensive epistemic credibility.

It is important to note that, on this formulation of the inversion thesis, the types of epistemic advantage posited by standpoint theory are localized and contingent. Standpoint theory need not and, here, does not presuppose an *essentialist* conception of the social kinds in terms of which standpoints are characterized, nor a claim that these standpoints confer categorical or comprehensive *epistemic privilege*.[13] Where the risk of essentialism is concerned, on this account it is an entirely contingent matter whether lines of social differentiation obtain that are robust enough to make a systematic difference to what epistemic agents are likely to know, or know well. That is to say, for social kinds to be epistemically salient, they need not approximate an implausible essentialist ideal of internal homogeneity, external boundedness, and stability. All that is required is that the structures of social inequality that create and sustain social kinds should establish similarities in social experience in a given context – for example, through patterns of socialization, education, and work (relations of production and reproduction) – make a systematic difference in the ways epistemic capacities are developed and epistemic resources distributed in that context. Moreover, even when there are grounds for recognizing that social difference is epistemically salient, this does not sustain any comprehensive assumption of credibility, superior insight, or 'incorrigibility' (Narayan 1988, 37). Any distinctive angle of vision, experience, or critical perspective associated with social marginality will confer advantage only contingently, and only with respect to specific epistemic problems.[14]

Epistemic advantage in this delimited, contingent and pragmatic sense can usefully be assessed on three dimensions.[15]

---

[12] See, for example, Sunstein's discussion of the epistemic costs of conformity to a dominant perspective or world view (2003, 5–9 and throughout).

[13] My use of the term 'advantage' is intended to mark a rejection of formulations of standpoint theory that presuppose (or advocate) a thesis of 'automatic privilege, (Wylie 2003, 28–30).

[14] See Fehr (Chap. 7) for discussion of how epistemic advantage of these kinds may be undermined by isolation or lack of uptake (forms of testimonial injustice), and by patterns of socialization that reinforce methodological conservatism or a disinclination to articulate dissenting perspectives.

[15] This analysis of epistemic advantage is developed in more detail in Wylie (2003, 32–39).

Evidence Those who negotiate social, legal, and economic institutions from a position of marginality come to know, indeed they often have to be attuned to dimensions of the social and natural world that can be ignored by those who are comparatively privileged or, indeed, that are systematically obscured (or inverted) by dominant world views that legitimate entrenched hierarchies of privilege. For example, they may know in intimate detail, how labor is exploited, how material conditions of life and social relations are sustained, how power inequities are reproduced and what their consequences are, especially for those who are subdominant.[16] They may also be attentive to aspects of the natural world that reflect their situated interests and needs, interacting with shared bio-physical environments in quite distinct ways.[17]

Inferential Heuristics and Explanatory Models Differential access to evidence is rarely an advantage on its own. Standpoint theorists often point to particular skills at discerning patterns in the available evidence that are associated with subdominant status. These include, most obviously, inferential acuity with respect to the power dynamics and mechanisms of oppression, and their systematicity (across contexts), that those living lives of relative privilege do not need to cultivate. It may also take the form of distinctive 'metaphors, models, analogies, and narratives' that enable the detection of a different selection of 'nature's regularities' than are captured by the conceptual resources of dominant culture knowledge systems (Harding 2006, 140). By extension, subdominant knowers may develop an expanded repertoire of explanatory hypotheses for making sense of experience that is unintelligible on, or indeed radically inconsistent with, dominant categories of sense making.

Critical Distance Finally, and crucially, standpoint theorists have particularly emphasized the kinds of epistemic advantage that arise when marginality enforces critical dissociation from a dominant world view, throwing into relief the parochial nature of conceptual categories and norms of credibility that are otherwise taken as a given and projected as universal.[18]

It is in connection with these last two factors – explanatory resources and critical distance – that epistemic advantage on the margins is most contingent and most potentially transformative. Born of epistemic injustice, it is in the struggle to take distance from a dominant world view, to critically scrutinize entrenched

[16] See, for example, Narayan on the reasons to cultivate epistemic humility (1988, 38).

[17] This is a point central to Harding's arguments for standpoint theory: 'even in "the same" environment, different cultures have different interests in the world around them' (2006, 140, 99).

[18] See Rooney (Chap. 1, this volume) for an analysis of just this kind of epistemic advantage: the meta-philosophical advantage, as she describes it, that puts feminists in a position to recognize background assumptions, to articulate critical analysis of the limitations they impose, and to develop alternatives to 'epistemology "proper."' Fehr (Chap. 7, this volume) also describes in general terms how 'alternative perspectives,' arising from situational diversity, 'can be fruitful in terms of providing alternative questions to ask, theories to test and methods with which to generate data' (147).

norms of credibility and formulate interpretive alternatives that a *standpoint on* epistemic agency can (sometimes) emerge from the resources of subdominant situated knowledge. When such conditions obtain, standpoint theory is a useful framework for understanding consequential patterns of epistemic exclusion or marginalization, and pivotal shifts in understanding that arise when insights from marginal standpoints throw into relief the partiality of a dominant world view.

8.2 Activist Research on the Academic Workplace Environment

Consider, then, the play of epistemic injustice and correlative advantage in the case of the grass-roots, activist research by which women documented what came to be known, in the 1980s, as the 'chilly climate' they were encountering in the male dominated disciplines and professions they were then entering in record numbers.

The catalyst for this activist research was growing concern, by the mid-1980s – two decades after legal guarantees of equal access had been instituted – that the demographics of college and university students had changed dramatically, but improvements in the representation of women in the professoriate, and their effective integration into the academy, seemed to have stalled; the pipeline was showing definite signs of leaking or, more accurately, of filtering and sluggishness. As Simeone described the situation in 1987, qualified women were still 'more likely than men to be unemployed, underemployed, or in part-time non-tenure track positions'; they were disproportionately concentrated in less prestigious institutions; they showed substantially higher rates of attrition, advanced through the ranks more slowly and, at the same rank, were paid less than their male counterparts.[19]

---

[19] By the mid-1980s the percentage of women receiving doctorates across all academic fields was nearly twice that of women in faculty positions (roughly 17% of faculty were women, compared to 33% of PhDs), and yet their distribution across the ranks conformed to the inverted pyramid pattern familiar from 20 years earlier. The percentage of *full professors who were women* remained tiny (roughly 11% in the 1980s in the U.S.; 7% in Canada) but, more telling, the percentage of *women who were full professors* was consistently a third or less than that of men and showed very little change over the previous 15 years. Women were slightly better represented in initial appointments to tenurable positions than in the relevant candidate pools – a function, it would seem, of equity and affirmative action policies – but they continued to swell the lower (most vulnerable) ranks of the professoriate, especially off-ladder ranks: they made up a third of assistant professors but 52–55% lecturers and instructors, and they were much more likely to hold nonladder positions or to be unemployed than men (the revolving door phenomenon). They were being tenured at lower rates than men (two-thirds of men compared to less than half the women), and they continued to be better represented less prestigious institutions and in smaller (non-graduate teaching) departments and universities, compared to men with comparable graduate training. These details are excerpted from Wylie et al. (2007), and from Wylie (1995a). Fehr (this volume, 134) provides a summary of current data on the representation of women and minorities in STEM fields (science, technology, engineering, and mathematics) that demonstrates the persistence of these patterns of attrition and the resulting "inverted pyramid" in the distribution of women and minorities by rank and institution.

It was at this juncture, in the early 1980s, that women who had successfully navigated the training pipeline, and who expected academia to be a model meritocracy – 'fair' in all the senses Cole defended for the sciences – found themselves increasingly frustrated by just the kinds of inequitable 'rates of exchange,' the lack of uptake and patterns of exclusion, that Rossiter brought into focus in her 1981 response to Cole.[20] When the occasion arose to compare their experience with that of other women in their local work environment, or in their disciplines and professions, they discovered, often to their surprise, that problems they had assumed to be idiosyncratic – to their personal situation, the peculiarities of their colleagues, the culture of their institutions or disciplinary subfield – were, in fact, widely shared.[21] This was a process of 'coming to consciousness' that has been described in a number of connections; for example, Fricker draws on accounts of revelatory insights generated by collective reflection on experiences of sexual harassment and post-partum depression to capture the contours and the harms of hermeneutical injustice (2007, 149, 153). In an academic context, Aisenberg and Harrington describe the 'shock of recognition' that galvanized members of the Alliance of Independent Scholars in Cambridge (Massachusetts) into action, convincing them of the need to more systematically document the strikingly consistent pattern of 'deflection from expected [tenure track] academic careers' that emerged when they compared their experience in a range of fields at their first meeting in 1980.[22] Working groups coalesced in academic institutions and societies across North America, undertaking local studies – often interview based, workplace ethnographies – in which they documented their experiences and, crucially, struggled to develop the conceptual resources necessary to capture emerging commonalities and patterns and to name the diffuse sense of alienation that so many described. The Association of American Colleges (Project on the Status and Education of Women) published a series of widely influential reports through the 1980s in which Hall and Sandler coined the term 'chilly climate.' Their growing awareness that, as Aisenberg and Harrington put it, they were 'hearing about a generalized experience' (1988, ix) is captured by the titles of the AAC reports. The first two, published in the early 1980s, posed a question: *The Classroom Climate: A Chilly One for Women*?; and *Out of*

[20]This turning point in thinking about the persistence of gender inequity in academic contexts is reflected in contributions to *Breaking Anonymity* (The Chilly Collective 1995), and described in more detail in Wylie (1995a).

[21]As indicated at the outset, although the STEM disciplines have been a particular focus of attention, especially in the public debate generated by the MIT report and by Summers's remarks, these patterns of marginalization are by no means unique to the sciences, and the processes of coming to terms with them that I describe here have taken shape across the social sciences and humanities.

[22]As Aisenberg and Harrington describe this initial meeting: 'The effect of that [initial] round of stories was electrifying. Women who had arrived with the sense that the drama and loss in their own academic careers was more or less unique, felt a shock of recognition, hearing their experience in the lives of others previously unknown to them. It seemed clear from that one meeting, as women of highly divergent backgrounds and fields told stories with strikingly similar plot turns, that we were hearing about a generalized experience' (1988, ix).

the Classroom: A Chilly Campus Climate for Women? (Hall and Sandler 1982; 1984). By 1986 Sandler shifted to the declarative: *The Campus Climate Revisited: Chilly for Women Faculty, Administrators, and Graduate Students* (Sandler 1986).

Twenty years after the first 'chilly climate' working groups had begun to coalesce, the authors of the MIT Report on the Status of Women retraced their steps. Nancy Hopkins describes the process by which they came to their conclusions about the insidious nature of 'post-civil rights' gender discrimination as entirely local and internal (Wilson 1999, A17). When she consulted other senior women in the School of Science about persistent problems she faced in attempting to secure additional laboratory space, they each discovered that they had been dealing with similar problems in isolation; Hopkins had asked for advice about a letter she'd drafted outlining her problems and, in the end, 16 of her colleagues redrafted and signed it as a collective 'letter of protest' to the Dean of the School of Science, Robert Birgeneau. The committee appointed by Birgeneau confirmed their suspicions about an overall pattern of gender difference in the distribution of resources, in professional recognition and compensation, and in institutional decision making and leadership roles. In addition, this report documented an age-graded pattern in these gender disparities. While men and women scientists at MIT start out on an equal footing (gender differences proved to be negligible at the junior ranks), the difference between them widens the more senior the comparison group.

What the 'chilly climate' reports and pamphlets of the 1970s and 1980s articulate, and what got public traction with media coverage of the MIT report in 1999, are two key insights that challenge the explanatory framework established by Cole. The first is that gender bias, like other forms of attributional bias, exists in the social fabric of everyday interaction; it takes the form of persistent, small-scale, but systematically gendered differences in recognition and response, evaluation and expectation. Crucially, as Sandler put it in the mid-1980s, this 'host of subtle personal and social barriers' often operates 'below the level of awareness of both men and women' (Sandler 1986, 17), unintended and unrecognized.[23] The grass-roots reports of the 1980s describe, in this connection, the following standard mechanisms by which the workplace is rendered inhospitable for women.[24] The most fundamental is an uncritical reliance on stereotypic assumptions about women's capabilities and

[23] Valian provides a comprehensive overview of the research on cognitive schemas that delineates the role of non-conscious gender schemas in generating gender-biased patterns of evaluation and interaction (1999). Although the grass-roots studies of workplace environment issues I describe here do often cite early studies of evaluation bias, they draw very little on this wider body of work on cognitive schemas that was taking shape at the time.

[24] I summarize here an analysis of central themes in the chilly climate literature that were evident by the late 1980s and have proven remarkably stable. I argued then that the types of practice documented by these reports fall into three broad categories – stereotyping, devaluation, exclusion – each of which may be reenacted in intensified form as reprisals against those who draw attention to these practices (Wylie 1995a, 38–40). Fehr (Chap. 7) describes how these mechanisms – specifically exclusion and various forms of evaluation bias – can systematically undermine the potential epistemic advantages of situational diversity.

(appropriate) roles in academia that translate into gender normative work assignments, with ramifying effects for recognition, compensation, and the allocation of resources. Women report being tracked into service and support roles: undergraduate teaching, student advising, heavy administrative assignments that typically emphasize 'housekeeping,' 'hostessing,' nurturing and facilitating roles. They find themselves disproportionately serving as 'associate' or 'assistant' positions rather than leadership roles with decision making power. By extension, these studies routinely describe instances in which women confront a double standard in response to character traits that are valorized for men (e.g., perceptions of 'assertiveness'; Valian 1999, 129), and in the way their credentials are assessed and projected. The cases described in these reports illustrate patterns of evaluation bias that are now well documented by experimental psychologists (Valian 1999, 127–133), and that constitute testimonial injustice on Fricker's account: men are assumed competent until proven otherwise, while women have to demonstrate their competence at every step; women's successes are seen as exceptional, attributed to the support of others or to luck, while their failures are treated as all that could be expected. As Sonnert and Holton describe these dynamics in their study of elite women scientists, women find themselves suspect, under 'heightened critical scrutiny,' with implications for their interactions with colleagues, research style, and publication patterns (1995, 156, see also Fehr (this volume, 151–152). Finally, a recurrent theme in these studies – as signalled by the 'chilly climate' metaphor – is that women often report a sense of isolation: they lack both formal and informal mentoring; they find they are cut off from key communication networks in their work units and disciplines (the 'sports buddy,' 'locker room' phenomena). As a consequence, they report being disproportionately affected by a lack of institutional transparency about performance expectations, resources, and procedures.

The second key insight, articulated with particular clarity by the authors of the MIT report, is a corollary to the first: small-scale differences in expectation, work assignment, recognition, and social integration, of the kind that chilly climate researchers had documented through the late 1970s and 1980s, can result in substantial and persistent gender differences in career trajectories and outcomes, manifest in everything from lifetime earnings profiles to striking age-graded differences in job satisfaction. Rossiter dubs this pattern of cumulative disadvantage the 'Mathilda effect' (1993), inverting Merton's model of cumulative advantage, the famous 'Matthew effect' (1968).[25] At the time that chilly climate researchers were exploring the micro-dynamics of workplace environments, a robust body of statistical analysis

[25] Merton's reference here is to Matthew (13:12): 'For whomsoever hath, to him shall be given, and he shall have more abundance; but whomsoever hath not, from him shall be taken away even that he hath.' Rossiter settled on the 'Mathilda Effect' after considering a number of possibilities; this is in honor of the nineteenth century sufferagette, Mathilda Gage who, Rossiter argues, developed a critical perspective on the impact of these differences on women's contributions to collective understanding (1993).

of the demographics of academia was taking shape that has documented these outcomes on a large scale.[26]

The response to these studies, when they first appeared in the 1980s, is a particularly telling indicator of what was at stake, epistemically as well as politically. Three recurrent themes in the public reactions of senior administrators, the media coverage, and the resulting public debate about chilly climate studies are particularly relevant here: denial of the facts as presented; denial that they stand as evidence of systematically gendered differences; and denial that, even if substantiated, instances or patterns of gender difference (e.g., in uptake, response, support, or outcomes) demonstrate unfairness in any sense that warrants intervention to change the situation.[27]

The first standard response was disbelief and indignation: the authors of chilly climate reports, and those whose experience they report, must be malicious or deluded (Wylie 1995b [1989], 159–160). The critics typically observed that they had never witnessed or heard of any incidents like those described in chilly climate reports, therefore they could not have occurred as described. Most striking are cases where, for example, the critics enacted, in their condemnation of these reports, precisely the patterns of gender normative stereotyping and evaluation bias the existence of which they were intent on denying.[28] The contours of both testimonial and hermeneutical injustice are clearly evident in these exchanges. It was more plausible to those who were not subject to or who benefit from such bias that the ad hoc working groups, who undertook the hours of interviews and the labor of assembling and distributing these reports, must have falsified the incidents they described, or must have deliberately sought the notoriety of a 'media event' (as one set of critics described it), than that the patterns of marginalization they reported could really be commonplace in the meritocratic culture of the academy. In the case of one such report (Backhouse et al. 1995 [1989]), aggrieved university administrators objected that the interviewees were 'hiding behind anonymity'; those whose accounts were reported anonymously must have 'made it all up' (Wylie 1995b [1989], 159). Despite entrenched conventions of research ethics in the social sciences that require interviewers to protect the identities of research subjects, the content of the report was so evidently unsettling, so threatening in its illegibility,

---

[26] Sonnert and Holton (1995) describe this pattern in the cohorts of high achieving women scientists they studied, making use of comparisons with a male control group. Xie and Shauman (2003) provide a detailed overview and assessment of large scale demographic studies, and Ginther's analyses of a persistent gender gap in compensation across academic and professional fields is an example of this work (e.g., 2004, 2009).

[27] I draw here on published accounts of the response to chilly climate reports that appeared in the 1980s, chiefly as described by the Chilly Collective (1995).

[28] In a discussion of the extremely hostile and high profile response of the President of the University of Western Ontario to a 1989 report, I observed that, in the end, this 'highly charged reaction to the *Chilly Climate Report*' did more to 'illustrate, graphically and publically, the problems we had hoped to document than any amount of 'anonymous' reporting could have done' (1995a, 51).

that anonymity was taken to be evidence of deceit and malicious intent.[29] Here the hermeneutical lacunae that rendered chilly climate phenomena inscrutable served to reinforce already entrenched suspicions about the testimonial credibility (the truthfulness and the competence) of the women who reported them. The irony is that it is precisely these patterns of credibility deficit – amplified and rendered explicit in public debate – that chilly climate authors strove to capture in their accounts of the persistent, demoralizing experience of finding their intellectual and professional contributions ignored, discounted, or attributed to others.

Even when the facts of chilly-making incidents and practices were accepted as reported, a second response was to deny their status as evidence of any systematic difference in the treatment of women as compared to their male peers: each instance must be explicable other terms, as idiosyncratic to the event, the individual, the situation, a localized conflict or misunderstanding, or to a generally hostile environment, such that no gendered pattern emerges in the details. Certainly it is challenging to demonstrate that there are robustly gendered patterns in small-scale, often unintended and unrecognized differences in response, recognition, inclusion. Chilly climate researchers typically started with individual women's stories, following well-established feminist practices of oral history and auto-ethnography that serve to create spaces, in collective discussion, interviews, and testimonial writing, in which women can begin to articulate, in their own terms, experience that does not fit normative expectations. But however powerful the 'shock of recognition' when striking and persistent similarities suggested that these stories embody systematic gender differences in response and recognition, the qualitative nature of these accounts rendered them suspect, especially for those who have no counterpart in their own experience to that which is reported.

Here again both hermeneutical and testimonial injustice configured the debate. Chilly climate researchers confronted a 'gap in collective interpretive resources' (Fricker 2007, 1) that put them at a disadvantage in at least two distinct but powerfully interconnected senses. The challenge they faced was to work against the grain of a set of presuppositions that both animate and obscure the very phenomena they were struggling to capture: presuppositions about the nature of cognitive authority and discrimination that privilege the role of deliberate intention both in individual action and as embodied in explicit policies. The methods of inquiry they relied on to do this – comparative enthnography, textual and qualitative analysis – were precisely what was needed to identify previously unrecognized 'microinequities' and the hermeneutical lacunae that rendered them inscrutable. And yet, in delivering insights that disrupted dominant expectations, these methods were further discredited. On these assumptions, even if the resonances evident in the reported experience of academic women proved to be widespread (a 'generalized experience'), any claim of systematicity would remain implausible so long as there

[29]The senior administrators in question quickly backed away from this line of critique, but it continued to be a recurrent theme in letters to the editor and public debate (Michell and Backhouse 1995, 138–141; Wylie [1989] 1995a, b, 160).

was no evidence of deliberate intent to discriminate on the part of individuals, or of explicit institutional barriers to the training, appointment, funding and advancement of women academics. The heuristic gap that rendered gendered patterns illegible in this domain was the lack of an explanatory mechanism conventionally recognized to be capable of generating systematic differences in women's experiences and academic career paths.

A final objcction to chilly climate studies draws out the normative implications of this last point. Even when systematic gender differences were successfully documented, either in a particular context or as a pervasive feature of academic life, the critics of chilly climate studies routinely denied that they reflect any unfairness on the part of individuals or institutions; if no harm was intended, and no intent to discriminate had been demonstrated, allegations of injustice were unfounded. Cole's argument in *Fair Science* depends on these presuppositions, and they routinely resurface in contemporary debate; they are evident, for example, in the arguments made by those who defended Summers's remarks (e.g., Pinker 2005). Two elements of this paradigm are relevant here. The first is that discrimination, as a form of injustice, is not just a matter of unintended consequences, or unfortunate inequities in the distribution of resources or rewards, however systematic they may be. It is, by definition, a consequence of intentional action that is morally culpable; only if harm is intended, and can be causally be attributed to the actions of an individual, is it morally or politically salient. The second assumption is a resolutely internalist conception of agency that sharply delimits an individual agent's moral accountability. Reasons for action that are introspectively accessible to an epistemic or moral agent – their conscious beliefs and intentions, and judgements that arise from deliberation on them – are the only relevant grounds for explaining their actions and the only legitimate basis for attributing responsibility for the outcomes of action. Each of these presuppositions has generated vast philosophical literatures and are certainly untenable as they stand, but in their vernacular form they are never far from the surface in public debate about the claims central to chilly climate reports, and are sometimes ardently defended by the critics of these reports.[30] So long as they frame discussion of women's workplace experience, they powerfully counteract the possibility that the persistent, often 'subtle' differences in treatment reported in chilly climate reports could be recognized as systematic, or as discriminatory.

Indeed, the problem with which chilly climate researchers and activists grappled was not just a lack of conceptual tools adequate to the task of capturing interpretively opaque experience – a function of gaps in the available hemeneutical resources, as Fricker describes it (2007, 148) – but the constraints imposed by dominant conventions of sense-making that foreclosed the possibility of recognizing the phenomena in question. On the presuppositions about agency that chilly climate researchers strove to make explicit, it was deeply implausible that systematically

[30]See, again, the review one such debate in *Breaking Anonymity* (Chilly Collective 1995) and the defenses of Summers that appeared a decade later.

gendered patterns of evaluation bias and interaction might arise from internalized cognitive schemas that operate 'below the threshold of consciousness' or that such 'micro-inequities' in interaction have the capacity to generate large scale differences in opportunity and outcome for women (Rowe [1973] 1990). Moreover, it was quite literally inconceivable that we might (collectively and individually) be accountable for the effects of these dynamics, given complementary normative assumptions about the nature of discrimination. In the case of climate studies that focus on academia, these hermeneutical barriers are compounded to the extent that the defining feature of these communities and institutions, and the cornerstone of their epistemic credibility, is a commitment to regulative ideals of intellectual meritocracy.[31] In short, the central claims of chilly climate studies about the cumulative effects of diffuse 'microinequities' were categorically implausible given the (resolutely externalist) self-understanding and epistemic authority of their own academic communities. The cost to those who attempted to name and to report forms of experience that called these assumptions into question was further erosion of their credibility as epistemic agents.

8.3 Epistemic Injustice and the Resources of Situated Knowledge

How did chilly climate researchers render intelligible to themselves the diffuse but persistent problems they continued to face even when anti-discrimination laws had long been in place and overt discrimination was (largely) a thing of the past? And what changed between the early 1980s and 1999 such that the central tenets of their analysis could get significant public traction with publication of the MIT report and, subsequently, through arguments for the extension of Title IX provisions to graduate training in the sciences (Munro 2006; Zare 2006)?

Where the first question is concerned, my thesis is that chilly climate researchers drew chiefly on the resources of their own situated, experience-based knowledge to develop what they could only describe metaphorically as an inhospitable 'climate.' In particular, apart from scattered references to early evaluation bias research and appeals to statistics on the representation of women in academia that demonstrated the need to look beyond 'civil rights era' barriers to access, their work was largely uninformed by a growing body of disciplinary research in the cognitive and social sciences that would ultimately vindicate their central insights. I return to this point shortly. The situated knowledge on which grass roots chilly climate researchers did depend incorporates each of the elements I identified at the outset as loci of epistemic advantage, but with some telling twists.

[31] Rooney (Chap. 1, this volume) offers a striking example, in the marginalization of feminist epistemology, of how critical challenges are deflected by appeal to entrenched epistemic ideals of objectivity and neutrality.

Evidence Social psychologists have articulated a principle of informational asymmetry that fleshes out the empirical detail of an insight that has long been central to feminist standpoint theory: it is that 'in any relationship defined by differential power (like gender), the dominant group (e.g., men) can afford to be oblivious to certain kinds of social cues, while the subordinate group (women) cannot'; consequently, 'dominants and subordinates have very different levels and kinds of information about each other' (Stewart and McDermott 2004, 529; citing Fiske 1993). In some cases this asymmetry ensures that subordinates and outsiders have access to evidence that the privileged do not in a quite literal sense. This is an insight routinely exploited by mystery writers, from Agatha Christie to Barbara Nealy, whose fictional investigators are discounted for precisely the attributes that put them in a position to learn crucial facts about means and motivation that elude normatively credible witnesses.[32] The case of chilly climate research is more complicated in several respects. Women academics are, in principle, insiders to the 'sacred grove' (Aisenberg and Harrington 1988), not sojourners or invisible outsiders; what mobilized (some) to take up chilly climate projects was typically dissonance between their expectations – rooted in a deeply held commitment to the meritocratic ideals of their chosen fields and, often, considerable privilege in other respects – and evidence from their direct experience of academic institutions. Chilly climate authors catalog moments of rupture in which a particular juxtaposition of responses or judgments throws into relief a double standard (e.g., the same credentials are read in very different ways), or a growing unease about gender inequity in the outcomes of deliberation (e.g., on appointments or admissions, publication or promotion) that alerts them to a persistent disconnect between the purportedly gender neutral norms of academic accomplishment and the highly gendered characteristics of the qualities that conventional working indicators tend to track.[33] They register shock and anger at what they slowly and often grudgingly came to recognize as pervasive, gendered patterns of credibility overspill and credibility deficit – testimonial injustice; evaluation bias – that had been invisible to them, and that remained largely inscrutable to those who better fit these dominant norms and benefit from them.

Inferential Heuristics and Explanatory Models Given the hermeneutic deficits facing the insiders who struggle to articulate this dissonant experience, chilly climate projects were explicitly and agonizingly exploratory. The process of coming to consciousness described by Aisenberg and Harrington, and by Hopkins, was an irreducibly collective and comparative undertaking. It was a matter of creating from the ground up the conceptual tools and interpretive heuristics necessary to reconceptualize systematic disadvantage in terms of the 'climate' of a workplace,

---

[32] Elsewhere I develop an analysis of evidential advantage in terms of Nealy's character Blanche White, a crime solving African American housekeeper (Wylie 2003).

[33] These are the mechanisms by which systematic forms of ignorance and underlying (taken for granted) assumptions become visible to those on the margins, and to insider-outsiders (Rooney, this volume, 10).

rather than its architecture, and to identify mechanisms that might generate large-scale discriminatory outcomes 'despite good will,' through inflicting 'a thousand cuts' or by suffocation under 'a ton of feathers' (Caplan 1993).

Critical Distance As this suggests, the situated experience that made chilly climate researchers aware of gendered 'micro-inequities' afforded them a critical advantage in discerning the ways in which the institutions and practices of academia fall short of its ideals. The ruptures created by finding that gender makes a difference to who counts as a credible knower – whose 'merit' becomes a medium of exchange in a meritocracy – enforced a critical dissociation from norms and conventions of academic practice which, in turn, made it possible to recognize the disconnect between what working indicators of epistemic credibility actually track and what they are claimed to track. It was asymmetries in extant 'working indicators' of epistemic credibility that chilly climate researchers both documented and exploited.

In recent years a number of high profile reports have appeared that systematically rebut the lingering presuppositions of Cole's paradigm, as revivified by Summers,[34] bringing to bear the results of research that probes the mechanisms responsible for evaluation bias and documents the cumulative effects of small scale gender biases in uptake and response. These reports demonstrate that fields as diverse as experimental psychology, sociolinguistics, and economics have substantiated many of the conclusions drawn, tentatively and quite independently, from the dissonant experience documented by grass roots 'chilly climate' researchers. It is now well established that factors operating below the threshold of conscious awareness – condition the adjudication of academic merit in the review of credentials, in assessing grant proposals, and in weighing the authority of publications. These effects are documented by studies of cognitive schemas, stereotype mobilization, and ascriptive bias in social and cognitive psychology (Steele 1997; Valian 1999); by the work by sociolinguists on small-scale interaction patterns that reproduce social hierarchy (Ridgeway 1992); and by sociological studies of institutional structures that can foster or counteract these dynamics (Reskin 2003).[35] In addition, since the early 1990s, models of the dynamics by which women and minorities are deflected from and marginalized within academia

---

[34]For example, the National Academy of Sciences report, *Beyond Bias and Barriers* (2007), and the AAUP report, *Faculty Gender Indicators* (West 2006). Although the NAS report frames its mandate in general terms without reference to Summers, it responds point for point to Summers's claims, demonstrating that the best empirical research available renders untenable the stereotypes and conventional assumptions he invokes.

[35]For example, Ceclia Ridgeway calls for attention to the micro-structure of interaction as the level at which gender stratification is generated; these include, for example, the double standards at an interactional level that Fosci describes as mechanisms by which structural inequities are maintained. And Barbara Reskin (in a 2002 Presidential Address to the American Association) inveighs against the continued focus on 'motives' and argues for focusing on 'organizational- and societal-level mechanisms' – patterns of practice; systems of accountability; degree of transparency and formalization in an organization – that allow cognitive schemas to operate and that perpetuate the double standards and patterns of ascriptive bias that underpin them.

(especially in the STEM disciplines[36]) recognize the interactive and cumulative effects of small scale disadvantage, giving the various forms of Rossiter's Mathilda effect a central place in their analyses (Cole and Singer 1991; Sonnert and Holton 1995). Finally, large scale quantitative analysis of national databases delineate, with growing precision, a persistent 'gender gap' in such measurable indices of recognition as salary, and document the age-graded patterns of cumulative disadvantage predicted by these models (Xie and Shauman 2003).[37]

No doubt a great many factors contributed to the sea change in the reception of chilly climate reports, from the early 1980s when they first began to appear to 1999 when the MIT report drew national attention. But certainly one key factor is a shift in the interpretive resources available in public discourse about 'post-civil rights' discrimination as a consequence both of the grass-roots chilly climate research and of these proliferating research programs in the social sciences and psychology.[38] By the turn of 2000 it was no longer radically incomprehensible that our judgments and behaviors might be substantially shaped by non-conscious cognitive schemas, or that large scale, morally and epistemically consequential inequities might arise from unintended and unrecognized differences in treatment of men and women.

8.4 Conclusion

In this analysis of the insights central to chilly climate reports on 'post-civil rights era sex discrimination,' I have argued that chilly climate researchers posited a set of generative mechanisms – elements of an alternative explanatory paradigm for understanding both their own localized experience and the patterns they discerned in this experience when they had occasion to compare it with other women in their fields and institutions. Although few directly engage Cole, the explanatory models they offer directly counter the conventional presuppositions made explicit by Cole and, 25 years later, by Summers; they show how gender normative behavior

[36] Science, technology, engineering and mathematics (STEM).

[37] These show, for example, that gender gaps in productivity are not as marked as Cole claimed and have been diminishing. Nonetheless, studies that control for an enormous range of factors cited as possible reasons for women's different employment and compensation profiles (various forms of self-selection, demographic or market factors) show a persistent gap that, it seems, must be attributed to residual gender discrimination, especially at higher ranks and more elite schools. They also suggest that the training and career advancement pipelines are more porous that assumed and leak at different places for different disciplines.

[38] In a response to Summers, Valian observes that 'although an abundance of research of this sort exists, it has not become part of our common understanding and thus has not yet redressed the imbalances between men and women in professional life' (2005). While I agree in general terms – certainly, there is much to be done to integrate insights about cognitive schemas into our understanding and our practices – the outcry generated by Summers's remarks suggests that the resources available for understanding the status of women in male dominated fields had shifted significantly since the time when Cole could take for granted that, with the exception of a few critics like Rossiter, his reading audience would accept terms of his analysis.

can give rise to systematic, large scale and ramifying differences in outcomes for academic women in the absence intentional discrimination or formal barriers to their participation.

At the same time, however, the more radical epistemic implications of this alternative chilly climate paradigm are routinely blunted by its staunchest advocates. As one outspoken advocate for gender equity in the physical sciences likes to put it, the institutions of her science are sexist to the core, but 'quarks have no gender'. Inequities in the application of epistemic norms, manifest in a reliance on working indicators of rational authority that relegate women to the margins, have no bearing whatsoever on the conceptual, empirical integrity of the science. The problem to be resolved is strictly a matter of testimonial injustice; for meritocratic ideals to be realized what must be rectified are systematic patterns credibility deficit and overspill, the misrecognition of epistemic agents. It is assumed that ideals of excellence – the substantive norms of credibility that define what counts as a well formed question and a credible answer – are impervious to the influence of factors that are recognized to distort their application when intellectual merit is adjudicated in the context of hiring, tenuring, promoting, awarding grants to, and publishing particular individuals. In fact, it is hard to see how systematic testimonial injustice could fail to entrench patterns of hermeneutical injustice, even in fields that deal with manifestly non-gendered subject matters. If, for example, patterns of workplace segregation obtain such that women typically work in a narrow range of subfields or on particular types of problems, and if the results of women's work gets less support and recognition than that of their male peers, given standard gender biases in citation and funding, then the evidence and insights women generate will have less impact on their field as a whole than the work done by men in areas where they dominate. Testimonial injustice thus translates into biases in the research agenda and in the epistemic resources available to the research community that, to varying degrees and in diverse ways, shape the trajectory of a discipline as a whole. It is the possibility that institutional inequity may have an impact on the content of their fields that many equity activists flatly refuse to consider. To take up these questions would require more than the resources of situated experience which brought testimonial injustice into sharp focus; it would require a well articulated critical standpoint on knowledge production.

It is perhaps unsurprising that equity activists would draw strong conclusions about testimonial injustice, delineating innumerable ways in which gender schemas determine (unfairly) who counts as a credible knower and who gets credit for contributions to the collective store of authoritative disciplinary knowledge, but deny that these injustices have any impact on the epistemic integrity and hermeneutical resources of the disciplines within which they work. The challenge of rendering the experience of epistemic injustice communicable is particularly acute in academic contexts, especially the sciences, because it calls into question the community norms of credibility to which chilly climate researchers are held accountable and to which they themselves subscribe. In many contexts, these community norms include a proscription against any appeal to idiosyncratic, personal experience, and yet it was women's dissonant experience (as scholarly or scientific insiders but gender outsiders) that threw into relief the contours of the cognitive schemas and localized interaction patterns that generate persistent patterns of testimonial injustice. The cost of relying on the epistemic advantages of situated knowledge, for many, was a resolve to circumscribe its import, sharply dissociating the claims they make about institutional inequity from any more probing critical analysis of the epistemic conventions of their fields.

Acknowledgments I gratefully acknowledge the support I received as a Research Fellow at the Michelle R. Clayman Institute for Gender Research at Stanford University when initiating this project in 2005–2006, and the generous feedback on earlier versions of this paper provided by colleagues – skeptical and encouraging alike – who heard it presented as a keynote address at the Central States Philosophy Association (2006) and at the second FEMMSS conference (Feminist Epistemology, Methodologies, Metaphysics and Science Studies, Phoenix, 2007), as well as in symposia at UCLA, the London Institute for Philosophy (2007), and the second biennial meeting of the Society for Philosophy of Science in Practice in Twente (2007). I hope this philosophical analysis does justice to the inspired and courageous work on which I draw, especially the Chilly Collective!

References

Aisenberg, Nadya, and Mona Harrington (eds.). 1988. *Women of academe: Outsiders in the sacred grove*. Amherst: University of Massachusetts Press.

Backhouse, Constance, Roma Harris, Gillian Michell, and Alison Wylie. 1995. The chilly climate for faculty women at western: Postscript to the backhouse report. In *Breaking anonymity: The chilly climate for women faculty*, ed. Chilly Collective, 97–132. Waterloo: Wilfrid Laurier University Press. 1989.

Bell, Derrick. 1992. *Faces at the bottom of the well: The permanence of racism*. New York: Basic Books.

Caplan, Paul J. 1993. *Lifting a ton of feathers: A woman's guide to surviving in the academic world*. Toronto: University of Toronto Press.

Code, Lorraine. 1995. *Rhetorical spaces: Essays on gendered locations*. New York: Routledge.

Cole, Jonathan R. 1979. *Fair science: Women in the scientific community*. New York: The Free Press.

Cole, Jonathan R., and Burton Singer. 1991. A theory of limited differences: Explaining the productivity puzzle in science. In *The outer circle: Women in the scientific community*, ed. Harriet Zuckerman, Jonathan R. Cole, and John T. Bruer, 277–310. New York: W. W. Norton.

Collective, Chilly. 1995. *Breaking anonymity: The chilly climate for women faculty*. Waterloo: Wilfrid Laurier University Press.

Fiske, Susan T. 1993. Controlling other people: The impact of power on stereotyping. *American Psychologist* 48: 621–628.

Fricker, Miranda. 1998. Rational authority and social power: Towards a truly social epistemology. *Proceedings of the Aristotelian Society* XCVIII(2): 159–177.

Fricker, Miranda. 1999. Epistemic oppression and epistemic privilege. *Canadian Journal of Philosophy* 25: 191–210. Supplementary.

Fricker, Miranda. 2006. Powerless and social interpretation. *Episteme: Journal of Social Epistemology* 3(1): 95–108.

Fricker, Miranda. 2007. *Epistemic injustice: Power and the ethics of knowing*. Oxford: Oxford University Press.

Ginther, Donna K. 2004. Why women earn less: Economic explanations for the gender salary gap in science. *AWIS Magazine* 33((1): 6–10.

Ginther, Donna K. 2009. Does science promote women? Evidence from Academia 1973–2001. In *Science and engineering careers in the United States*, ed. Richard B. Freeman and Daniel F. Goroff. Chicago: University of Chicago Press.

Hacking, Ian. 1999. *The social construction of what?* Cambridge: Harvard University Press.

Hall, Roberta M., and Bernice R. Sandler. 1984. *Out of the classroom: A chilly campus climate for women?* Washington, DC: Association of American Colleges.

Hall, Roberta M., and Bernice R. Sandler. 1982. *The classroom climate: A chilly one for women?* Washington, DC: Association of American Colleges.

Harding, Sandra. 2006. *Science and social inequality: Feminist and postcolonial issues*. Chicago: University of Illinois Press.

Hartsock, Nancy C.M. 1997. Standpoint theories for the next century. In *Politics and feminist standpoint theories*, ed. Sally J. Kenney and Helen Kinsella, 93–101. New York: Haworth Press.

Hartsock, Nancy C.M. 1983. The Feminist Standpoint: Developing the Ground for a Specifically Feminist Historical Materialism. In *Discovering reality: Feminist perspectives on epistemology, metaphysics, methodology and philosophy of science*, ed. Sandra Harding and Merrill B. Hintikka, 283–310. Boston: D. Reidel Publishing Company.

Lewin, Arie Y., and Linda Duchan. 1971. Women in academia: A study of the hiring decision in departments of physical science. *Science* 173: 892–895.

McConkey, Jane. 2004. Knowledge and acknowledgement: "Epistemic Injustice" as a problem of recognition. *Politics* 24(3): 198–205.

Merton, Robert K. 1968. The Matthew effect in science. *Science* 159(5): 56–63.

Michell, Gillian, and Constance Backhouse. 1995. Epilogue: The remarkable response to the release of the chilly climate report. In *Breaking anonymity*, ed. Chilly Collective, 133–169. Waterloo: Wilfrid Laurier University Press.

MIT, Committee on Women Faculty in the School of Science. 1999. A study of the status of women faculty in science at MIT. *The MIT Faculty Newsletter*, 11(4). http://web.mit.edu/fnl/women/women.html.

Moya, Paula M.L. 2000. Reclaiming Identity. In *Reclaiming identity: Realist theory and the predicament of postmodernism*, ed. Paula M.L. Moya and Michael R. Hames-García, 1–26. Berkeley: University of California Press.

Munro, Neil. 2006. Title IX: Not just for athletes. *National Journal* 37(12): 50–51.

Narayan, Uma. 1988. Working together across difference. *Hypatia* 32: 31–48.

National Academy of Sciences, National Academy of Engineering, and Institute of Medicine Committee on Maximizing the Potential of Women in Academic Science and Engineering, and Committee on Science, Engineering, and Public Policy. 2007. *Beyond bias and barriers: Fulfilling the potential of women in academic science and engineering*. Washington, DC: The National Academies Press.

Pinker, Steven. 2005. Psychoanalysis Q-and-A. *The Harvard Crimson Online*, January 19, 2006. http://www.thecrimson.com/article/2005/1/19/psychoanalysis-q-and-a-steven-pinker-in-an/?print=1

Reskin, Barbara F. 2003. Including mechanisms in our models of ascriptive inequality. *American Sociological Review* 68: 1–21.

Ridgeway, Cecilia R. (ed.). 1992. *Gender, interaction, and inequality*. New York: Springer.

Rossiter, Margaret W. 1981. Fair enough?: Review of J. R. Cole, fair science, and L. S. Hornig, climbing the academic ladder. *Isis* 72(1): 99–103.

Rossiter Margaret, W. 1993. The [Matthew] Matilda effect in science. *Social Studies of Science* 23: 325–341.

Rossiter Margaret, W. 1995. *Women scientists in America: Before affirmative action, 1940–1972*. Baltimore: Johns Hopkins University Press.

Rowe, Mary. 1990. Barriers to equality: The power of subtle discrimination. *The Employee Responsibilities and Rights Journal* 3(2): 153–163.

Sandler, Bernice R. 1986. The campus climate revisited: Chilly for women faculty, administrators, and graduate students. In *Project on the status and education of women*, ed. Bernice R. Sandler. Washington, DC: Association of American Colleges.

Simeone, Angela. 1987. *Academic women: Working towards equality*. South Hadley: Bergin and Garvey Publishers.

Sonnert, Gerhard, and Gerald Holton. 1995. *Who succeeds in science? The gender dimension*. New Brunswick: Rutgers University Press.

Steele, Claude M. 1997. A threat in the air: How stereotypes shape intellectual identity and performance. *American Psychologist* 52: 613–629.

Stewart, Abigail J., and Christa McDermott. 2004. Gender in psychology. *Annual Review of Psychology* 55: 519–544.

Summers, Lawrence H. 2005. *Remarks at NBER conference on diversity in the science and engineering workforce*. Cambridge: Harvard University, Office of the President, [cited January 14, 2005].

Sunstein, Cass R. 2003. *Why societies need dissent*. Cambridge: Harvard University Press.

Tosi, Henry L., and Steven W. Einbender. 1985. The effects of the type and amount of information in sex discrimination research: A meta-analysis. *Academy of Management* 28(3): 712–723.

Valian, Virginia. 1999. *Why so slow? The advancement of women*. Cambridge: MIT Press.

Valian, Virginia. 2005. Raise your hand if you're a woman in science. *Washington Post*, January 30, 2005, B01. http://www.washingtonpost.com/wp-dyn/articles/A46421-2005Jan29.html

Weeks, Kathi. 1996. Subject for a Feminist Standpoint. In *Marxism beyond marxism*, ed. Saree Makdis, Cesare Cesarino, and Rebacca Karl, 89–118. New York: Routledge.

West, Martha S., and John W. Curtis. 2006. AAUP faculty gender equity indicators. Washington, DC: American Association of University Professors. http://www.aaup.org/AAUP/pubsres/research/geneq2006.

White, Harrison C. 1982. Review essay: Fair science? *American Journal of Sociology* 87(4): 951–956.

Wilson, Robin. 1999. An MIT professor's suspicion of bias leads to a new movement for academic women. *The Chronicle of Higher Education*, December 3 1999, A16–18. http://chronicle.com/article/An-MIT-Professor-s-Suspicion/21165

Wylie, Alison. 1995a. The contexts of activism on 'Climate' issues. In *Breaking anonymity: The chilly climate for women faculty*, ed. Collective Chilly, 29–60. Waterloo: Wilfrid Laurier University Press. 1989.

Wylie, Alison. 1995b. The remarkable response to the chilly climate report. In *Breaking anonymity: The chilly climate for women faculty*, ed. Chilly Collective, 155–165. Waterloo: Wilfred Laurier Press. 1989.

Wylie, Alison. 2003. Why standpoint theory matters: Feminist standpoint theory. In *Philosophical explorations of science, technology, and diversity*, ed. Robert Figueroa and Sandra Harding, 26–48. New York: Routledge.

Wylie, Alison, Janet R. Jakbosen, and Gisela Fosado. 2007. Women, work and the academy: Strategies for responding to 'Post-Civil Rights Era' gender discrimination. In *New Feminist Solutions*. New York: Barnard Center for Research on Women. http://www.barnard.columbia.edu/crow/newfeministsolutions/index.htm

Xie, Yu, and Kimberlee A. Shauman. 2003. *Women in science: Career processes and outcomes*. Cambridge: Harvard University Press.

Zare, Richard N. 2006. Sex, lies, and title IX. *Chemical and Engineering News* 2006: 46–49.

Zuckerman, Harriet, Jonathan R. Cole, and John T. Bruer (eds.). 1991. *The outer circle: Women in the scientific community*. New York: W. W. Norton.

Part III
Contexts of Oppression: Accountability in Knowing

Chapter 9
More Than Skin Deep: Situated Communities and Agent Orange in the Aluoi Valley, Vietnam

Nancy Arden McHugh

Abstract I build upon feminist arguments for situated knowledge and pragmatist arguments for experimental inquiry to articulate and argue for an approach that I refer to as situated communities. This approach seeks to generate effective and ethical scientific research practices by asking that researchers focus on communities in their complex environment as subjects of study instead of relying primarily on clinical trials and laboratory research. Communities should be recognized as situated epistemic agents and as changing, evolving centers of life. Doing so requires that these communities are understood in their materiality through bodies that are aged, gendered, abled/disabled, raced, classed, colonized, bordered, materially advantaged and disadvantaged, engaged in particular daily practices within a complex environment.

To illustrate my argument I analyze the effects of Agent Orange on communities in the Aluoi Valley, Vietnam and the accompanying research on Agent Orange. I argue that when studied through the situated communities approach instead of in the isolation of the laboratory, it becomes much more obvious why Agent Orange can cause the congenital anomalies, cancers, and other diseases the Vietnamese claim it does. I focus especially on women in this region because they carry the largest social burden of the effects of Agent Orange due to their role in agriculture, housework, childbearing, breastfeeding, and caring for children and adults affected by Agent Orange.

Keywords Agent Orange • Communities • Dioxin • Pragmatism • Situated knowledges

Mainstream philosophy of science and mainstream science obscure the practical social and political significance of scientific knowledge practices by idealizing the

N.A. McHugh (✉)
Department of Philosophy, Wittenberg University, Springfield, OH 45501, USA
e-mail: nmchugh@wittenberg.edu

H.E. Grasswick (ed.), *Feminist Epistemology and Philosophy of Science: Power in Knowledge*, DOI 10.1007/978-1-4020-6835-5_9,

laboratory and the clinical trial as models for objective knowledge acquisition. Our daily lives are mediated by magazine ads, commercials, and news blips that report the latest clinical trial of the latest drug or the most recent results of experiments on the toxin that happens to top our list of social concerns. We absorb this knowledge rather passively because we are taught to believe that this data tells us something, something about what our lives would be like if we take this drug or whether we are safe from the effects of this chemical. Few of us question whether the methodology used to gather this information is the best or only methodology to give us the knowledge that we need to live well and act effectively. We rarely think about how this particular knowledge-making practice drives policy and action. Nor do we consider that there may be other, equally effective or more effective methodologies to generate scientific knowledge and action.

Feminist philosophy of science and pragmatist perspectives have challenged the orthodoxy of scientific practice on numerous levels, such as the neutrality and objectivity of scientific methods, practitioners, and knowledge. They have also challenged claims of epistemic individualism, the fact-value distinction, and the qualitative distinction between the natural and social sciences.[1] In this paper I build upon feminist arguments for situated knowledges and pragmatist arguments regarding experimental inquiry to formulate a position that I am calling situated communities. This position entails a move back to some of the values endemic to scientific practice by calling for a reorientation of contemporary science. It first requires an engagement with the everyday world in which we live to generate scientific knowledge and action instead of relying on the primacy of laboratory experimentation and clinical trials. Second, situated communities requires an increased awareness and attention to the ethical consequences and social outcomes of scientific methodologies. Both of these are practices that pragmatists argue were and should continue to be an intrinsic part of the values of science. I argue that the situated communities approach not only provides us with a better epistemic lens and a more effective methodology, it also provides the knowledge that we need to practice responsibly. I use the case of Agent Orange in Vietnam as an example of the inability of predominant scientific methods to provide substantive knowledge about the effects of Agent Orange in these Vietnamese communities, thus reflecting an inability of these methods to responsibly address the health and social needs of Vietnamese victims of Agent Orange. I point to more situated methods that are employed in examples of research that work outside of the predominant model. This situated approach provides a better understanding of the effects of Agent Orange and directions to act responsibly.

I begin by recounting an experience I had in Vietnam that led me to the situated communities argument and finish by coming back to Vietnam to use the situated communities approach to assess the evidence for Agent Orange causing congenital anomalies, cancers, and other serious health effects in the people living in the Aluoi Valley.

---

[1]See for example, Dewey (1925, 1929), Harding (1986, 1991, 1998), Longino (1990, 2002), Nelson (1993), and Seigfried (1996).

9.1 Being in Vietnam

The Tu Du hospital in Ho Chi Minh City is Vietnam's largest women's hospital. When I went Vietnam summer of 2004 with a group to study how the change to a market economy, Doi Moi, affected the Vietnamese, we were asked to visit the hospital and to tour its Peace Village. I knew little about Peace Villages, little about Agent Orange, and little about the destruction that I was about to see.

As I walked out of the offices, a clinical space that revealed very little about what I was about to experience, I was troubled that two generations after the American war children were being born with an alarming rate of congenital anomalies in the communities having the highest levels of exposure to Agent Orange. The director of the hospital said they suspect there are genetic changes occurring at the somatic level, in utero, as well as the at the germ cell level, the level of the sperm and egg. What little I knew about research done on Agent Orange and U.S. Vietnam veterans indicated that dioxin could not have genetic effects on these levels; dioxin was supposedly unable to bind with or alter the structure of DNA.[2] Yet, I didn't know how else to explain the effects they were describing to me.

As I approached the Peace Village housed within the hospital I began to be challenged in a new way. I saw what was literally a village, set up with the goals of community interaction in mind, nothing like the sterility and false sense of safety generated in U.S. hospitals' common space. This was a space that reflected the needs of an impoverished community. Most of the patients housed in the Peace Village came from rural areas; many were from the Central Highlands and were poor. The village reflected energy and life, unlike U.S. hospitals that feel lifeless, literally and metaphorically.

As we walked through the village, we acquired an escort. He was a young man who several years earlier had been separated from his conjoined twin. He was dynamic, spoke to us in English, (typically American, none of us spoke Vietnamese) and did not seem overly hampered by the loss of the leg he shared with his twin. His vibrancy did not prepare me in anyway for what I was about to experience as he escorted us up the elevator into the rooms that housed the other children in the Peace Village. What I saw can't be described well. The best I can say is I saw bodies and lives destroyed in a way that was beyond my experience, beyond the experience of most westerners. This was a war zone, but 30 years after the American war ended. It hit me at a gut level that is hard to describe.

A Vietnamese-American woman came walking out of a room I was about to enter. She was carrying a child who suffered from hydroencephalitis, a swelling of the brain and cranium. The little girl also had no eyes, her eye sockets were fused shut, her mouth and palate were severely deformed, as were her arms and legs.

---

[2]Like many people in the U.S. who heard about Agent Orange, my knowledge came primarily from the 1978 lawsuit, settled in 1984, by U.S. Vietnam veterans against the manufacturers of Agent Orange. Until recently the claims by the U.S. government and the chemical manufacturers dominated the public and scientific opinions on the effects of Agent Orange.

The woman's name was Trinh Kokkoris. The name didn't mean much to me and it wouldn't have to most U.S. citizens, but it should have. The name Kokkoris meant a lot to the Vietnamese. In January of 2004 her husband Constantine Kokkoris had filed the first class action lawsuit against 37 chemical companies on behalf of the Vietnamese victims of Agent Orange. Though the children in this room were not named in this legal suit, these children and countless children and adults like them would be beneficiaries if damages were awarded.

The physical evidence that I was seeing and the claims of the hospital's doctors didn't fit with the physical laboratory claims made by researchers who worked on the effects of dioxin. I want to emphasize that for both claims there was evidence, but they were different types of evidence, from different settings. One was in the clinical setting of the laboratory and one was here, in the living, situated environment of Vietnam, a physical and social environment in which Agent Orange existed and has existed as part of daily life since it was first sprayed in 1961. Scientists tend to dismiss evidence from the 'wild' nonclinical setting because this evidence doesn't accord with predominant scientific methodologies, like toxic risk assessment or randomized control trials, which rely upon isolating substances to understand their effects or isolating organisms to understand how they are affected. These methods are thought to ensure a more purely objective body of evidence because of their isolation from the complexity of the everyday world, which ironically is the setting in which life takes place and we actually experience things.[3] My visit to the Tu Du Hospital helped me to recognize a gap in what many scientists and lay people want from science, for it to generate knowledge to improve human living, and, in this case, its ability to do so. I began to question whether our current scientific methods could meet the needs of communities that are situated outside of dominant culture and experience multiple impacts, such as from poverty, poor access to medical care, environmental contaminants, stress, war, racism, colonialism, and sexism. From this experience I began to formulate the argument for situated communities.

9.2 Knowing One's Place: Situated Knowledges and Concrete Engagements

The view that all knowers and knowledge are situated is one of the most important and tangible insights generated in feminist science studies. It has resulted in epistemological and methodological reframings of scientific practices and has led to ongoing critical work in feminist science studies and feminist epistemology. Though Donna Haraway was the first to use the term 'situated knowledges' in her 1991 essay 'Situated Knowledges: The Science Question in Feminism and the Privilege of

[3]Code also works to drive this point home in *Ecological Thinking*. This is especially evident in her discussion of the knowledge Rachel Carson generated from turning to the world for her knowledge instead of to reports from laboratory testing.

Partial Perspective,' it has been developed more fully in feminist epistemology in recent years by such writers as Sandra Harding, Patricia Hill Collins, and Lorraine Code.[4] At its most basic level, situated knowledge is the claim that all knowledge is generated from a knower's particular location, which consists of the complex unfolding of one's social, material, epistemological, gendered, lived bodily experience. There is no purely objective knowledge in the sense that there are no individual purely objective knowers, i.e., epistemic subjects who are totally free from values, biases, and background assumptions.[5] All knowers are situated within and through their experiences and this bears upon the knowledge they acquire. Though Haraway's argument for situated knowledges was primarily epistemological, methodological implications were implicit. With the development of the epistemological components of situated knowledges in feminist science studies, there also came an increasingly explicit emphasis on its methodological import.

The move toward viewing situated knowledges as both epistemic and methodological is important. It provides the tools for not only understanding how knowledge functions, but also what we can do with knowledge that is situated. In other words, this move provides a way for knowledge to be transformative. In the second edition of *Black Feminist Thought* Patricia Hill Collins is particularly careful to distinguish between epistemology and methodology. As she points out, epistemologies give us accounts of truth, standards for knowledge, and tools to assess that knowledge. Methodologies are the 'principles of how to conduct research and how to interpret' the frameworks we use to understand the world (2000, 252). Methodologies are also those means we use to interact with the world and people in the world.

'Situated knowledges' is both epistemological and methodological. It is epistemological in that it is a theory about how to gain more accurate or more objective knowledge and it explains how individuals and groups have particular epistemological worldviews based on their material location. It is methodological because, as I argue below, it becomes a tool to generate knowledge about the world and to provide better means for our interactions with the world.

Sandra Harding's standpoint epistemology also has argued for both the epistemological and methodological salience of location. As early as her 1986 book *The Science Question in Feminism* and in more detail in her 1991 book *Whose Science, Whose Knowledge*? Harding argued that all knowledge is generated from a standpoint, i.e., from a particular social and historically mediated perspective. All knowledge is generated from a location, and some knowledge and starting points for generating knowledge, those of women, are better at developing objective

---

[4]Among other feminists that have influenced the direction of situated knowledge arguments are Chela Sandoval (2000), Sarah Hoagland (2001), and Chandra Mohanty (2003). In this paper I don't take up their work on situated knowledges because I am focusing on feminists whose work has most directly influenced discussions in feminist philosophy of science. In my project, *Actions Which Change the Face of the World*, I develop and utilize a broader range of work in feminist epistemology to address the situated nature of knowledge.

[5]See for example Collins (1986), Longino (1990), Haraway (1991), and Harding (1991) for nuanced discussions of the subjective nature of individual knowledge.

knowledge because of the standpoint from which they originate. Harding argued that not only should we recognize the situated nature of all knowledge, we also need to employ location or situation as a methodology from which to start thinking. Just as standard scientific methods ideally sought to gain objective knowledge, Harding's standpoint approach sought to maximize scientific objectivity. She argued that because '[w]omen are valuable "strangers" to the social order,' scientific questions should be initiated from women's perspectives (1991, 124).

In *Science and Social Inequality: Feminist and Post-Colonial Issues* (2006) Harding continued to develop the situated nature of standpoint theory by drawing on the work of feminist ethnoscientists, such as Vandana Shiva (1989a, 2000), and feminists working on gender and development, such as Rosi Bradiotti (1994) and Drucilla Barker (2000). The work in these areas provided significantly more context for the experiences that shape particular women's lives and knowledge; among these are gendered practices, governmental involvement, economic influence, environmental conditions, and women's access to basic resources, including food, water, fuel, and medical care. Harding's attention to the particularities of situation allowed her to employ standpoint theory to more directly address issues of gendered, global injustice. She asks her readers to '[c]onsider for example, the different interests of women concerned with the relation between apparent increases in cancer and living "downstream" from toxic industries and, in contrast, tribal or peasant women living on the edge of the expanding Sahara desert, who experience decreasing supplies of water, food, and fuel, which they must supply to their communities' (2006, 99). The standpoints and needs of these women differ, but so do the methods necessary to address these related but particularly different examples of injustice. We can't engage these issues by thinking of women's lives only collectively. Instead we must also understand that women's experiences, standpoints, and needs differ based on the material conditions of their lives. This recognition puts situated knowledges in a more effective position to tackle issues of global injustice.

The increased focus on materiality and concrete nature of situation and its epistemological and methodological importance is especially apparent in Lorraine Code's argument for ecological thinking. In *Ecological Thinking: The Politics of Epistemic Location* (2006) Code emphasizes the importance of place, as habitat and as an epistemological location. She argues that a significant aspect of situated knowledge is that it is not just a place from which to interrogate knowledge or from which to generate knowledge; location – social location and physical location/ habitat – is a place to be interrogated. Code views situation or 'habitat as a place to know' (2006, 37) and emphasizes that, like all living things, humans are ecological subjects. Just as ecology must take into account all the interactions an organism engages in, experiences, and is affected by, ecological thinking 'builds on the relations of organisms with one another and with their habitat, which comprises not just the physical habitat or the present one, but the complex network of locations and relations, whether social, historical, material, geographical, cultural, racial, sexual, institutional, or other, where organisms – human or nonhuman – try to live well, singly or collectively' (2006, 91). Code's work generates an epistemology from the methodologies in ecology. She then uses this epistemology to generate a methodological approach to knowing the world. Thus, ecological thinking offers a way to know 'us', humans, in the world, and provides a fuller accounting and direction for engaging in the world than less situated modes of philosophical practice have provided.

Given the analysis generated in this section, the following claims can be attributed to arguments for situated knowledges:

1. Situation is an epistemic location, i.e., a place from which to know.
2. It is a vital location, that is socially, materially, and historically salient to its members.
3. It is a place to know or a place to interrogate.
4. It is also a methodological location from which to initiate critical, transformative practices, practices that are informed by location.
5. It is a place whose conditions are transformed by its own methodology as well as a place that methodologically transforms epistemology.

9.3 Pragmatism as Methodology: Experimental Inquiry and Practice

Situated knowledge arguments provide an important framework for thinking about the ways that gender and material location shape epistemologies, methodologies, and needs. The power of situated knowledges can be honed by combining it with the insights of classical pragmatism, which, through the work of John Dewey, championed experimental knowing/inquiry as the most promising method for doing philosophy as well as for doing and critiquing science. By tying situated knowledges to pragmatism and experimental inquiry I ground situated knowledges in a practice that was intrinsic to the rise and success of scientific practice. Thus, as I argue below, because of its emphasis on goal driven, physical and ethically responsible engagements with the world, this pragmatic direction is able to provide a more critical method by which to assess if science has been able to meet values and goals that were set out in science's development.

In *The Quest for Certainty* (1929) Dewey argued against the passive, distanced epistemological inclinations endemic to philosophy, which he labeled the 'spectator theory of knowledge,' in favor of the epistemological practices developed in the rise of science in the early modern period, what he called 'experimental knowing' or 'experimental inquiry.' Experimental knowing, which served as a model for all knowledge acquisition, 'is [a] mode of doing, and like all doing takes place at a time, in a place, and under specifiable conditions in connection with a definite problem' (1929, 102). Furthermore, unlike the passivity of the spectator theory of knowledge, experimental inquiry is directed. It gains knowledge by varying conditions and directing its inquiry toward a goal, not passively receiving information (Dewey 1929, 123). Dewey argued that what really marks the difference between the methodologies of philosophy and experimental inquiry is the emphasis that experimental inquiry places on physical doing or activity; in other words, it emphasizes a targeted, physical engagement with the world to create change and understanding through *altering* conditions and relations. Thus experimental inquiry intentionally opens the door for further engaged inquiry and transformation, within the physical and social world, where philosophy had, effectively, shut the door.

Experimental knowledge is concerned with the materiality of the world, or as Dewey put it 'with the world in which we live, the world which is experienced' (1929, 102). Experimental knowledge initiates its inquiry from the 'things of the environment experienced in our everyday life, with things we see, handle, use, enjoy and suffer from…' (1929, 103). As a practice or 'mode of doing' it situates its activities in terms of a specific problem or question, within a specific location, set of conditions, and time, and sees the everyday world as offering opportunities for inquiry and challenge. Dewey argued that the problem with which we are working determines what particular methodologies or operations we are to use, unlike philosophical methods that determine what kinds of questions to ask given what kinds of methods are at our disposal. We know that our ideas and thoughts are well-founded when they direct our activity toward what is required, i.e., what we hope to solve, achieve, or change. Our ideas matter in the sense of how they can help us to 'rearrange and reconstruct in some way, be it little or large, the world in which we live' (1929, 138).

For all Dewey said about the efficacy of experimental inquiry coming out of science as a model for inquiry, he was concerned that science was not living up to its own standards. Not only did science fall back into some of the same problems that the spectator theory did, it also pushed aside those sciences – the biological sciences – that were concerned with human needs. Through relying upon physics as a model for all science, resulting from the prominence of logical positivism and unity of science programs at the turn of the twentieth-century, science distanced itself from the needs of the everyday world. This resulted in an approach to science that was 'remote from any significant human concern' and 'at the expense of all that is distinctly human' (1929, 196).

The biological sciences then began to model the physical sciences by moving away from experimental inquiry, toward the model of the physical sciences that was becoming more dominant. They sought to limit the sphere of inputs for knowledge, which resulted in generating knowledge that did not necessarily reflect human living nor could guide us in changing the conditions of living. The biological sciences, thus, now rely upon an isolated mode of laboratory experimentation and clinical trials as norms instead of experimental inquiry. For example, the rise of Evidence-Based Medicine (EBM) since 1992 and its emphasis on randomized control trials as the 'gold standard' (Sackett et al. 1996, 71) for generating evidence is indicative of the relevance of Dewey's target of criticism in contemporary science and medicine. Dewey argued that the move away from experimental inquiry is a significant loss for experimental human sciences and results in a distancing of research from the concrete situations pertinent to it, and thus from useful knowledge that can create change. As Kravitz et al. argue in 'Evidence-Based Medicine, Heterogeneity of Treatment Effects, and the Trouble with Averages,' the standard methods of EBM indicate treatment for the average person, but patients 'who deviate far from the average trial participant…may behave differently' when treated (2004, 675).

Thus the people whose lives are the most complex may be those the least likely to benefit from therapies indicated by EBM. On the contrary, embracing experimental inquiry as a framework for scientific and medical investigation has the potential to do the kind of work that the human sciences seek to do: to improve the material conditions of people's lives by providing an accurate and full understanding of human living. Thus, through his pragmatic philosophy, Dewey asked scientists to return to the methodology of experimental inquiry and reintegrate it into scientific practices.

Embracing experimental inquiry does not make 'soft' science nor is it anti-science. It is the very foundation of how we came to do science. However, it does illustrate that contemporary scientific methodology, which holds as its predominant methods laboratory research and randomized control trials, is not the only way to engage in scientific practice or to engage the world. Dewey's insights recognize that science, good science, can be done through the concrete practice of experimental inquiry and that good science has always been deeply tied to the achievement of human good and human values. Dewey is not asking for anything new to be added to science. He is merely asking researchers to reengage the practices that initially gave science its distinctive methodological power.

9.4 Situating Communities: A Pragmatist Feminist Approach to Scientific Research

Situated knowledge arguments focus on the situated nature of the epistemic agent. Thus, the philosophical import of these arguments lies in recognizing the situation of the knower, not what is being known. I characterized situated knowledges as moving toward increasingly particularized engagements and becoming increasingly methodological in import. I build from these insights to provide another way of thinking about situation by linking it to experimental inquiry. I argue that if our goals are to acquire accurate knowledge that we can act upon to improve human living, then we need to emphasize the situatedness of the communities that are studied by scientists and recognize that communities need to be studied in this complexity.

Dewey isn't telling us anything new when he articulates his despair that the natural and human sciences now rely primarily on methods that are 'most remote from any significant human concern.' Though we recognize that clinical trials are designed to be distant from how we actually do live, they are part of our daily vocabulary and we treat them as if they give us certain knowledge. We are inundated with reports of the efficacy of the most recent weight loss pill or depression medication. Whether we listen to these carefully or not, what has become normalized in our culture is the assumption that these tell us something, something significant and that if we too take pill X, we will experience similar results. Laboratory experimentation, though not as much a part of popular press, too instills us with the same confidence. When we find through toxic risk assessment that, with all other variables eliminated, that chemical Y couldn't harm humans in any context, we tend to believe it, because this type of methodology has come to signal to us the epitome of pure, accurate knowledge.

Because the laboratory environment intentionally is distanced from the outside world and free from complex factors that are thought to complicate results, the knowledge generated also is thought to be free from bias, accurate, and universal.

It shouldn't take Dewey to make us see how relying primarily on these two modes of investigation is problematic, but those within the sciences, and the public who has been habituated see these methods as the mark of good science, view data from the 'wild' non-clinical, non-laboratory setting with suspicion. I am not arguing that we should do away with laboratory experimentation or clinical trials, but that we need to study living situated communities *also* if our goal is to acquire knowledge that is accurate, effective, and ethical. We require knowledge that helps us to understand the complexity of the world, and knowledge that helps to better direct our engagements with this world, both epistemologically and with an eye toward social justice. These are values that we should return to in scientific inquiry. Experimental trials and laboratory experimentation do provide a certain degree of epistemological success, i.e. they provide us with some information about the world and guidance for action. For example, randomized control trials of birth control pills indicate that birth control pills have a 99% effectiveness in preventing unwanted pregnancies. What these trials don't indicate is that women taking birth control pills are not exercising the same input control that women in clinical trials are screened for and directed to employ. Studies that were initiated because some women more typically conceived while on birth control have now found that birth control pills are less effective for overweight and obese women (OBGYN and Reproductive Week 2005). Furthermore, women don't necessarily live in a world where birth control pills can be taken at the same time every day, thus diminishing their effectiveness. Trials do indicate how birth control pills work in a controlled setting, but they don't indicate how birth control functions in the complexity of women's lives. If the goal is to help women prevent unwanted pregnancies, then we need to know not only how birth control pills function in an idealized setting, but how they function practically in the lived, complex uncontrolled lives of those women that are meant to benefit from them. Thus, these trials do generate knowledge; they just don't provide the complex array of knowledge that is needed to help women live well. Starting from the everyday world of women's lives would have initiated a more complex array of questions and a more complex mode of study.

Furthermore, the communities that are studied by researchers and are impacted by science and medicine develop and are situated by the pertinent conditions and social heritages the members share.[6] Location, health, environments, histories of marginalization, race, gender, sexuality, class, ability, and age shape and form communities. These same factors serve to situate communities, not only by generating specific knowledge or ways of being, but also by marking them in historically, socially, and physically distinct and significant ways. Not only do people come to see the world in a way that is mediated by their situation, i.e., they occupy specific epistemological perspectives, but people live in transaction with these pertinent

[6]See Anderson (1983), Dewey (1954), Du Bois (1995), Mohanty (2003).

conditions. In other words, these conditions continually shape communities and communities continually shape their conditions. Communities are occupied by humans who are embodied and in transaction with their environment. Communities so understood are thus the subjects and objects of knowledge; to borrow from Dewey, they are the knowing and the known.

Experimental knowing starts from the conditions of a community, the conditions of 'the world in which we live, the world which is experienced' (1929, 102) and initiates its inquiry from the 'things of the environment experienced in our everyday life, with things we see, handle, use, enjoy and suffer from' (1929, 103). It thus starts inquiry from the situatedness of the community. This means that we need to ask different sorts of questions than the sciences have been asking. My case example of Agent Orange in the Aluoi Valley will provide an illustration of the type of questions that need to be asked according to a situated communities approach, along with some of the answers.

If, according to a situated communities approach, research is to be initiated from the complexity of the everyday world, researchers must be in intimate contact with the communities whose lives they are intending to benefit. Though researchers do come with a knowledge-set about their study and the subjects of their study, depending on how they are situated they may not necessarily have insider knowledge of the intricacies of a community and the lives of its members that comes from being situated in and through a community.[7] Nor are researchers likely to have the knowledge of what it is like to live with a particular contaminant or illness. They lack the embodied, authoritative knowledge that can only come from direct lived experience. Only through prolonged conversation, careful listening, and recognition of members of the community as epistemic agents can researchers learn what kind of questions they need to ask, obtain, and understand the answers to these questions, and observe factors that may not come up through dialogue. Scientific research has long functioned through an epistemology of distance. Through the study of situated communities it will need to employ an epistemology of intimacy.[8]

Within the current climate of mainstream philosophy of science and mainstream science, the approach I am advocating here is likely to be cast as anti-science.[9] Yet given the historically held scientific goal of understanding and improving human living (knowing and doing), especially in the case of the biological sciences, and given my argument's foundation in experimental inquiry, it is difficult to cast it

[7] See Collins (1986, 2000).

[8] See Lugones (2003) and Frye (1983) for insightful arguments on arrogant perception and loving perception. In a different version of this paper, I address how these relate to science and my argument.

[9] Arguments that seek to create change in science frequently are cast as reactionary and designed to denigrate science when their actual goal is to improve how science is practiced. For example, feminist science studies was 'feminist critiques of science,' but 'critique' was viewed by mainstream science studies, scientists, and popular press as anti-science even though these early analyses were largely generated by female scientists whose goal was to develop better scientific knowledge, not to dismantle science.

in this manner. Dewey admits that we don't know if in every case experimental inquiry will give us sure results. Because this approach is built upon experimental inquiry, which focuses on the outcomes of our actions, we can't determine what the result will be each time we approach something from the perspective of situated communities. Yet, this also is the case with clinical trials and laboratory experimentation. And, as Dewey, points out, this is the very point of experimental inquiry – it is to be tried (1929, 271).[10]

From this situated communities perspective, I am going to provide a case analysis of the effects of Agent Orange on a particular set of communities in Vietnam. The argument I have set out for situated communities becomes more apparent when it is put in to action. This should not be surprising considering this exactly what feminist and pragmatist arguments indicate: it is in the doing that we see the import.

9.5 Agent Orange in the Aluoi Valley

9.5.1 From Operation Ranch Hand to Dioxin Reservoirs

In 1961 the U.S. government launched Operation Ranch Hand, formerly called Operation Hades, on the land and people of Vietnam. From 1961–1971 the U.S. government sprayed areas of southern and central Vietnam with chemical defoliants to eliminate forest cover hiding Vietnamese soldiers and food sources for soldiers and civilians. Spraying continued by the South Vietnamese military, at lower quantities, until 1975. Most of these chemical defoliants contained a type of dioxin labeled TCDD (2, 3, 7, 8-tetrachloro-dibenzo-para-dioxin), which is the most toxic human-made substance.[11] The Vietnamese consider the use of Agent Orange by the U.S. government chemical warfare and to quote Dr. Tran Xuan Thu, the 'first war

[10] I consider implications of the situated communities approach in my manuscript *Actions Which Change the Face of the World*. Among these are whether the situated communities approach is time consuming and therefore burdensome, issues of epistemic authority, the limits and extent of our knowledge, the ethical and epistemic consequences entailed with speaking for, with and to marginalized groups, issues of epistemic honesty and humility, the challenges of pluralism, and structural constraints with funding more situated projects.

[11] Agent Orange was not the most toxic of these chemical defoliants sprayed in Vietnam, Agent Purple was. However, Agent Orange is the most referenced of these and of the most concern because it was the most heavily sprayed defoliant through aerial and hand spraying as well as the highest source of contamination through leakage in and around former U.S. military bases, areas that are in immediate proximity of Vietnamese hamlets. Data collected in 2003 increases the U.S. government's post-war estimate defoliant spraying by seven million liters (Stellman et al. 2003, 1) and contamination by dioxin from an estimate of less than 170 kg to greater than 600 kg (Dwernychuk et al. 2005, 998). This does not include containers leaks at U.S. bases, which are the location of the most heavily contaminated dioxin sites in Vietnam and the rest of the world.

of eco-destruction in the world history' (2006, 1).[12] It is estimated that 4.8 million Vietnamese were exposed to Agent Orange and three million have or are experiencing the effects of Agent Orange. Thu reports that '[a]ccording to data collected from certain provinces, among victims, half were civilians and 85% of households had two or more victims, 3% [had] five [victims]' (2006, 10).

Most research on Agent Orange's effects on humans has relied heavily on animal testing in the laboratory setting. Until relatively recently, the consensus among scientists has been that the evidence to support health effects experienced by U.S. Vietnam veterans and their children from Agent Orange are at best inconclusive if not just plain false.[13] However, recent assessments of animal studies have questioned the denial of the mutagenic toxicity of dioxin. The National Institutes of Health report on Agent Orange, *Veterans and Agent Orange, Update* 2004 provides a meta-analysis of research on Agent Orange. They report that the toxicity of TCDD results from two manners through which it interacts with the body (NIH 2004, 44, 55). The first is through the way TCDD is routed in the body, absorbed, distributed through tissues, transformed, and eliminated (NIH 2004, 44). The second manner is through its ability to bind with and alter the action of AhR (aryl hydrocarbon receptor), a gene regulatory protein. It is speculated that cell cycle control is affected through TCDD-bound AhR, which leads to cell death as well as the 'inappropriate' cellular and hormonal responses and increased oxidative stress (2004, 66). The NIH report finds that human developmental effects and cancers resulting from TCDD exposure are 'biologically plausible,' yet cannot be determined with anything near certainty because of the differences in the way various species are affected by TCDD (2004, 340, 397). That is, because all studies assess the effects of TCDD on nonhuman animals, no claim can be made about its effects on humans. Though this distinction may seem trivial, this argument was used in dismissing the suit brought forth by the Vietnamese against the manufacturers of Agent Orange.[14]

The situated communities approach that I am advocating does not suggest that we throw out these studies, but reasons that we need an additional approach for generating effective knowledge, knowledge that can better develop our understanding of the effects of dioxin on humans in a particular community. It seeks to understand the ways that TCDD interacts with and affects human bodies in the manner described by the NIH as biologically plausible and in, perhaps, other ways not considered by this research. But, unlike the studies that initiated their inquiry from the conditions of the laboratory, this approach starts inquiry from the

[12]This sentiment is echoed by an article in *Nature*: 'In 1961, for the first time in the history of mankind, large-scale chemical warfare was started in South Vietnam by the Kennedy Administration' (1982, 114).

[13]See, for example, Lathrop (1983), Gough (1986), American Council on Science and Health (1981).

[14]See the court documents Memorandum, Order and Judgment: Agent Orange Product Liability Litigation, 10/3/2005 and the epilogue to this paper.

situatedness of the community, directly considering how particular communities and particular members of these communities can be affected.

Like all inquiry, the situated communities approach begins inquiry from a series of questions. Unlike approaches that are more common to the sciences that intentionally limit contextual influences and multiple inputs when beginning inquiry, these questions start from the conditions of the community – its particular location, the lives of community members, the local environment, the social and historical context – and seeks to gain knowledge from this situated approach. The questions a researcher should ask in the case of Agent Orange in the Aluoi Valley in the Central Highlands of Vietnam would be: How long were/are the members of this community exposed? How long would it exist in the ecosystem of the Central Highlands of Vietnam that has particular rainfall patterns, soil, vegetation, and animal life? How does spilled and leaking Agent Orange interact in the environment differently than Agent Orange that was sprayed four decades ago? What practices and occupations of this community are conducive to exposure to dioxin? Practices include diet, food preparation, length of infant and child nursing, bathing, recreation, transportation, and home construction. Furthermore, what does it mean to live with a toxin? Unlike U.S. soldiers who had acute exposure, the Vietnamese have experienced generations of lived exposure. What role does gender play in exposure? Does the higher body fat of women make them more susceptible to dioxin than men? Does the dioxin in their body fat affect ovum, fetal growth, and nursing infants? How are children, the elderly, and the infirmed differently affected by dioxin than healthy adults? Each of these questions is united by the need to understand the situatedness of these communities in order to develop an understanding of the effects of dioxin with the goal to improve lives and the lives of future generations.

9.5.2 Situating Agent Orange in the Aluoi Valley

The material I use to address these questions is generated by Vietnamese, U.S., and Canadian researchers. Some projects were collaborative efforts, others were not. All of the scientific research is from 2001 to 2006. The researchers rely heavily on working with the communities not only to get the samples needed but to understand their ways of living. Operating outside of the predominant scientific paradigm, these research projects, when considered together as a whole, provide one model for the situated communities approach that I advocate.

The Aluoi Valley has been the subject of study because it was heavily sprayed by U.S. forces and there were three U.S. military bases in the valley with large amounts of Agent Orange leakage from barrels left at the end of the American War. Though it has been labeled an Agent Orange ‘hot spot’ because of heavy aerial spraying, at this point, overall, the region does not contain high levels of dioxin in the soil (Dwernychuk et al. 2002). This is attributed to ‘tropical rains, erosion, and chemical degradation’ (Dwernychuk et al. 2005, 998). But in this area there are hot spots that are the result of heavy hand spraying, spillage and leakage from containers of Agent Orange stored on U.S. bases during the war (Dwernychuk et al. 2002, 118). Such hot spots are marked by contamination of the soil, the ground water, the food, including fish (which are the most highly consumed protein source and the most contaminated), frogs, ducks, chickens, pigs, various greens and root vegetables, and breast milk, which feeds Vietnamese children well into and through their toddler years. Contamination includes the dirt that makes of the floors of the house, the wood used to make cooking tools, houses, boats, sleeping pallets. The majority of people living the Aluoi Valley belong to one of three Vietnamese ethnic minority groups, the Pa Co, Ca Tu, and Ta Oi. Like many of the ethnic minority groups in Vietnam these groups are physically and socially isolated, poor, and live as a subsistence society through farming. The isolation of these communities and their status as minority groups in Vietnam makes living with dioxin even more dangerous, because, as I will argue later in the paper, they lack access to medical and informational resources, in addition to having to rely upon their immediate environment for subsistence.

When dioxin contamination is studied in this valley, in a living environment, it is hard to not see the destruction that the U.S. caused with its use of Agent Orange. Most soil samples from the studied area show high levels of dioxin, levels that exceed the U.S. EPA guidelines for safe residential housing (Dwernychuk et al. 2002, 123). Considering that the members of these communities live in houses with dirt floors, frequently walk barefoot, and through farming and general food production are in close contact with the soil, U.S. guidelines for safety are inadequate to ensure the safety of these people (Dwernychuk et al. 2002, 125). If the soil in this area exceeds U.S. safety guidelines, it is probably significantly more dangerous for this community considering the key differences between their daily lives and those of Americans.

In terms of food, the highest levels of contamination are in fish and duck fat, two significant sources of protein for these communities. This is not surprising considering these are fish cultivated in human-made ponds, dug out of contaminated soil and filled with water that is contaminated (Dwernychuk et al. 2002, 125). The bodies of the ducks and fish accumulate and magnify the dioxin that they ingest, which then through consumption is concentrated and magnified in human bodies (Dwernychuk et al. 2002, 127). Furthermore, animal fat is a delicacy as well as a necessity in Vietnam because of its high caloric content. In a community that physically labors to meet subsistence standards and is still frequently undernourished, high caloric foods represent an important part of the diet. But, due to the bioaccumulation and biomagnification of toxins in fat, these important food sources pose a serious risk for the Vietnamese.

Studies show that older people had higher concentrations of dioxin in their blood, accumulated through years of lived exposure and bioaccumulation, as well as exposure at particularly acute times during and following the American war. If women bear children, they have lower levels of dioxin than men. This can be attributed to the 'offloading' of toxins that occurs when women breastfeed.[15] Dioxin leaves the mother's body through nursing and goes directly to the infant's body.

[15] So, though women's bodies accumulate more dioxin because of their higher body fat (dioxin is lipophilic, i.e. accumulates in fat), they also lose the dioxin from breastfeeding.

Firstborn infants are more greatly affected than latterborn, though all receive contaminated breast milk. In the Aluoi Valley the firstborn infants of women receive an average daily intake of dioxin up to 27 times the amount considered safe by the World Health Organization (Dwernychuk et al. 2002, 130). In one particular hamlet it exceeded this guideline by 47 times.

Compared to men, women socially and physically experience the effects of Agent Orange to a greater extend. This is due to their role in the household and reproduction. In rural areas, Vietnamese women do 76% of the agricultural labor (Longino 2002, 6), such as rearing livestock and fish, working in the fields, gathering foodstuff, tending the family garden (Ha 1997, 66) and almost all of the housework, including washing clothes, preparing food, caring for children and ill members of the household, house cleaning, and educating children. They also are responsible for aiding in the care of sick and elderly members of the community (Ha 1997, 66).

In the Aluoi Valley these activities situate women in such a way that they are more likely to be exposed to dioxin. For example, fish carry high loads of dioxin because they live in water that is contaminated with it. It follows that the women who are farming the fish are not only exposed by consuming fish, but also by coming into contact with the dioxin in the water.[16] Similarly, if the dirt flooring in homes is contaminated by dioxin, not only are women exposed to it by walking on it and living with it, as are all members of their household, they are also exposed through the dust generated by sweeping and cleaning. Most of the activities the women engage in, because they are in such immediate contact with soil and water contaminated with dioxin, increase women's risk of exposure and increase the risk of fetal exposure. Furthermore, women are the caregivers to those who are ill from the effects of Agent Orange, both inside and outside the family, again increasing the effects, in this case the social effects, of Agent Orange.

Vietnam has the highest abortion rate in the world. 'It is estimated that 40% of pregnancies are terminated' (Asian Development Bank 2002, viii). Thus, there are 2.5 abortions per women (Asian Development Bank 2002, viii). When I was speaking with the director of the Tu Du hospital in Ho Chi Minh City she attributed the high rate of abortion partly to selective abortions for infants with congenital anomalies that are attributed to dioxin exposure.[17] In the Central Highlands when infants are born suffering from the effects of Agent Orange, not only do women have the primary responsibility in caring for them, they are frequently doing so without adequate medical care and information. Though health care in rural areas has improved in recent years, village clinics in remote areas are not equipped to deal with the types of illnesses that can result from exposure to Agent Orange. Urban hospitals receive a vast majority of the government funding allocated for

[16] Dioxin is hydrophobic so it rests on top of the water. Thus people working on or in the water easily come in contact with it.

[17] There certainly are other reasons for this high rate, but the director of the hospital was clear that selective abortion because of congenital anomalies was an important contributor to their high abortion rate.

medical care (Binh 1997, 10). This means that adults and children in rural and remote areas can't be cared for well.

Even with this brief bit of data I have presented, it is not hard to see how in a living, situated community, Agent Orange can cause the kind of damage the Vietnamese claim it does, at the same time as results in the laboratory, the space Dewey describes as most remote from human concerns, are found to be inconclusive. What I presented thus far shows for specific communities in the Aluoi Valley a history of lived exposure in locations that have been designated Agent Orange hot spots. We know that in the valley there are high levels of soil, water, and food contamination that have resulted from documented Agent Orange leaks. We also know from the empirical data that there are high levels of Agent Orange in human tissue, blood, and breast milk. Furthermore, in areas of Vietnam where there is very low dioxin contamination in the soil, they also do not have high levels of dioxin in human tissue, blood, and breast milk (Schecter et al. 2001). So thus far I have established a consistent level of exposure and pattern of bodily toxicity.

A meta-analysis of Agent Orange studies that was published in the *International Journal of Epidemiology* in 2006 found that the rate of congenital anomalies among civilian families in this region compared with nonexposed civilian families was 3.27% greater (Ngo et al. 2006, 1220). This number does not include miscarriages – 47.03% in the Highlands compared to 5.77% in Northern samples – or stillbirths, which occur at a higher rate in this area, or abortions (Vietnam Courier 2003, 76). Thus there are 3.27% more congenital anomalies in live births among the exposed people than there are among unexposed civilians in other areas of Vietnam. Furthermore, adult and childhood cancers and skin diseases occur in these areas at a significantly higher rate than in noncontaminated areas of Vietnam (Thu 2006, 14). Thyroid cancer, ovarian cancer, and Hodgkin's disease occur at a statistically higher rate in this region as do multiple, but seemingly unrelated cancers (Thu 2006, 14). Furthermore, developmental problems in children occur at a higher rate, which may be the result of exposure to contaminated breast milk or exposure in utero (Ngo et al. 2006, 1224). When this data is viewed in light of the above evidence, one can see a correlation between exposure to dioxin, bodily toxicity, and disease.

9.5.3 From Evidence to Obligations

It is important to ask ourselves what kind of evidence we need in order to make the claims that dioxin has caused disease in this community and that the U.S. has a responsibility to these communities. Do we need to understand the specific mechanisms of dioxin's interaction with human bodies or does this more situated approach provide a robust enough series of connections to allow us to make correlations between dioxin exposure and significant health effects? When studying Agent Orange in a living situated environment, we see a correspondence between exposure and disease, a pattern that can't be made apparent in the laboratory. Though I am unable to show the direct biological mechanisms that lead from exposure to dioxin to disease, I am able to show a correlation between disease patterns and exposure. This correlation emerges from the ability to critically track dioxin from the soil, water, and food to the body through samples of blood, skin, and breast milk, and then trace correspondingly high levels of disease in these specific communities, diseases that we would expect from a toxin that alters DNA leading to cellular death as well as 'inappropriate cell activation.'

Experimental inquiry indicates that the problems that we work with determine the methods we ought to use. In the case of Agent Orange in Vietnam, laboratory experimentation cannot deal on its own with the problems that need our work, and it constrains our ability to answer the question whether Agent Orange causes the health effects described by the Vietnamese. But with a pragmatist feminist focus it becomes obvious that if we employ experimental inquiry and the evidence that can be generated by the situated communities approach, then the data pointing to a correlation between dioxin and disease is enough to determine that we do have scientific knowledge about the health effects of Agent Orange. This knowledge ties the U.S. government and the 37 chemical manufacturers to obligations, both moral and legal, to improve the lives of those affected by Agent Orange and to change conditions to mitigate exposure to dioxin in the future. Again, knowledge to improve human living is an important goal of science. I am not asking for anything more than goals that science had already set for itself.

9.6 Conclusion

In this paper I have argued that our epistemologies tie us to methodologies that allow scientists to engage communities more or less responsibly. The methodologies that are indicative of predominant scientific practice rely on understanding conditions that are remote from how communities actually do live. The situated approach that I am advocating not only provides a fruitful way of understanding communities in a more nuanced and complex manner, it also allows researchers to recognize how these communities' health needs may not be met by the type of evidence that results from their methods. When researchers look to the everyday world to understand how communities live with diseases, toxins, poverty, and oppression, it presents opportunities for them to ask questions that not only provide greater epistemic success, but also point in the direction of how to responsibly engage these communities.

In the case of Agent Orange in Vietnam, the stakes are high. One type of epistemological worldview indicates that in fact we have no obligations to the Vietnamese. Because researchers can't make claims about dioxin beyond what is found in toxic risk studies on animals, they can't make any claims about what dioxin does to humans. But the situated communities approach asks us to look at the lives of the people of the Aluoi Valley and to engage science from the perspective of their lives, in their place. In doing so it becomes apparent that there is a significant correlation between high disease rates and high levels of environmental and bodily dioxin contamination. In this situation it is difficult to deny that the U.S. does have ethical obligations to the Vietnamese to, at minimum, assist in cleaning up the local environment and providing means to meet their health and social needs. The U.S. government and the 37 chemical manufacturers also may have obligations that reach beyond these basic ethical obligations to include financial remuneration, much like that awarded to U.S. Vietnam war veterans. As Dewey so adamantly asserted, with the responsibility of engaging in scientific practice, comes obligations that go far beyond the practice of science. It includes the responsibility of helping others to live well in their situation.

9.7 Epilogue

It has been 6 years since I was in Vietnam. I am still haunted by what I saw and by the intentional and ongoing harm created by the U.S. and U.S. chemical companies. I was horrified all over again in March of 2005 upon learning that Judge Weinstein had thrown out the petition filed on behalf of the Vietnamese victims of Agent Orange. In the 233 page legal decision Weinstein cited numerous references to toxic risk assessments and other forms of laboratory experimentation. These showed no or dubious connections between dioxin and illness other than chloracne. Very little evidence came from people working in Vietnam, studying Agent Orange in its environment. He argued that there was insufficient evidence that Agent Orange caused birth defects, miscarriages, stillbirths, and cancers. In reality what was insufficient was the methodology used to gather the information. How we study Agent Orange and dioxin matters to three million people. Because predominant scientific methods prioritizes evidence gathered in settings 'most remote from any significant concern' over evidence that comes from a living, situated community, the people of the Aluoi Valley will lack the social, medical, and financial resources to care for themselves and their communities, and they will not have the resources to change the physical structure of their community to reduce their exposure. Thus, the victims of Agent Orange continue to be victims of how we do science.

Acknowledgments This material is based upon work supported by the National Science Foundation under Grant No. 0541512. Any opinions, findings, and conclusions or recommendations expressed in this material are those of the author and do not necessarily reflect the views of the National Science Foundation. Thank you to Heidi Grasswick, the reviewers of this volume, and the 'Writing Group,' Tammy Proctor and Molly Wood, for their invaluable criticism.

References

American Council on Science and Health. 1981. *The health effects of herbicide 2, 4, 5-T*. New York: American Council on Science and Health.

Anderson, Benedict. 1983. *Imagined communities*. New York: Verso.

Asian Development Bank. 2002. *Country briefing: Women in Viet Nam*. Manila: Regional and Sustainable Development Department and Mekong Department Asian Development Bank.

Barker, Drucilla. 2000. Dualisms, discourse, and development. In *Decentering the center*, ed. Uma Narayan and Sandra Harding, 177–188. Bloomington: Indiana University Press.

Binh, Do. 1997. Some problems concerning social policies toward rural women in the period of economic restructuring. *Vietnam Social Science* 2(58): 3–13.

Braidotti, Rosi, et al. 1994. *Women, the environment, and sustainable development*. Atlantic Highlands: Zed Press.

Charlene, Seigfried Haddock. 1996. *Feminism and pragmatism: Reweaving the social fabric*. Chicago: University of Chicago Press.

Code, Lorraine. 2006. *Ecological thinking*. Oxford: Oxford University Press.

Collins, Patricia Hill. 1986. Learning from the outsider within. *Social Problems* 33(6): 14–32.

Collins, Patricia Hill. 2000. *Black feminist thought*, vol. 2. New York: Routledge.

Dewey, John. 1925. *Experience and nature*. Carbondale: Southern Illinois University Press.

Dewey, John. 1929. *The quest for certainty*. New York: Capricorn Books.

Dewey, John. 1954. *The public and its problems*. Athens: Swallow Press.

Du Bois, W.E.B. 1995. *W.E.B. Du Bois: A reader*. New York: Henry Holt and Company.

Dwernychuk, L.Wayne, et al. 2002. Dioxin reservoirs in southern Vietnam – A legacy of agent orange. *Chemosphere* 47: 117–137.

Dwernychuk, L.Wayne, et al. 2005. Dioxin hot spots in Vietnam. *Chemosphere* 60(7): 998–999.

Editorial Staff at the Vietnam Courier. 2003. *Herbicides and defoliants in war*. Honolulu: University Press of the Pacific.

Frye, Marilyn. 1983. *The politics of reality*. Freedom: Crossing Press.

Gough, Michael. 1986. *Dioxin, agent orange: The facts*. New York: Plenum.

Ha, Nguyen. 1997. Evaluation of the housework of rural women. *Vietnam Social Science* 2(58): 65–72.

Haraway, Donna. 1991. *Simians, cyborgs, and women: The reinvention of nature*. New York: Routledge.

Harding, Sandra. 1986. *The science question in feminism*. Ithaca: Cornell University Press.

Harding, Sandra. 1991. *Whose science? Whose knowledge?: Thinking from women's lives*. Ithaca: Cornell University Press.

Harding, Sandra. 1998. *Is science multicultural?* Ithaca: Cornell University Press.

Harding, Sandra. 2006. *Science and social inequality*. Chicago: University of Illinois Press.

Hoagland, Sarah. 2001. Resisting rationality. In *Engendering rationalities*, ed. Nancy Tuana and Sandra Morgen, 125–150. New York: SUNY Press.

Kravitz, Richard, et al. 2004. Evidence-based medicine, heterogeneity of treatment effects, and the trouble with averages. *Milbank Quarterly* 82(4): 661–687.

Lathrop, David, et al. 1983. *Operation ranch hand II*. San Antonio: USAF School of Aerospace Medicine.

Longino, Helen. 1990. *Science as social knowledge: Values and objectivity in scientific inquiry*. Princeton: Princeton University Press.

Longino, Helen. 2002. *The fate of knowledge*. Princeton: Princeton University Press.

Lugones, Maria. 2003. *Pilgrimages/peregrinajes theorizing coalition against multiple oppressions*. New York: Rowman & Littlefield.

Mohanty, Chandra. 2003. *Feminism without borders*. Durham: Duke University Press.

National Institutes of Health Committee to Review the Health Effects in Vietnam Veterans of Exposure to Herbicides. 2004. *Veterans and agent orange update 2004*. Washington, DC: National Academies Press, 1982. *Nature* 298(5870): 114.

Nelson, Lynn Hankinson. 1993. Epistemological communities. In *Feminist epistemologies*, ed. Linda Alcoff and Elizabeth Potter, 121–160. New York: Routledge.

Ngo, Anh, et al. 2006. Association between agent orange and birth defects: Systematic review and meta-analysis. *International Journal of Epidemiology* 35: 1220–1230.

OBGYN and Reproductive Week. 2005. Getting Pregnant While on the Pill. January 24, 2005.
Sackett, David, et al. 1996. Evidence-based medicine: What it is and what it isn't. *British Journal of Medicine* 312: 71–72.
Sandoval, Chela. 2000. *Methodology of the oppressed.* Minneapolis: University of Minnesota Press.
Schecter, Arnold, et al. 2001. Recent dioxin contamination from agent orange in residents of a Southern Vietnam City. *Journal of Occupational and Environmental Medicine* 43: 435–443.
Shiva, Vandana. 1989a. *Staying alive: Women, ecology, and development.* London: Zed Press.
Shiva, Vandana. 1989b. *The hijacking of the global food supply.* Cambridge: South End Press.
Shiva. 2000. *Stolen harvest: The hijacking of the global food supply.* Cambridge Massachusetts: South End Press.
Stellman, Jeanne, et al. 2003. The extent and patterns of usage of agent orange and other herbicides in Vietnam. *Nature* 422: 681–688.
Thu, Tran. 2006. 'Consequences of chemical warfare in Vietnam.' Unpublished manuscript.

Chapter 10
'They Treated Him Well': Fact, Fiction, and the Politics of Knowledge*

Lorraine Code

Abstract In this paper I show how fact and fiction, collaboratively, can inform a moral epistemology that moves toward deriving principles for understanding difference; responding well to alterity. Specifically, I examine impediments to knowing, from positions of white privilege, how it is to live racial inequality. Starting from Nadine Gordimer's novel, *July's People*, written when South African apartheid was moving violently toward its dissolution; yet where polite, concepts/ideals integral to liberal enlightenment discourse, such as emancipation, equality, and welfare, were under strain, I examine the phrase 'they treated him well' for how it permits the novel's white protagonists to ignore the extent of an Otherness that is allegedly erased in the provisions they make for the comfort and welfare of July, their black servant. The language is neutral, well-intentioned, self-confessedly liberal, and oblivious to the barriers and exclusions it sustains. Yet contains the "white folks" within an epistemological-ethical imaginary of sameness where they cannot understand the need to relinquish taken-for-granted distinctions, taxonomies and assumptions about "natural kinds" through which they know "their" world, even when those distinctions lose their pertinence. July knows their world and their ways far better than they know his, yet their failure to recognize the extent of his epistemic privilege ultimately leads to disaster. The paper will elaborate the epistemological consequences of this apparent incommensurability.

Keywords Difference and alterity • Epistemology of ignorance • Epistemic responsibility • Moral epistemology • Social imaginary

---

*An earlier version of this paper is published under the title 'The Power of Social Imaginaries: Fact, Fiction, and the Politics of Knowledge' in *Trans-Humanities*, (June 2010) 2(1): 9-31 (Seoul Korea: 2010 Ewha Institute for the Humanities).

L. Code (✉)
Distinguished Research Professor Emerita, York University, Room Number Ross 442S, Ross Building, 4700 Keele Street, Toronto, ON, Canada, M3J 1P3
e-mail: codelb@yorku.ca

H.E. Grasswick (ed.), *Feminist Epistemology and Philosophy of Science: Power in Knowledge*, DOI 10.1007/978-1-4020-6835-5_10,

> *You see, they are not in the least like ourselves. They don't need and can't use the luxuries that you and I must have. The have the animal capacity to endure the pain of, shall we say, domestication.*
>
> Chinua Achebe,
> Anthills of the Savannah

10.1 July's People

My thinking in this paper gains purchase from the structural injustices elegantly represented in Nadine Gordimer's (1981) novel, *July's People*, written when South Africa was in turmoil as the old apartheid order with its clear, cruel, and patently unjust epistemological-moral-political boundaries was moving, violently, toward its dissolution; yet when polite, orderly concepts and ideals integral to liberal enlightenment discourse, such as emancipation, equality, and welfare, were seriously under strain. Evidence of how this conceptual frame blocks the recognitions integral to knowing people and their circumstances responsibly and well exposes the epistemic limitations of the frame itself, highlighting its local and temporal specificity. For it contains the 'white folks' in the novel within an imaginary of human sameness where they cannot understand why it has become impossible to rely on the taken-for-granted taxonomies, distinctions, and assumptions about human 'natural kinds' through which they know 'their' world.

July, in Gordimer's novel, is a black servant, a 'boy', who has spent 15 years in the service of a middle-class white Johannesburg family, from whom he has travelled home only every second year to his wife and children in his village. In the novel, it is he who enables the white folks – a man, woman, and their three children, whom Gordimer ironically dubs 'his people'–to flee the riots, looting, and burnings that threaten their lives: he takes them on a long journey to his village, commandeers a hut for them – his mother's hut – gives them sanctuary. The novel tells of the complexities of fragile relationships that disintegrate and shift in this new situation: shows what the book's jacket description names the 'terrifying, tacit understandings and misunderstandings between black and white'; the knowings and unknowings that become more impossibly tangled than they always, imperceptibly, had been.

In this reading, I take the phrase 'they treated him well' as emblematic for how the white folks see their long association with July, fraught as it is with mis-recognitions lurking beneath the apparent recognitions it assumes. It is an intriguing phrase, commonly uttered thoughtlessly; yet closer analysis exposes layers of suppressed meaning and epistemological assumptions beneath its surface. When or where would anyone use it? Surely not in or about what Iris Marion Young (2004, ms. 1) calls relationships of 'equality and inclusion'. Most commonly, it might be said of people's dealings with their pets; and, of course, their servants, underlings, prisoners, slaves. Discursively, it gestures toward profound differences in social position, even as it permits those who 'treat others well' to ignore the extent of an Otherness – an alterity – allegedly erased, obliterated, in the provisions they make – as did July's white people – for the comfort and welfare of those thus treated: the 'decently-paid and contented male servant, living in their yard since they had married, clothed by them in two sets of uniforms... given Wednesdays and alternate Sundays free, allowed [sic] to have his friends visit him and his town woman sleep with him in his room' (Gordimer 1981, 9). '...[H]e wasn't kept short of anything' (155). The language is neutral, well-intentioned, self-confessedly liberal, and oblivious to the barriers and exclusions it sustains. In a different register, it recalls Marilyn Frye's comment about women in a male world, who indeed count as people because they are 'biologically human ... and have certain linguistic capacities and emotional needs' in virtue of which they 'should be treated *humanely*, as we are enjoined to treat the very ill, the elderly and members of whatever race(s) we take to be below our own in the pecking order,' but whose 'personhood' is denied (1983, 48 italics added). Without equating or conflating these markedly different Otherings, note that 'humanely' does similar work, with an analogous extension, to the idea that 'they treated him well.' Clearly, to be treated 'humanely' is, at the very least, not to be treated with the respect due to a person as an end in her– or himself.

10.2 Moral Epistemology

My project in this essay is to outline a conceptual frame for thinking about the moral-epistemological issues Gordimer's novel raises. Moral epistemology figures in what follows in two senses with different, but overlapping, implications. The first sense derives from my thinking about epistemic responsibility over the two and more decades since the publication of my book by that title (Code 1987). Although I focus here on the ethics and politics of knowing other people responsibly and well, singly and/or collectively, the extension of responsibility injunctions is much broader. They pertain to knowing events, phenomena, facts, and situations; they imply that there are often choices about how and what to know, some of which are more responsible than others, according to the 'quality' of epistemic conduct they attest to on the part of would-be knowers attempting to know as well as possible, where it might be easier, more expedient, not to. Responsible epistemic conduct involves being accountable to, responsive to the putative known, to subject(s) and circumstances that become(s), or is/are positioned as 'object(s)'of knowledge: to subjects who are to be known, morally.

The second, interconnected sense of moral epistemology is more familiar: it is about guidelines and normative principles that shape the knowing integral to, yet often invisible in, moral judgements. When ethics and epistemology are conceived as distinct domains, as they often are in professional philosophy, such knowing falls below the threshold of visibility. A tacit conviction that circumstances and evidence are self-presenting leaves little space for evaluating the quality of knowledge that may function as the unexamined basis of the condemnation or commendation that are properly the tasks of ethics. Hence, and germane to my reading of *July's People*, moral deliberation in Anglo-American philosophy frequently relies on uncontested assumptions about human sameness, about a matter-of-course ease of 'putting oneself in someone else's shoes' and knowing his/her/their circumstances accordingly. Such assumptions impede possibilities of knowing people and their situations well enough to recognize when and how sameness can in fact *not* be assumed: they make it difficult to see how differences make a difference. In consequence, moral deliberation may start from an irresponsible epistemic 'take' on an issue, read as understood from the start and simply presented for moral evaluation. By contrast, I shall argue, explicit, active recognition of the specificity of differences – to the extent that is realizable – is an epistemological prerequisite for countering inequitable social practices, compensating for disadvantages, revaluing devalued attributes, positions or actions, or making space for members of marginalised groups to achieve the acknowledgement on which their epistemic and moral credibility depends. Thus the cognitive underpinnings of moral deliberation have to be moved onto the same investigative plane as the deliberation itself.

This appeal for explicit, active recognition of differences owes its formulation, in part, to Annette Baier's reminder about the epistemic complexity of recognition. Maintaining that there 'could not be a *private* practice of self-criticism which was not parasitic on a public one', for assessment by fellow practitioners always involves 'not just policies but actual *practices* of criticism...', Baier maintains: 'Cognition (including knowing how) requires *re*cognition' (1985, 13–14, italics added). She is in effect contending, rightly I suggest, that people have to *know well* in order to recognise (re-cognise) effectively: that a complex web of experiences and knowing, socially deliberated and negotiated, is a *sine qua non* for adequate recognition, perhaps on a distant analogy with Meno's paradox. Practices of criticism, in this sense, involve public and private deliberation about the degree to which recognition can be claimed. Yet, both epistemologically and in everyday practice, as I shall show with respect to *July's People*, such requirements are difficult to fulfill when they are impeded by an entrenched *social imaginary* that tells against the very possibility of knowing what must be known if recognition is to be good enough to counteract presuppositions generated, countenanced, and enforced by an epistemology of ignorance and a politics of unknowing that sustains it, the best intentions of well-meaning white citizens notwithstanding. Nor is recognition ordinarily a one-way street, although my argument here is largely unidirectional in its (emblematic) focus on white failures to recognize salient intricacies of a black African life.

Epistemic responsibility invokes a cluster of prescriptions, cautionary practices, and prohibitions: it echoes Judith Butler's thought, in a different context, about 'doing ... [epistemic] justice to someone' (2004a, 57), hence being aware and wary of possibilities of enacting epistemic violence, intentionally or inadvertently. A cautionary warning underpins her analysis, to the effect that careless, too-swift, insensitive classifying, summing-up; thoughtless claims to know on the basis of minimal cues or evidence, of too-slight resemblance to other putatively 'like' people or practices, too-glib assumptions about 'natural kinds' can generate oppressive, paternalistic, and/or condescending claims to know (an)other person(s) better than people can know themselves. Such putative 'knowing' paves the way for unjust, harmful acts, attitudes, and practices. Yet my purpose here is broader than the negative, quasi-conservative one of devising principles and strategies for avoiding error. In its positive, creative dimensions, epistemic responsibility advocates projects of promoting just epistemic conduct that are creative in generating innovative, revisionary knowledge projects with the social-political transformations, renewals, and disruptions they may animate.

My reading of the effects of irresponsible epistemic conduct is consonant in many respects with Miranda Fricker's (2007) analyses of the damaging personal and structural effects of epistemic injustice in social understanding. She shows how epistemic injustice, whether testimonial or hermeneutical, is a source of harm to subjects as knowers and as knowable. Testimonial injustice, in its capacity to deny 'access to what originally furnishes status as a knower' reduces a testifier to 'less than a full epistemic subject' (145). Hermeneutical injustice, in drawing on socially embedded interpretations and understandings which are differentially available across relations of power and privilege, excludes certain people from communal interpretive discourses. Fricker notes: 'relations of power can constrain women's [and members of other marginalized social groups'] ability to understand their own experience' (147); '...relations of unequal power can skew shared hermeneutical resources so that the powerful ... have appropriate understandings of their experience ready to draw on ... whereas the powerless ... [may have] at best ill-fitting meanings to draw on in the effort to render [their experiences] intelligible' (148). A woman unable to gain a hearing for damage inflicted by persistent sexual harassment, owing to 'a lacuna in the collective hermeneutical resources' (150); a man unable to own his 'nascent identity as a homosexual' in 1950s America (163–5), are just two of her examples. When 'some social groups are unable to dissent from distorted understandings of their social experiences' owing to lacunae in 'collective hermeneutical resources' generated by prejudice or bias, they can find no way of claiming recognition for what happens to them. Thus when 'sexual harassment', 'racism', 'homophobia', or 'sexism' had not been named, when their implications had not achieved common currency in the rhetorical spaces of western societies, testimony 'about them' could not claim a hearing or expect communal uptake, in a strong sense of 'could'. People whose experiences can now be understood and responded to under these descriptions were hermeneutically marginalised, systemically and systematically, their testimony routinely dismissed as mere reminders of 'how things are for a woman in a man's world/a gay man in a heterosexist world/a black person in a white world', in the absence of a structural rubric to address and contest ensuing harms. The conceptual resources such terms make available have opened the way toward reconfigured epistemic engagement with the practices they name, and created spaces for revised social-political intervention.[1]

Collective hermeneutical resources such as Fricker invokes come into being, and are shaped and sedimented within what Cornelius Castoriadis (1994) calls an *instituted social imaginary*, loosely comprised of the normative social meanings, customs, expectations, assumptions, values, prohibitions, and permissions – the

[1] I discuss Fricker's analysis of epistemic injustice at greater length in Code (2008).

habitus and *ethos* – into which human beings are nurtured from childhood. As they make sense of their place, options, responsibilities within a world, both social and physical, people internalize, affirm, challenge, contest, or refuse these social meanings and imaginary significations, whose enactment is integral to their identification, and self-identification, as members of that society or group. Entrenched and interlocking metaphors, images, rhetorical assumptions and devices silently but pervasively govern the thinking of a social group at any period in its history, making some ideas, practices, and projects plausible, intelligible, and others not. While biases, stereotypes, prejudices, and received dogmas are indeed part of any imaginary, the idea itself is more encompassing, and enabling than any of these, singly, are. As social creatures, people think and live within or against a going social imaginary which broadly defines and constitutes the objects and kinds, activities and projects they can know and care about, and offers guidance about how to be. It sets boundaries on the credibility of knowers and of institutions of knowledge production, even as it also opens interpretive possibilities. A social imaginary, then, is a loosely integrated system of metaphors, images, assumptions, ways of thinking, with powerful if tacit features that generate and underwrite possibilities of knowledge production, interpretation, uptake, and implementation.

As I show in *Ecological Thinking* (Code 2006a), dominant social imaginaries legitimate or preclude certain kinds of epistemic and other relations between and among people and to the physical-natural-conceptual world; and philosophical systems instantiate and reinforce these imaginaries. Yet despite their time- and place-bound hegemony, they are neither rigid nor permanent, but open to interrogation and displacement. Thus, to the instituted imaginary Castoriadis juxtaposes the *instituting* imaginary, which he conceives as the critical-creative activity of a society that exhibits its autonomy in its capacity to put itself in question: in the ability of (some of) its members to act from a (collective for some collectivity) recognition that the society is incongruous with itself, with scant reason for self-satisfaction. Imaginatively initiated counter-possibilities can interrogate the social structure to destabilize its pretensions to 'naturalness' and 'wholeness,' thus initiating a new making (a *poiesis*). For Castoriadis, as for Michèle Le Dœuff in a different context, 'there is no intellectual activity that is not grounded in an imaginary' (2003, xvi).

Briefly, then, within the social imaginary a patriarchal society holds in place, a predatory sexual relationship between men and women is just how things are, as in a racialized society, an oppressive white/non-white hierarchy attests to how things naturally are, and the (intersecting) injustices, tyrannies, and misinterpretations such imaginaries hold in place work to claim recognition, acknowledgement, for the exclusions, harms and oppressions they silently countenance. Framed in the language of instituted and instituting imaginaries, epistemic responsibility is thus more complex than when I first articulated it, for within an uncontested instituted imaginary, sexist, patriarchal, racist or myriad other exclusionary practices might appear to enact quite responsible ways of knowing people marked as 'Other' (from a white male norm) according to the going wisdom about 'women' and 'non-white' and other Others as members of 'natural kinds'. The explanatory power of social imaginaries, then, is in their demonstration that it is not only individual acts of epistemic injustice that are the focus of critical inquiry, but the systemic, sedimented, interlocking character of sexist, racist, paternalistic, and other oppressive thought and action that make those individual acts possible. Yet seeing why a range of practices had seemed to be responsible under the sway of a displaced imaginary does not straightforwardly exonerate those practices or the attitudes they betray: this is where deliberation begins. When the imaginary itself comes under critical scrutiny, space for understanding and contesting the structural complexity of irresponsible epistemic conduct opens out. By contrast, interventions that treat acts of hermeneutic or epistemic injustice as discrete, isolated moments may fail adequately to address either their reciprocally rationalizing character, or the structures of power and privilege that sustain them.

Endeavouring to know – to understand – well enough how it is/what it means to live specifically embodied, racialized subjectivities from positions of white privilege is central to articulating a moral-political epistemology capable of disrupting what Charles Mills, in *The Racial Contract* (1997), famously calls 'an epistemology of ignorance' which has infiltrated and sustained white western racism. Not only, borrowing Mills's phrases, does such ignorance excuse white signatories to 'the racial contract' from any need to understand 'the world they themselves have made,' it also allows them to exist in an 'invented delusional world' held in place by a social-epistemic imaginary of self-deception and 'structured blindnesses'(18–19). The contract, in effect, generates and thrives on systemic cognitive failures, whose effects are to 'naturalize' multiple patterns and practices of inequality and oppression in an interlocking structure of immoral beliefs that owes something to Enlightenment humanism's core assumption that 'only Europeans were human' (27). For Mills, the social contract also 'tacitly presupposes an 'epistemological' contract', which naturalizes contingent social orders and configures as 'natural kinds' both those advantaged and those disadvantaged by its hegemonic explanatory power. A silent 'agreement to misinterpret the world'(18)[2] filters out empirical evidence that would unsettle or counter any suspicion that these beliefs might, indeed, be held together by webs of distortion and error: of irresponsible knowledge production and circulation. Just such a configuration of 'natural kinds' is discernible in the actions and utterances of July's people.

10.3 'Fact', and Fiction

In its critique of social-institutional policies and practices that engage with matters of equality or inequality and analogous justice issues from positions of principled or indifferent *ignorance* of the detail and lived implications of human differences,

[2]See also Mills (2007), where (citing David Roediger) he refers to 'the fundamental epistemic *asymmetry* between typical white views of blacks and typical black views of whites: these are not cognizers linked by a reciprocal ignorance but rather groups whose respective privilege and subordination tend to produce self-deception, bad faith, evasion, and misrepresentation on the one hand, and more veridical perceptions, on the other hand' (17). This asymmetry is central to my reading of *July's People*.

Iris Young's 'structural inequality approach' is helpful. She shows how practices engaged out of ignorance/inadequate local knowledge are ill-equipped to generate sensitive, well-informed ways of addressing inequalities integral to, embedded in social and institutional structures, with the epistemic injustices they breed: indeed such practices may, unintentionally, reinforce exclusions and harms. Prompted by Young's analysis, I return to thinking about *July's People*, in order to show how fact and fiction can collaborate to advance projects of understanding difference and responding well to alterity, from positions of unmarked privilege.

Striking in Gordimer's novel is how the white woman, Maureen, takes human sameness for granted in matters both personal and material: sameness of relationships and feelings, of conjugal arrangements and everyday expectations, of the significance of places and objects, of how gendered divisions of labour and material things are not merely structured, but lived. Striking too is how she persists in such assumptions, despite her newly-affirmed commitment to acquiring a sense of how it is for him, for July, and for the people who are more plainly 'his': the people of his village. According to her, and this is one of the troubling assumptions: 'The human creed depended on validities staked on a belief in the absolute nature of intimate relationships between human beings. If people don't all experience emotional satisfaction and deprivation in the same way, what claim can there be for equality of need?' she wonders (64). She is unable to recognize (in Baier's sense) the particularity of her own conceptions of sexual loyalty, 'suburban adultery' (103), and love to the white middle-class society where she has learned them; unable to move to a critical or self-critical point from which to ask whether even these seemingly universal human verities might not count as universal, after all. So in the village her attempts to achieve 'equal' exchanges with July are predicated on imagined arrangements and understandings she cannot know would have been his – as when she assumes, 'He's afraid I'll tell about his town woman' (103). When she tries to become part of the women's work in the village, to establish a commonality with them, using their names and insinuating herself into the rituals and routines of their daily work, she cannot understand their resistance except by imagining that 'Martha', July's wife whose English name she uses glibly, without a by-your-leave, is too 'shy' to talk woman-to-woman with her.[3]

The extent to which naming is itself emblematic of relations between 'his kind' and 'their kind' (cf. 1) is evident in the white folks' surprise when, in their hearing, July is first referred to by his name, Mwawate. '"July" was a name for whites to use: for 15 years they had not been told what… [he] really was called' (120). They had not considered the implications of calling him by the name of a month in their English calendar. They have been ignorant even of his name. (Analogously, writing of Sojourner Truth's famous cry – 'Ain't I a woman?' – Susan Babbitt (2005, 7–8) shows that the white ignorance to which Truth speaks is no small matter, for it invokes something much greater than a 'difference' from other [=white] women, to

[3]Raimond Gaita (2003, 265), in a related context, reminds his readers that words 'can mean different things in different mouths'.

expose the structural injustices of Truth's life. Not having a name, a place to live, the capacity to care for her children: these quotidian expectations that assumed human sameness conjures up are starkly absent from her life, contrasted with their status as simple, routine expectations in lives lived outside the boundaries of such oppressions and deprivations. Undermining the power of an intransigent epistemology of ignorance might begin with recognizing and acknowledging the systemic workings of these seemingly small unknowings.)

According to Young, a 'commitment to substantial equality requires attending to rather than ignoring such differences' (2004, ms. p. 2). Somehow, it requires being attentive to discern their implications even where they are hardest to see and least expected, while learning to recognize the effects of the ingrained politics of ignorance that perpetuates their invisibility. July's white people would likely aver just such commitments, informed by the liberal politics they try to live, enacted in those practices of treating him well. Yet theirs is a small-'l' liberalism, conceived not in a sense indebted to John Stuart Mill or any other classical political theorist, but to 'liberal' as a term of self-identification, community-identification, in how they and/or their friends conceive of their positioning vis-à-vis apartheid, situating themselves on the side of 'the good'. It is a liberal stance that enlists the polite terminology of tolerance, which too readily descends into indifference – especially epistemic indifference. It relies on a cluster of careless assumptions about 'them', in how they are and are not just like 'us'. Although the white folks know that they would not want to live as they require/allow July to, they assume it is fine for him, hence that they have treated him well. Back there, the condescension such an approach entails was integral to, built in to how they, following the available scripts, could be with him: they might not even have seen it as condescension. Thus they 'treated him well' according to how they conceived his circumstances – they did not ask him. He was someone to 'treat', not to consult. It was not their intention to renounce their privilege but to use it responsibly, according to their lights. Indeed, according to the laws and customs of that old world order, there may have been a certain justice at work in their practices.

Yet their new situation, in his place and among the people of his tribe, presses urgently for their asking again, and more deeply, how could they attend well enough within the dominant imaginary in which they had lived, albeit critically and self-consciously, which they carry with them, intact, to this village? How could they attend well enough to differences to which they thought they were 'sensitive' yet which, structurally, systemically, they could not recognize well enough? How could they know that the discursive and material structures of their lives had inoculated them against what Rae Langton (2000, 128) calls 'doxastic shock'; had suppressed, subjugated – in a quasi-Foucauldian sense – the very knowledge they require to live this situation responsibly, and well?[4] Admittedly, a viable

[4]Langton observes: 'Many a woman has experienced vividly at first hand that demolition, that shaking of established belief, which Descartes thought necessary for the acquisition of knowledge… Foundations… are shaken, not by reflections on demons and sensory delusion, but by a life under inequality or oppression – a life which suddenly reveals for what they are those many falsehoods one had accepted as true' (2000, 127–128).

political epistemology of difference can neither derive from nor countenance a suggestion that such knowledge is impossible, that all exhortations are futile, or even that theirs is an innocent ignorance, so they need not address it. 'Ought implies can' functions, appropriately, as a fundamental moral-epistemological precept. But this is the conundrum: how can these people know well enough to fulfill these requirements, to honour this unspoken 'ought'? Fricker, for example, suggests that 'in the absence of a critical awareness of gender prejudice' those whose practices and utterances betrayed such prejudice 'were not *culpably* at fault until the requisite critical consciousness… became available to them' (2007, 100, her italics). Racial prejudice is analogous in this respect, and her claim makes good sense. But is it enough? *Should* the white folks have recognized their cognitive failures once they moved away from the familiarity of their white urban lives?

Questions about the scope and limits of knowing across radical differences are both about logical possibility and situated epistemic practice. They are among the most urgent questions facing philosophers at the beginning of the twenty-first century, following the collapse of androcentricity, eurocentricity, and the other 'centricities' that supported and sustained a hegemonic epistemology of sameness. But because they are quite newly on the table, and so much in process, it would be premature to propose a ready-made answer, a formula, a putatively worked-out solution that would risk homogenizing differences, all over again. At the very least – again quoting Fricker – 'the requisite collective… consciousness is … likely to dawn only gradually' (100). Recent work by feminists and critical race theorists on epistemologies of ignorance, which is my focus and inspiration here, has begun to map the terrain, articulate the issues, expose and engage epistemic injustices, set out certain cautionary principles (cf. especially Tuana and Shannon Sullivan 2006; Sullivan and Tuana 2007). But this is indeed only the beginning. Nonetheless, some interim diagnostic and prescriptive conjectures are in order, and it is to these I now turn.

Epistemology after Thomas Kuhn (1970) has been haunted by the prospect of an incommensurability that would lock scientific and other practitioners working under different paradigms into radically separated, incommunicable worlds. Kuhn's efforts to dispel anxieties spawned by his apparently hard-edged paradigm conception are well known (cf. Kuhn 1977). Moral epistemology faces analogous problems, often articulated as an imperative to avoid moral relativism above all. Yet responsibly understanding what is at issue when July's white people find themselves 'blocked by an old vocabulary', the vocabulary of 'back there' (Gordimer, 127) even from within a commitment to recognizing differences, requires 'us', whoever we are, to revisit the vexed issue of incommensurability, perhaps to draw its sting so as to acknowledge a certain promise concealed beneath the worries prompted by its absolute articulations.

Consider Ofelia Schutte's (2000) plea in favour of acknowledging 'incommensurable speaking positions': her insistence on the need to refuse the imperialism, indeed the epistemic violence, involved in superimposing the familiar upon the 'strange' and suppressing the leftovers. She reminds 'us' of how those outmoded, tacitly presumed 'centricities' work, in their imperialistic assumptions of human and circumstantial sameness, to standardize and naturalize ways of knowing, thereby blunting the effectiveness of such public and private (self)-criticism as Baier advocates. The 'data' thus taken as given feed into and are fed by an imaginary of neutrality, objectivity, and replicability: epistemic values imagined to be so incontestably universal, truth-preserving, and hence impersonal and formal, that they 'could not possibly be oppressive' while, paradoxically, both detached neutrality and a liberal egalitarianism insensitive to difference become excuses for avoiding recognition, thereby mutating into tools of oppression; of epistemic injustice. Yet Schutte rightly (in my view) contends: '...what we hold to be the nature of knowledge is not culture-free but is determined by the methodologies and data legitimated by dominant cultures' (47). Focussing her analysis on questions of alterity, Schutte looks to existential-phenomenology for a nuanced and radical approach with the power to show that it is 'the other' who 'makes it possible for the self to recognize its own limited horizons in the light of asymmetrically given relations marked by sexual, social, cultural, or other differences' (45–47). With its commitment to 'bracketing' hitherto fixed ideas and beliefs in order to clear (conceptual) spaces for developing descriptive experiential analyses of the Life-world, the existential-phenomenological conception of alterity makes available a concept of '*the other*' (46, italics original) which, she contends, 'combines the notion of the other as different from the self with the acknowledgement of the self's decentering that results from the experience of such differences' (46). Such a decentring can begin to effect a release from the tenacious fetters of presumed human sameness and the classical unified self, and toward ethical possibilities of understanding across radical diversity. Hers is a heuristic suggestion which, as I read it, catches something of what *July's People* achieves in the imagined encounter with alterity and incommensurability Gordimer creates.

Yet when racism 'epidermalizes'attributes of otherness (borrowing Iris Young's word), it can legitimize uncritical practices, premised on just such an allegedly benign recognition as Gordimer's Maureen might claim: that 'we' are all alike beneath 'our' merely accidental surface differences. It condones practices of treating him well from a conviction that his cultural imaginary translates readily into 'ours' and ours into his, so that all this unselfconsciously unmarked 'we' must do is provide for him circumstances like ours, scaled down to a level appropriate to his station, with scant thought for how it might be, for him, to live them. When we – whoever 'we' are – try to carry these practices from here to there, however, 'there' becomes a place curiously unmarked by its own specificities, to the extent that the *habitus* and *ethos* of here accompany us as extensions of our own white skin; both the literal epidermis, and the skin of custom and expectation, while preventing both him and us from acknowledging the differences that divide us.

These thoughts are not new for feminist and anti-racist theorists; and I have often addressed them. Yet the issue remains urgent even through repeated attempts to know across differences responsibly enough to think and act well. Neither July nor his white people can discern the subterranean detail of how white/black racism has made both him and them; in their case, because they do not know well enough to stand back from the incommensurables in this new-found 'public' world of the village, and to approach them well; in his, perhaps, because he has been trained to find safety in not knowing. This problem sits at the heart of the politics of difference. Thinking further with Schutte's appeal to existential-phenomenology, stepping onto this dangerous territory amounts to performing a variation on a Nietzschean transvaluation of values, where some of the most cherished values – 'beyond good and evil' – that have held white western self-styled liberal democratic societies together have to be radically revalued in consequence of their negative underside: a negativity consigned to the realm of unknowing by the very smoothness of these societies' polite surface. It involves taking very seriously her proposal that the 'incommensurable', which may not be 'subject to perfect cultural translation' (54), may yet be the most vital part of exchanges that attempt to reach across relationships of alterity. So the vexed question of what 'recognition' can achieve, how far it can go, whether it can animate a better politics and generate change, remains on the table, but unresolved.

10.4 Narrative Understanding

My reading of Gordimer's novel suggests how an artifactual narrative can interrupt, engage with, even displace more traditionally 'factual' empirical-experiential accounts. As I have suggested, a novel invites its readers into a life world – indeed, in Heidegger's sense – it *throws* them into that world, with its factical given-ness, which unsettles thought, yet has to be addressed. The novel develops what amounts to a quasi-phenomenological descriptive analysis of people and events consonant, I believe, with Schutte's recommendations. Hence, it can open possibilities of intelligibility in naturalized epistemology, showing rather than stating how 'the other' might indeed bring 'the self' up against its own limited horizons. Here, I cast July as 'the other', and Maureen as 'the self', while simultaneously positioning the reader as both self and other. Reading human lives through narratives that make new sense of them, if never definitively or completely – whether they present themselves as straightforwardly fictional, as autobiographical, as ethnographical investigations – opens the way for readers to re-imagine forms of life that cannot have been theirs, challenges their expectations in matters of recognition and oppression, and can animate strategies of thinking toward a critically creative instituting social imaginary where the occupants of positions of privilege might begin to take responsibility for how that privilege has made them.

This task is delicate: it needs to be performed in awareness of the dangers of too swift optimism about intelligibility achieved. It requires enlisting fictions in knowledge production without becoming entangled in what amounts to a narrative miasma: a term that captures something of the (justifiable) wariness Susan Babbitt expresses about a faulty logic at work in simply 'collecting stories from appropriately diverse sources' as a putative route toward acquiring 'an idea of the central properties of humanness' (2005, 11). It is as though without further analysis, stories could speak, unequivocally, for themselves: as though they simply accumulate to add more knowledge, or count as definitive experiential moments, immune to negotiation, hermeneutic engagement or critique, rather than as the points of – contestable – entry they afford.[5] In *Epistemic Responsibility* (1987), I address such questions by proposing that well-achieved narratives offer what I, pretentiously, call 'knowledge-by-second-hand-acquaintance', tipping my hat to Bertrand Russell. Now, I am less sanguine about the immediacy and the consequent scope of such putative knowledge; hence similarly troubled by simple practices of 'collecting stories', more inclined to suggest that stories remind us of 'our' (whoever we are) relative opacity to one another, captured in Schutte's reminder that there is always a residue of meaning that will not be reached in cross-cultural endeavours; and to honour that opacity. Alongside this sense that 'we' must trim our expectations of knowing one another well, however, run multiple lines of potential intelligibility – and it is to this project of achieving a just, if negotiable, measure of intelligibility across differences that divide us even more radically than our most cautious deliberations suggest, that such perceptively crafted, situationally grounded narratives as Gordimer's novel can contribute. (Yet even acknowledging the differences that divide 'us' is fraught, for the injunction admits of inhumane readings as readily as it holds space open for responsible circumspection, for reserving judgement. Consider how assiduously the Nazis learned that they must 'appropriately acknowledge' the differences dividing the pure Germanic Volk from the Jews. The example points to the dangers in my project and terminology, showing how the recognition I have argued for is always intertwined with larger issues of responsibility and justice.)

I return briefly to Judith Butler, whose cautionary reminder of the effects of taking entrenched modalities of human intelligibility for granted, thereby failing 'to think critically – and ethically – about the consequential ways that the human being is produced, reproduced, deproduced', bears crucially on matters of epistemic injustice. Both the white folks and July are produced, reproduced, and ultimately deproduced in the politics of the town and, subsequently, of the village. Butler 'cannot imagine a "responsible" ethics or theory of social transformation operating without such critical inquiry' (2004b, 222). Preparing the way for 'resignification as politics' (223), her challenges expose a fragmented society that is indeed incongruous with itself, where hitherto apparently settled ontological questions erupt with new urgency. Questions about who and what counts as 'real', Butler reminds her readers, are questions of knowledge and power, for 'power dissimulates as ontology' (215) 'keeping our notion of the 'human' open to a future articulation is essential', she argues, 'to the project of a critical international human rights discourse and politics.' Hence, substituting 'black Africans' for 'women' in Butler's text, 'the ways in which... [a Black "boy" is] said to "know" or to "be known" are already orchestrated by power precisely at that moment in which the terms of "acceptable" categorization are instituted' (215). I have pointed to naming as one such categorization. One of the cognitive failures, for the white folks, is that they do not/cannot shift ground to the point of coming to know what or how July can

---

[5] I have also benefited in thinking about these issues from Susan Babbitt's (2006).

know, cannot conceive of him as occupying a black South African standpoint, if not in the fullest (standpoint theory-derivative) sense of his being a politically informed, collectively occupied and enacted standpoint, but as nonetheless claiming epistemic privilege that they themselves cannot. He in fact knows them and their situations, the intimacy of their lives and their vulnerabilities among 'his' people, far better than they know him. In the village, by contrast with how he was in his town life, July now 'chose what he wanted to know and not know. The present was his; he would arrange the past to suit it' (96). Indeed, as the namings and epistemic situations shifted it began 'to amuse July to be the mentor, as if he didn't take too seriously a white's wish to comprehend … what he had never needed to know as a black had the necessity to understand, take on, the white people's laws and ways' (112). Practising a version of what Alison Bailey (2007) aptly calls 'strategic ignorance',[6] July was, then, positioned to manipulate these people in ways that Maureen came too slowly and too late to realize. Hence, at the end of the novel, 'She ran…'.

Continuous with my thinking, together with Schutte, about incommensurability, opacity, alterity, and the politics of knowledge, Lewis Gordon, in an essay tellingly titled 'What Does It Mean to Be a Problem?', cautions against an epistemic closure in knowing other people, which amounts to reducing a feature of reality to absolute reality with which one ends an inquiry, presents it as knowledge accomplished, proclaiming, in effect, 'say no more' (2000, 88). He, by contrast, advocates an epistemic openness in learning about Others, which issues in the judgement 'there is always more to be known'. Gordon, too, works within an existential-phenomenological framework consonant in many respects with Schutte's. He deplores epistemic-political acts of declaring closure by naming a social role or racialized identity, presenting it as all a knower needs in order to legitimate a plethora of other judgements in which the humanity of the subjects of study is erased in an aggregating process that sees them as a group. Such acts of closure sustain ignorance of the multiple modalities of lives lived under those ill-fitting labels, those putatively natural kinds. They require principled refusals of a retreat to (Sartrean) bad faith that would deny/look away from the 'anonymity' that undergirds social groups, just as July's white people have been accustomed to do.

10.5 Conclusions

As my scavenger approach to epistemic resources in this essay will have shown, I incline toward an epistemological pluralism whose sources are multiple, often reciprocally enriching, sometimes challenging intransigent assumptions or imaginings.

[6] Bailey writes: 'Strategic ignorance is a way of expediently working with a dominant group's tendency to see wrongly. It … uses dominant misconceptions as a basis for active creative responses to oppression. It seeks out resistant paths through the logic of purity that turn white ignorance back on the oppressor jiujitsu style' (2007, 88).

My thoughts about literary contributions to knowledge enlist the conceptual framework of the ecological thinking I develop in my book by that title, in the sense that when subjectivities and situations are drawn well enough to open out new places of critical-interpretive engagement, a novel can offer ways of understanding habitus and ethos, patterns of situated moral-epistemic response akin to those existential-phenomenology at its best also constructs. These diverse textual resources point toward an epistemic stance from which to think away from the taken-for-granted structural inequalities, whose lived meanings a viable politics of difference and post-colonial resistance has to understand better than ready-made formulas or accumulations of information – of facts – can allow. The goal, as I have indicated, is not merely the conservative one of avoiding the errors such intransigent imaginings condone, but a creative and constructive one of seeking new understandings. It is to such a positioning that Gordimer contributes in *July's People*: a way of engaging with 'difficult knowledge' that is sensitive enough, and with the necessary humility, to recognize something of how it is, phenomenologically, for people caught in the palpable moral and epistemic chaos she portrays, where the touchstones Maureen (the white woman) tries to hold on to in order to maintain a stable moral-epistemic frame are somehow the wrong ones. She cannot get beyond, and in her isolation from an interpretive community, has trouble seeing that she cannot get beyond, the language and assumptions of back there,[7] conceived not just social-structurally but ecologically. She is ignorant of her own ignorance, and confirmed in her isolation by a tacit assumption that a viable, reliable interpretive community can be comprised only of 'one's kind'.

In its unerring sense of place and personae the novel straddles resources available from bio-regional narratives and some versions of philosophical anthropology, which I read, loosely as inquiry that seeks draw together diverse investigations of 'human nature' in order to understand human beings as creatures of their environment and creators of meaning.[8] (Hence, and germane to this point, Charles Mills observes: 'Often, for their very survival, blacks have been forced to become lay anthropologists, studying the strange culture, customs and mind-set of the "white tribe" that has such frightening power over them' [2007, 17]). It structures the politics of difference so as to discredit any principled ignorance of individual or group differences, while advocating an epistemic openness to situation, place and subjectivity, a circumspect engagement with alterity. Like any novel, it admits of diverse readings. It could be read as an example of participatory ethnography, casting

[7]Schutte writes of 'nodes in a linguistic interchange or a conversation in which the other's speech, or some aspect of it, resonates in me as a kind of strangeness, as a kind of displacement of the usual expectation.' She continues: 'Cultural alterity requires that one not bypass these experiences or subsume them under an already familiar category' (2000, 49).

[8]Raimond Gaita (2003, 277), for example, suggests that 'if the discursive is not longer restricted to the exercise of the kind of thought in which form and content are separable, then, in roughly those parts of philosophy which the Europeans call philosophical anthropology, there will be no marked distinction between the narratives that must to some degree nourish inquiry and philosophical engagement with them.'

Maureen as a participant observer living among Others more or less as they live, attempting to assimilate by emulating their ways, engaging in the minutiae of their everydayness. Such a reading would leave it open to the contestations such practices invite, where the anthropologist may be blissfully ignorant of the ploys her 'subjects' deploy to assert their agency in constructing what she takes away as 'their story', or where she unwittingly reads them through lenses ground to enable her to see only what she already knows, is oblivious to incommensurability. Hence again, given that July knows their world and their ways so much better than they know his, their failure to recognize the extent of his epistemic privilege ultimately contributes to social disintegration. Yet a novel moves beyond detached observation and the conceptual poverty of thought experiments with 'possible worlds' or otherwise, and of truncated, contrived examples. It eschews the presumptions endemic to beliefs in an uncomplicated ease of 'putting oneself in someone else's shoes' (La Caze 2002); and exposes some of the structural and conceptual limitations even of such self-consciously engaged practices as the white woman's attempts to know.

The contrast that interests me, then, is not between outright ignorance and knowledge achieved or complete, but between that 'little knowledge'–snippets of half-truth – that folk wisdom sees as 'a dangerous thing' – and knowledge responsibly achieved. Thus, in my first sense of 'moral', knowledge and recognition move away from views, biases, prejudices received or ingested, and toward increasingly creative and self-critical engagement with practices, forms of life, and situations which, responsibly engaged, can issue in recognitions of bio-regional complexity 'at home' and 'away', and of folk whose ways cannot, before the fact, be assumed to be 'just like ours'. In my second sense, it exposes some of the egregious harms and epistemic injustices enacted when decisions and judgements are based on uncontested assumptions to the effect that the world, both human and other-than-human simply presents itself to 'us' as we have always taken it to be. The goal is not epistemic autonomy, and closure may never be warranted; but taking on the intellectual and emotional challenge of staying with indeterminacy, with ambiguity, where premature closure risks performing epistemic violence, can be a more rather than a less courageous epistemological stance. Ambiguity and indeterminacy are indeed at odds with the goals of much orthodox epistemology, but ecological thinking thinks with them.[9]

Methodologically, I am proposing no rules for the redirection of inquiry, no route toward perfect knowledge, but re-examining some of the pieces and practices that could allow knowledge claimants to recognize human differences, more responsibly than otherwise. The message is not 'take a novel, add a bio-regional narrative, do some phenomenological anthropology, and think about how they can be stirred together, and what else you might need'. It is about linking lines of inquiry, weaving them into established forms of empirical evidence, for their capacity to unsettle patterns of ignorance and incredulity that, ironically, have structured

[9]See in this regard Code 2006b.

western epistemology and everyday knowing for too long. This way of thinking complicates the notion of epistemic responsibility in its insistence that knowers do not escape the pressures and limitations of their embodiment, but that these conditions thwart, animate, perhaps enable processes of inquiry. Rather than declaring a responsibility to transcend 'epidermalized' and otherwise embodied specificities, epistemic responsibility ecologically configured requires showing how lived embodiment and a strong sense of place and circumstance figure centrally among conditions for the possibility of knowledge, morality, and epistemic-hermeneutic justice. This piece suggests how a fictional narrative can perform an act of intervention into seamless imaginaries held in place by layers of ignorance of the very things on which they base their thought and action.

Acknowledgments My thanks to Heidi Grasswick and Sonia Kruks for their careful readings of this paper, and for helpful comments and suggestions. I presented earlier versions at Dalhousie University, the University of Dayton, McMaster University, the Ontario Institute for Studies in Education, Twente University in the Netherlands, the Institute of Advanced Studies (IAS) at Durham University, EWHA Women's University in Seoul, and as the first of my 2008 Hanna Lectures at Hamline University. I am grateful to colleagues present at those events for insightful responses and recommendations.

References

Babbitt, Susan. 2005. Stories from the south: A question of logic. *Hypatia: A Journal of Feminist Philosophy* 20(3): 1–21.

Babbitt, Susan. 2006. Review of reading across borders: Storytelling and knowledges of resistance, by Shari Stone-Mediatore. New York: Palgrave Macmillan, 2003. *Hypatia: A Journal of Feminist Philosophy* 21(3): 203–206.

Baier, Annette. 1985. Mixing memory and desire. In *Postures of the mind: Essays on mind and morals*. Minneapolis: University of Minnesota Press.

Bailey, Alison. 2007. Strategic ignorance. In *Race and epistemologies of ignorance*, ed. Shannon Sullivan and Nancy Tuana, 77–94. Albany: State University of New York Press.

Butler, Judith. 2004a. Doing justice to someone: Sex reassignment and allegories of transsexuality. In *Undoing gender*. New York: Routledge.

Butler, Judith. 2004b. The question of social transformation. In *Undoing gender*. New York: Routledge.

Castoriadis, Cornelius. 1994. Radical imagination and the social instituting imaginary. In *Rethinking imagination: Culture and creativity*, ed. Gillian Robinson and John Rundell, 136–154. London: Routledge.

Code, Lorraine. 1987. *Epistemic responsibility*. Hanover: University Press of New England.

Code, Lorraine. 2006a. *Ecological thinking: The politics of epistemic location*. New York: Oxford University Press.

Code, Lorraine. 2006b. Skepticism and the lure of ambiguity. *Hypatia: A Journal of Feminist Philosophy* 21(3): 222–228.

Code, Lorraine. 2008. Advocacy, negotiation, and the politics of unknowing. *The Southern Journal of Philosophy* 46(S1): 32–51.

Fricker, Miranda. 2007. *Epistemic injustice: Power and the ethics of knowing*. Oxford: Oxford University Press.

Frye, Marilyn. 1983. The problem that has no name. In *The politics of reality*. Trumansburg: The Crossing Press.

Gaita, Raimond. 2003. Narrative, identity and moral philosophy. *Philosophical Papers* 32(3): 261–277. Special Issue on 'Narrative Understanding'.
Gordimer, Nadine. 1981. *July's people*. New York/London: Penguin.
Gordon, Lewis R. 2000. What does it mean to be a problem? In *Existentia Africana: Understanding African existential thought*. New York: Routledge.
Kuhn, Thomas. 1970. *The structure of scientific revolutions*, 2nd ed. Chicago: University of Chicago Press.
Kuhn, Thomas. 1977. *The essential tension: Selected studies in scientific tradition and change*. Chicago: University of Chicago Press.
Langton, Rae. 2000. Feminism in epistemology: Exclusion and objectification. In *The Cambridge companion to feminism in philosophy*, ed. Jennifer Hornsby and Miranda Fricker, 127–145. Cambridge: Cambridge University Press.
Le Dœuff, Michèle. [1998] 2003. *The sex of knowing* (trans: Kathryn Hamer and Lorraine Code). New York: Routledge.
Marguerite, La Caze. 2002. *The analytic imaginary*. Ithaca: Cornell University Press.
Mills, Charles W. 1997. *The racial contract*. Ithaca: Cornell University Press.
Mills, Charles W. 2007. White ignorance. In *Race and epistemologies of ignorance*, ed. Shannon Sullivan and Nancy Tuana, 11–38. Albany: State University of New York Press.
Schutte, Ofelia. 2000. Cultural alterity: Cross-cultural communication and feminist theory in north-south contexts. In *Women of color and philosophy: A critical reader*, ed. Naomi Zack, 44–68. Oxford: Blackwell.
Sullivan, Shannon, and Nancy Tuana (eds.). 2007. *Race and epistemologies of ignorance*. Albany: State University of New York Press.
Tuana, Nancy and Shannon Sullivan eds. 2006. Feminist epistemologies. Special Issue of *Hypatia: A Journal of Feminist Philosophy* 21:3.
Young, Iris Marion. 2004. Structural Injustice and the Politics of Difference. Unpublished lecture delivered at the IAPh conference in Göteborg.

Chapter 11
Wrongful Requests and Strategic Refusals to Understand

Gaile Pohlhaus, Jr.

Abstract In *The Alchemy of Race and Rights* Patricia Williams notes that when people of color are asked to understand such practices as racial profiling by putting themselves in the shoes of white people, they are, in effect, being asked to, 'look into the mirror of frightened white faces for the reality of their undesirability' (1992, 46). While we often see understanding another as ethically and epistemically virtuous, in this paper I argue that it is wrong in some cases to ask another to attempt to understand certain positions or lines of thought. In developing my argument I draw on the work of María Lugones to argue for a view of agency that is epistemically interdependent. I examine the case described by Patricia Williams to demonstrate specifically how the understanding requested in this case unfairly undermines both epistemic and non-epistemic agency. I distinguish appropriate requests for understanding from inappropriate requests so as to make clear that I am not suggesting that it is wrong to make such requests when the understanding sought after is difficult, painful, or even when it forces one to reconsider the meaning of one's actions. Finally, I examine an example from Susan Brison to show how strategic refusals to understand may provide a pathway toward new ways of knowing and being in resistance to oppressive regimes.

Keywords Epistemic ethics • Intersubjective agency • María Lugones • Patricia Williams • Understanding

G. Pohlhaus, Jr. (✉)
Department of Philosophy, Miami University, Oxford, OH 45056, USA
e-mail: pohlhag@muohio.edu

H.E. Grasswick (ed.), *Feminist Epistemology and Philosophy of Science: Power in Knowledge*, DOI 10.1007/978-1-4020-6835-5_11,

> In *The Alchemy of Race and Rights* Patricia Williams recounts how during debates in the 1980s over the use of buzzer systems by store owners in New York City, there were 'repeated public urgings that blacks understand the buzzer system by putting themselves in the shoes of white storeowners – that, in effect, blacks look into the mirror of frightened white faces for the reality of their undesirability' (1992, 46).

> In *Aftermath*, an extended reflection on her survival from sexual assault and near death strangulation, Susan Brison writes 'When I started telling people about the attack, I said, simply, that I was a victim of an attempted murder. People typically asked in horror, "What was the motivation? Were you mugged?" and when I replied, "No, it started as a sexual assault," most inquirers were satisfied with that as an explanation. I would have thought that a murder attempt plus a sexual assault would require more, not less, of an explanation than a murder attempt by itself' (2001, 3).

In the first of these scenarios, legal theorist Patricia Williams focuses attention on a situation in which persons are asked to understand others in a way that seems utterly inappropriate. In the second, philosopher Susan Brison engages in what I will call an appropriate and strategic 'refusal to understand,' expressed through her confusion over why a murder attempt plus a sexual assault makes more sense to her interlocutors than a murder attempt alone. While we often view understanding others, and particularly understanding those with whom we do not share a common viewpoint, as ethically and epistemically virtuous, in this paper I consider: (1) whether and when it is wrong to ask another to attempt to understand certain positions or lines of reasoning and (2) whether and how explicitly refusing such requests might be ethically[1] and epistemically productive.

By 'understanding others' I mean attending to the sense of another's reasoning so that one is able to follow and to feel the possible force of that reasoning. We ordinarily expect understanding in this sense from responsible knowers for a number of reasons. First, it seems unreasonable to reject another's position before one has attended to her argument or to the possible reasons that could support her position. As a matter of responsible inquiry we are often implored to 'consider all sides.' Second, genuine disagreement arises only when one has seriously considered another's reasoning, and engaging in genuine disagreement can be epistemically productive. For one, we may find out that we are wrong to disagree. In cases where we do not find that we are wrong to disagree, such engagement often results in more clarity about where and why we disagree, thereby deepening self-understanding. Finally, attempting to understand another in a way that would be recognized by the speaker as capturing what she means or is trying to say, as opposed to 'twisting her words' or dismissing her out of hand, is something we generally think we owe our interlocutors as a matter of respect. Explicitly refusing to do so could even amount to a refusal of an interlocutor's position as an epistemic subject insofar as it preempts her from making a particular contribution to an important and common knowledge practice, the giving and receiving of reasons. Given the importance of

---

[1] I use 'ethically' rather than 'politically' here to indicate that the kind of productivity involved, as I will show in the second half of this paper, is not toward any particular politically identifiable end but rather productive toward opening possibilities for a more ethical life together for which there are, as of yet, no defined ends.

this type of understanding, why might it be wrong to ask for it? Moreover, why might it be not only warranted to refuse a request for this type of understanding, but even ethically and epistemically productive to do so?

While merely considering someone's reasoning may seem innocuous, particularly when one is not required to concede the conclusion, there is a deeper problem operative in cases like that which Williams describes. Understanding another's reasoning requires one to do more than hold a particular set of claims in the mind. It requires one to follow the sense of those claims, so that the claims may be evaluated for what they mean. The meanings of words and our ability to discern those meanings, however, is not something that exists independently of human practices and ways of being in the world. To follow the sense of a claim is to comport oneself toward the world in particular ways and to participate within the 'grammar' which structures the sense of the claim. Our words and language practices situate us in relation to the world and one another not simply by pointing us toward objects waiting to be discovered but rather by providing socially established patterns from which we interact in and with the world. In Sara Ahmed's language, the senses of our words 'orientate' us.[2] When we accidentally misunderstand another and then subsequently correct ourselves, there is a shift both in comportment and response from disorientation to 'following the sense' of what was said. This reorientation manifests itself in various ways. For example, it may shift our affect, as happens when we suddenly understand a joke. Or it may bring some things into focus while (or by) placing others out of focus as might happen when we realize that our interlocutor is talking about a dream as opposed to an actual event. In the former, attention might focus on images and emotive details, whereas in the latter, both images and emotions might be disregarded altogether with attention paid to the logistics that could render such an event plausible.[3] Once oriented to what was meant, we find that we can continue our engagement with the person instead of talking past one another, we are now 'on the same page.' When we are already attuned to the sense of what another is saying, the propositional and non-propositional attitudes that give sense to what is said and the range of responses (both physical and verbal) that indicate proper uptake of that sense go largely unnoticed as the field within which our claims figure as prominent. And when seeking to understand another whose claims we do not yet quite understand or see the point of, we try to find out where exactly she is 'coming from' so that we can 'find our feet' in what she says.

Importantly, the various attitudes and practices within which our claims make sense are held in place intersubjectively. To use one of Wittgenstein's examples,

---

[2]For more on 'orientating' see Sara Ahmed's *Queer Phenomenology* (2006), particularly Chaps. 1 and 2.

[3]The connection between understanding, practice, background assumptions and range of sensible responses is also demonstrated in the following anecdote from Wittgenstein's *On Certainty*: 'I am sitting with a philosopher in the garden; he says again and again "I know that that is a tree," pointing to a tree that is near us. Someone else arrives and hears this, and I tell him: "This fellow is not insane. We are only doing philosophy" '(1969, #467).

the shape of an arrow points in a particular direction not because that is how arrows in some predetermined way direct attention, but rather, the direction we follow just happens to be the way 'we' do things (1965). It is this 'doing' along with the place(s) it holds within our various practices that maintains the normative force of the arrow's 'directedness.' Of course, history is rife with ways of doing things and practices that have been called into question, rightfully criticized, and challenged. Among the things we do is to direct our attention to the ways in which things are done, to consider the implications of doing things in that way, and to attempt to change them. Just because our language and practices have a somewhat arbitrary nature (we do not *have* to follow arrows in the way they 'point'), does not mean that we can change them at will from a place outside them.[4] We change our ways of proceeding from where we are and with the engagement of others who may or may not follow us. Once followed, our ways of approaching the world and one another become a new background within which we make and evaluate claims. The practices that comprise this new background (as with all practices) are maintained through our continued use of them. Consequently, following the sense of another's reasoning is not wholly neutral, but requires one's participation in, and so maintenance of, that which gives sense to her claims. In some cases, as I will argue, that which gives sense to another's reasoning can alter and even curtail the listener's range of possible significant action. The curtailing of the listener's agency and the request that the listener participate in that which curtails her agency is what makes such cases as the first I quote at the outset of this paper so offensive.[5] In contrast, explicitly refusing to think within certain structural contexts, as I contend is happening in the Brison passage, can expand agency in a way that brings listener and speaker into a more ethical epistemic relation.

In what follows, I first clarify the type of actions with which I am concerned and make the case for the claim that one's range of agency with regard to these types of action is determined intersubjectively by practices that structure the senses of what we say and do. I then return to the case described by Patricia Williams to demonstrate specifically how asking for understanding in this case unfairly undermines the agency of certain persons. I distinguish appropriate requests for understanding from inappropriate requests so as to make clear that I am not suggesting that it is wrong to request understanding that is difficult or painful, or that forces one to reconsider the meaning of one's actions. Finally, I examine the Brison quote to show how strategic refusals to understand may provide a pathway toward new ways of knowing and being in resistance to oppressive regimes.

---

[4]In 'Forms of Life: Mapping the Rough Ground' (1996) Naomi Scheman argues cogently that this fundamentally Wittgensteinian insight need not lead to relativism or caprice.

[5]It should be noted as well that part of the insult involved in cases like the one presented by Williams is that marginalized people are more often than not the ones being asked to understand. I owe thanks to Alison Bailey for reminding me of this point. What I aim to show in this paper goes further, arguing that even if this history of asymmetry were empirically absent there is something wrong in asking for some positions to be understood.

11.1 Meaningful Action and Agency Shortchanged

Feminists have argued in numerous ways that one's ability to act in the world depends upon other agents. Here I draw on insights from a number of those arguments to demonstrate the degree to which meaningful action is enabled and constrained by intersubjective coordination with other agents. By intersubjective coordination I mean the manner in which agents recognize and respond to the world (including each other) within a background of propositional and non-propositional attitudes that determine what in that world is significant and how. This coordinated recognition and response is maintained (at least in part) by the way agents habitually comport their bodies over time within and toward the world in complex patterns that support the continuing of those very patterns.

To demonstrate simply what it means to habituate one's body in coordination with others, consider the example of driving a car. When one learns to drive a car, one must keep in mind various rules and practices, but also learn to use one's eyesight in particular ways (paying attention, not only to what is ahead, but also behind, and to the side through the use of mirrors). Once one has familiarized oneself with the 'rules' and has sufficiently habituated one's movements and one's attention accordingly, those rules and comportments become largely unconscious. Habituating the body and attention in this way greatly enhances one's ability to traverse long distances on roadways. At the same time, however, this very expansion of agency can constrict others' agency. If, in response to this expansion of agency, the placement of buildings and services is adapted solely to those bodies that can readily use and access motor transit, then some parts of the social world and the interactions that take place within them may become out of reach for people who can neither use nor access motor transit. Social constructionist disability theorists have long argued in this manner that one's ability to act in the world is something that does not reside in the individual as such, but rather is a function of the relationship between bodies and an environment that is shaped to fit particular bodies and not others.[6] Consequently, how the world is shaped by those who move within it has an effect on who can do what. For these reasons, the ways in which we coordinate our bodies and attention with others can expand *and* constrict agency; moreover, they can do both at the same time.

Our coordinating capacities, however, go far beyond habituating our bodies to particular patterns so as to facilitate movement at high speeds without running into one another. Specifically, how we understand and respond to particular movements of the body and the practices within which those movements have significance is an important component of certain actions. These kinds of action are typically ones that make life something more than a mere biological existence, as in the case of the action of 'eating the first bite of cake on one's birthday' as opposed to simply

---

[6] For a good articulation of the social constructionist position on disability see Chap. 2 of Susan Wendell's *The Rejected Body* (1996).

consuming food.[7] To clarify, certain actions depend upon what those actions *mean* and the way in which actions mean depends on far more than the way my individual body does or does not move. This is because the significance of my actions requires uptake, or recognition of the act that I intend, and whether this recognition occurs is not something within my complete control. For example, as a female, I cannot, no matter how hard I try and no matter what I do, perform a Roman Catholic wedding ceremony. I could perform something like it, provided that there were a couple who would ask me to perform such an action, but even if all parties present should wish me to bestow matrimony on a couple, this would not make the ceremony Roman Catholic. In a similar vein, agency is at stake in the current debate over same sex marriage. While a same sex couple wishing to enter into a marriage could perform a ceremony that they, their friends and family recognize as a marriage, in most of the United States (to the extent that national government and most local governments in the U.S. refuse to recognize their union by denying them all the rights and responsibilities of married couples) the couple is prevented from marrying and being married. In neither case is the action something that I am merely told that I cannot do, leaving open the possibility that I could transgress the law, perform the action, and suffer the consequences. Instead, our social set up and the meanings made available within it prevent even the possibility of my performing the action in question. These two examples show that my agency with regard to meaningful action depends upon particular social systems and the responses of others within them. Just as the arrangement of material space can enable and disable whole groups of people, so, too, can the arrangement of 'rhetorical space' (to use Loraine Code's apt phrase).[8]

The examples above show that the possibility of a specific action can be foreclosed by a particular set of social practices. Of course, with regard to the second example, there are a whole range of actions that are either made difficult or illegal for a couple who wishes to be but is not recognized as married. Still, the case that opens this paper is one where what appears to be at stake is a less formally, yet equally binding constraint, and a constraint that directly affects a whole range of actions. This kind of constraint is like that found in an example María Lugones gives in her book *Pilgrimages/Peregrinajes* within the context of demonstrating what she means by the term 'world.' Understanding both the example and how Lugones uses the term 'world' can help to clarify further the kinds of cases with which I am concerned.

In the chapter of *Pilgrimages/Peregrinajes* entitled 'Playfulness, World-Traveling, and Loving Perception,' Lugones explains how she came to understand and use the term 'world' when she encountered a confusion about the nature of her own

---

[7] The point here is not limited to the sociological claim that different societies have different things they take to be significant for living a life. Rather, to build and to have a life that is sustained over time regardless of one's particular society requires sets of social practices within which actions take on significance.

[8] See Loraine Code's *Rhetorical Spaces* (1995).

character and ability to act. Specifically she found that she was able to say truthfully about herself that she is fundamentally a playful person and that she is not playful at all. In other words, she could remember occasions in which her character was that of a fundamentally playful person and could call on persons she knew who were able to confirm that she indeed is playful even while she could remember occasions in which she was a person incapable of being playful at all and could call on other people to confirm that fact (2003, 86–87). To resolve the contradiction of being both playful and not playful, Lugones develops the notion of a 'world.' Before turning to that concept, however, I would like to dwell a bit on what Lugones means when she says that in some cases she is not playful at all even though she remembers herself as being fundamentally playful elsewhere.

One possible way of resolving the contradiction of both being and not being playful that is considered by Lugones is to say that while she *is* a playful person, she just has difficulty expressing that playfulness in particular situations, for example, because she is not at ease in them (87). However, Lugones stresses that this does not accurately describe her experience, for in the hypothetical case that would resolve the contradiction, Lugones notes she 'could work on it' (87), suggesting that in the actual cases where she is not playful she finds that no matter how much work she could possibly do, it would never result in an ability to be playful. This case resembles a description Lugones gives in an earlier chapter concerning agency under oppression, where she notes that often an oppressed person can remember being able to perform an act, but then finds that in the present situation she, 'cannot do so because the action does not have any meaning or has a very different sort of meaning than the one it has in the other reality [she remembers]' (57). If indeed this description fits Lugones's own experience with not being playful, then 'playfulness' and all actions associated with being playful are simply not an option for Lugones in a way that bears resemblance to a woman's inability to perform a Roman Catholic wedding ceremony. No matter how her body moves, no matter what she says, 'doing something playful' is not among the possible descriptions for her actions.

To explain how it can be possible to remember being able to do something one now finds it in principle impossible to do, Lugones posits that there is more than one 'world' and that she is not the same person across different 'worlds.' While Lugones indicates some characteristics of what she means and does not mean by 'world,' she does not stipulate an exact definition of it since, as she says, 'the term is suggestive and [she does] not want to close the suggestiveness of it too soon' (87).[9] Nonetheless, what she does say about the term and her manner of using it are enough, I think, to help clarify the way in which possibilities for action are intersubjective and can be unfairly constrained in some cases due to the nature of that intersubjectivity. Of 'worlds,' Lugones says they are spaces inhabited by people

[9] The fact that Lugones leaves the term open ended and 'suggestive' allows her to show rather than just say one of the key insights she develops in the essay in which she introduces it. Specifically, Lugones's treatment of the term 'world' exemplifies the kind of attitude she advocates as a way of being with and loving others.

containing 'a description and construction of life, including the constructions of the relationships' that sustain them (89). Worlds are 'not autonomous, but intertwined semantically and materially, with a logic that is sufficiently self-coherent and sufficiently in contradiction with others to constitute an alternative construction of the social' (20). In other words, within any individual world, material life is infused with semantic life and semantic life is animated by material relationships. As Lugones notes, a world must 'be inhabited at present by some flesh and blood people,' so it is not an abstraction that exists independently of human interaction. Rather, worlds are actual 'lived social arrangements' (25) that exist in tension with one another due to relations of power that are imbedded in and made possible by human intersubjective relating.

Lugones's use of the term 'worlds' is similar to (if not the same as) her modification and use of Victor Turner's term 'structures'(Turner 1974) so we can supplement what we have said so far with what Lugones says about structures. Lugones describes structures as '"patterned arrangements"... [that] construct or constitute persons ...in the sense of giving them emotions, beliefs, norms, desires, and intentions that are their own' (2003, 60). In contrast to Turner, Lugones does not see the enacting of roles within these patterned arrangements as personae that are put on and taken off by an individual who transcends the roles and relations, since even the ability to take on a persona depends on some sort of structure (60). In discussing this issue, Lugones uses the example of moving from the role of 'husband' to the role of 'doctor,' which she notes is itself structure dependent (60). In other words, that the same person can move from the role of husband to doctor requires a particular set of patterned social arrangements which enable this possibility and within which we can make sense of one person's taking on these two roles. To highlight this point, we can contrast Lugones's example with the example of being a 'mother' and being a 'hospital chief of staff.' Sexist structures do not allow fluid movement between these two roles, which is to say: the dominant set of practices and meanings within which mothers are mothers and within which chiefs of staff at hospitals are chiefs of staff does not support the possibility of being both. To the extent that the replacement of sexist structures with non-sexist structures changes the situation, allowing more possibilities for movement between these two roles, what it means to be such things as a 'mother' and 'chief of staff' has also changed. Importantly, these changes in structures and meanings do not come about because of an individual act of will or an individual way of understanding what it means to be a mother or chief of staff. Individual intentions, actions, and understanding are made possible by structures and not the other way around. Nonetheless, the patterns and practices that comprise structures exist only insofar as they are animated in material and semantic relationships.

Keeping in mind what Lugones says about 'structures' we can return to the notion of 'worlds' to say that within the heterogeneity of any given world, resistant practices and meanings can arise among persons that bring new worlds into being. This is not a matter of thinking about oneself however one happens to choose, but rather it is a matter of coherently and creatively inhabiting meaningful relations with others that give rise to agential possibilities. When this happens, it is possible for there to be more than one world in any given present. Because these worlds exist simultaneously, anything that happens in one of them will potentially have effects in all present worlds. What those effects will be depends upon the set of patterns and practices that maintain the given world in which those effects happen.

Returning to Lugones's example of being playful, to say there are worlds in which she is not playful attests to there being historical practices and meanings that construct her possibilities in such a way that make playfulness impossible for Lugones in principle. To say that there are worlds in which she is indeed a playful person, attends to alternate historical practices and meanings that construct Lugones in ways that make playfulness not only a possibility, but a fundamental part of who she is. These 'worlds' exist in tension and are embodied in her interactions with others.

In sum, Lugones's notion of 'worlds' helps us to see that who a person is and what she can do occur within a living context; the meaning or significance of our thoughts and actions is something that cannot be separated from the circumstances in which they occur. These circumstances are connected to a number of aspects of human social life, including the history of ways we understand and interact with one another. This history does not determine how we *must* proceed, but it does set constraints on our range of possible action insofar as it is the place *from which* we proceed. When such histories include categorizing certain persons as members of particular groups constrained by such things as stereotypes and institutional prejudice, the range of meaningful action available can be unfairly stratified in ways that are both material and semantic. Nonetheless, the possibility of multiple 'worlds' shows that our sustained relating with one another can form creative agential resistance that exceeds the bounds of worlds of oppression, even while standing in precarious tension with those oppressive worlds. Importantly, this resistance is not individual and the possibility of resistance does not make or guarantee an *actual* world of resistance. Rather, worlds must be animated interactively among agents. Insofar as our practices of understanding are integral to the way in which our material relations have semantic life and to the way our semantic life directs us materially, whom we understand and how we understand them can be a matter of continuing worlds of domination or animating worlds of resistance.

11.2 Wrongful Requests

Returning specifically to the first scenario with which we began, we can now detail why some requests for understanding are epistemically and ethically wrong. As Williams's essay reveals, in the case she describes, persons are being called to understand something that only makes sense from within patterns and practices that hold oppressive power relations firmly in place and that actively prevent those asked to understand from calling attention to this fact. To use Lugones's language, in this kind of case, persons are being asked to inhabit worlds that oppressively constrain their agency, including their epistemic agency, so that it is impossible to fill the request of understanding without simultaneously foreclosing the ability to demonstrate the harm perpetuated by the world that sustains the understanding. In these cases, demonstrating the harm that the requested understanding does can only be done from worlds that actively resist the sense of the world one has been implicitly asked to inhabit.

The piece from which the Williams's example comes is, as its subtitle announces, 'a commentary on the genre of legal writing' (Williams 1992, 44), investigating the relationships between communication, understanding, and agency. Specifically, one of the main insights conveyed by the essay is that the call for neutrality and for 'understanding all sides' in our communication with one another is anything but neutral and can make certain 'sides' of the situation invisible without appearing to do so. While Williams focuses on what can and cannot be conveyed via legal writing, her essay sheds light on the question of what understanding another can *do* with regard to one's possibilities for action and communication generally speaking. The essay invites the reader to think about calls for understanding within the context of at least two infringements on Williams's own agency: (1) being refused entrance to a Benetton clothing store and (2) being thwarted on multiple occasions and in a variety of contexts from her attempts to convey that story. Williams's rendering of the first of these infringements demonstrates how a 'world' can unfairly constrain certain persons' range of possible action while simultaneously making invisible the way in which it does so. Williams's attention to the second of these infringements shows how the call to neutrality and to understand 'all experiences' can reassert the very 'world' analyzed in the first of these infringements.

Williams's detailing of the initial infringement begins with an explanation of the use of buzzer systems to allow entry into stores during the 1980s in New York City and some examples of the reasoning given at that time in support of their use. Of note is that some of the arguments, specifically those that contrast the infringement on agency in being refused entrance to a store with that of murder and assault, were persuasive enough that 'even civil rights organizations backed down' (44). It is important to note here that Williams does not analyze these arguments directly nor does she ask her reader to follow their logic to see if they are cogent. Instead she indicates what those arguments *do* (put those who follow them in a place from which they cannot respond) and then proceeds to offer a way of thinking about how they do it. Demonstrating how the arguments have this effect requires us to see the arguments from a world that explicitly resists the commitments that maintain the sense of those arguments themselves. Williams accomplishes this task by writing from a field of sense that is maintained by anti-racist gestures and assumptions. Williams's own account of the use of buzzers to profile shoppers incorporates details and moves that are actively suppressed and/or deemed insignificant according to the sense of the public debate. This shift has the effect of bringing the background of the public debate into focus thereby allowing her readers to analyze the 'ways of seeing' that maintain the sense of the public debate. Using Lugones's language, Williams's own account asks us to see the debate on buzzers from within a world that is resistant to the one in which that debate took place. This resistant world emphasizes what racial profiling does with attention to the concrete (as opposed to abstracted) agency of the one profiled. It also shows how the world from which the practice of racial profiling is understood 'only as an abstract precaution' (46) is structured by racist assumptions concerning what it means to be Black and by racist omissions concerning the history of race in the United States.

Williams presents to the reader a specific example from which to think about the use of buzzers, her own experience of being refused entry to a store. She begins by indicating why she wanted entrance to a particular store on a particular day: Benetton Clothing store on the Saturday before Christmas to purchase a sweater she saw in the window as a gift for her mother. These details call the reader's attention to Williams as a unique source of valuing, a person whose experiences are imbued with meanings generated from within her own life and relations with others. This attention foregrounds the 'fullness' of Williams's 'public participatory self' (46), as one who has a life from which the significance of our social interactions can be understood. In contrast, these kinds of details are actively prevented from mattering within the world that structures the public debate, as Williams reminds us: within that debate there are people, 'who approve of those who would bar even as they deny that they would bar *me*' (46 emphasis in original). Of course, *every* person barred from entry to a store is a '*me*,' but this point is actively discounted in the public debate by figuring the one profiled as an abstract individual whose defining characteristic is to be either a purchaser or a criminal. Reducing the one profiled to such a narrow description invites the use of crime statistics (actual or imagined) as the most significant aspect to be considered among groups divided according to racialized categories with no attention whatsoever to the history from which those categories came to be. Under such a framework, race matters in racist ways (generalizing action across a group of individuals as though being a member of that group caused members to act in particular ways) while engaging in racist omissions that do not allow race to matter in anti-racist ways (the profiled is imagined to be one who 'happens' to belong to a group devoid of any attention to the historical and institutional inequalities that created the group to begin with).

Actively resisting the move to consider herself primarily in terms of a statistic, Williams details the character of her encounter with the store clerk to whom her own meaningful relations in the world brings her. What is considered a 'mere inconvenience' from the perspective that framed the public debates now comes to light as a fundamental disrespect of another's personhood that leaves open very little possibility of response. The structure of power in the situation literally constrained the range of sensible action available to Williams. In Williams's words, when she was denied access to a store for which access was clearly being given to those whose skin was not racialized Black like her own:

> There was almost nothing I could do... No words, no gestures, no prejudices of my own would make a bit of difference to [the white store clerk]; his refusal to let me into the store...was an outward manifestation of his never having let someone like me into the realm of his reality. He had no compassion, no remorse, no reference to me; and no desire to acknowledge me even at the estranged level of arm's-length transactor. (45)

Within this light, namely a light that makes sense of Williams as a full agent within the context of racism in the U.S., the use of buzzers is neither necessary nor merely inconvenient; rather it is an enactment of power relations with a long tradition that maintains white privilege and dominance. As Sara Ahmed writes, '[being] "stopped"...does not simply stop one from getting somewhere, but changes one's relation to what is "here"' (2006, 160).

Moreover, the details of the clerk himself, a 'white [male] teenager wearing running shoes and feasting on bubble gum' (Williams 1992, 45), reverberate with the arguments of the public debate, cited by Williams just four sentences earlier, that 'it is not all blacks who are barred, just "17-year-old black males wearing running shoes and hooded sweatshirts"' (44). The contrast asks the reader to consider how it is that things like 'wearing running shoes' and 'chewing bubble gum' appear differently when inhabited by white teenage males vs. black teenage males, when worn by those inside New York City boutiques vs. those left standing on the streets outside them. In other words, the contrast asks us to consider not only what kind of worlds make it possible for some to bar entrance to others, but also that make it possible to 'see' this barring as a 'necessary evil' rather than a fundamental infringement on the rights of those barred.

In highlighting the way in which profiling via buzzers sets up an asymmetrical relation between persons, both of whom are capable of being sources of value, Williams reveals the public debate to be grounded in white privilege in ways that reduce Black persons to types of bodies considered nothing more than potential criminals. Given the orientation of the debate, the 'repeated public urgings that blacks understand the buzzer system by putting themselves in the shoes of white storeowners' is a demand that Blacks inhabit a world (that is, operate within a set of patterns and orientations) in which their very own bodies are regarded as 'kinds' of bodies whose movements take on predetermined meanings in ways that white bodies are not subjected to. Such an understanding makes sense only within a world that is maintained by 'whiteness' as an invisible ground within which that which is not white appears as a particular type about which one can make generalizations.[10] As Williams points out, when white men engage in all kinds of acts, most people, 'not only [do] not claim but actively resist [believing] that [they represent] any kind of 'white male' norm' (243). Within a world that is oriented in this way, those whose bodies are identified as Black cannot simply move through the world as those whose bodies remain unmarked (in their whiteness) can, but must anticipate and negotiate within a context that already finds them suspect. Moreover, because these assumptions are built into the sense of the debate, it curtails the ability of people of color to convey this fact from within that debate. In other words, if Williams were to approach the debate directly, understanding it on its own terms, she would have to consider herself as less than a full epistemic agent to begin, as one in the presence of whom it is reasonable to fear for one's life. From within such a world, it is hardly possible to call attention to the outrageousness of being asked to engage from this position. Asking Williams (or any Black American) to understand

[10] See Ahmed (2006) Chap. 3.

the arguments of those who support racial profiling is to ask her to follow (and subsequently maintain) the sense of a context in which her agency, including her epistemic agency, is automatically curtailed for no good reason.

The remainder of Williams's essay details her attempts at creative resistance to the world that structures her agency in such oppressive ways. In each case, she highlights how calls for a 'neutral' presentation of the information implicitly require one to participate in worlds or habits of thought that undermine her own epistemic authority and/or agency in racist ways. For example, in attempting to publish an article about this incident in an issue of a law review journal on inclusion and exclusion, Williams was asked to remove all references to her race from her account because the journal had a 'race neutral' policy that forbade references to one's race. Williams notes that the removal of race from her story made it lose all sense. Why on earth was she barred entrance to a store in New York City on the Saturday afternoon before Christmas? And given that there must have been some bizarre occurrence that would help us make sense of the refusal to allow her to shop, why is she so upset about it? Alternately, if one is to make sense of the story, the account becomes one, 'in which the reader [has] to fill in the gap by assumption, presumption, prejudgment, or prejudice… [one in which the reader is made] to participate in old habits of cultural bias' (Williams 1992, 48). In other words without the information that Williams is Black clearly stated in the story, it becomes a story which can only be understood by a habit of thought in which one fills in the detail that Williams must be Black so that the store clerk's actions can 'make sense.' However, it is precisely this kind of 'making sense' that Williams is working against. Instead, Williams insists on conveying her story on terms in which the sense of the story is not dependent on filling in an assumption about who she is based on what is done to her, but rather in seeing and understanding the store clerk for who *he* is based on *his* actions, namely, racist. This example highlights yet again that the sense of our claims depends on a set of background attitudes, comportments, and habits of thought, or what Lugones would call a 'world,' and that asking another to understand one's claims is in effect asking another to inhabit the world that gives those claims their sense. In this particular case using the assumption that 'black persons are frightening to some people' constrains Williams in a way that the assumption 'a white person who uses his power in a situation to deny access to a Black person is a racist' does not.

In sum, while understanding an argument, position, or viewpoint need not require one to agree with what is understood, it does require one to participate within the world that gives sense to what is to be understood. Continued participation in (or cooptation into) the practices and habits of thought required to understand from within particular worlds sustains the life of those worlds. Because material and semantic relations are infused with power, some worlds are arranged in ways that unfairly constrict some of its members' possibilities for meaningful action and epistemic participation. For that reason, it is wrong, both ethically and epistemically to ask another to understand when the conditions for the possibility of that understanding systematically and asymmetrically constrict her agency.

11.3 Some Caveats

Here I would like to clarify a few things that I am *not* saying. First and foremost, my argument does not imply a condemnation of requests for understanding others who have something to tell us that we simply do not want to hear. There are many cases in which we ought to understand what others have to say even though it is troubling or even damning, for example, when we have hurt another through our actions. Notice, however, that understanding that my actions are unjust need not limit my agency in the way that understanding the white storeowner's fear in Patricia Williams's essay limits the agency of Black subjects. Consider for example that the white storeowner ought to understand how his use of a buzzer system is unjust. An attempt on the part of the white storeowner to understand or follow reasoning that supports the view that his actions are wrong might be difficult or painful for him, since he may like to think of himself as someone who acts rightly. That this is difficult or painful, however, in no way unfairly limits his possibilities for action. He may find himself no longer able to make sense of the thought that he is innocently trying to protect himself and his business by barring certain persons from shopping at his store. But there are many actions he can perform that would meaningfully embody 'protecting my business;' the understanding does not require him to forfeit all possibilities for action that can be recognized in this way. Nor are there significant actions available to others that are absolutely unavailable to him due to a prior categorization as a type of person.

In saying that it is wrong to request others to understand in cases that unfairly limit those others' agency, I am not suggesting that agency is, should, or could be infinitely unlimited. As indicated above, the meaning of my words and actions is not, nor should it be, determined solely by what I think or want them to mean. This is an important point, since often persons in dominant positions say and do things that effectively harm others without expressly thinking that is what they are doing; the fact that they do not consciously intend for their actions to do harm does not take away the harm that those actions do. To wit, the case of wrongfully 'asking another to understand' may not be consciously *intended* to be asking another to limit her agency, but that is what these requests do. What one thinks or wants one's words to mean is not automatically equivalent to what they mean. We often make mistakes and are corrected for those mistakes, and in some cases we are even held accountable for those mistakes. Instead, I am arguing that we should approach others and have the right to expect to be approached by others in ways that enable an equitable range of possibilities for meaningful action. This range is not infinite; nonetheless, it ought not to be systematically asymmetrically bounded.

Lastly, my argument does not necessarily imply that one ought to cut off all engagement with others who proceed in ways that limit one's range of meaningful action. First, it can be quite dangerous and in some cases impossible to refuse to engage with those in relation to whom one is materially vulnerable. Navigating power relations safely may force one to attend to how others understand the world, even when that understanding leaves little or no room for one's own agency.

So I do not condemn those who *do* understand others in ways that shortcut their own agency. My argument concerns only the wrongfulness, both ethical and epistemic, of requesting that type of understanding from others. The ethical wrong consists in asking another to participate in a set of meanings that constitute her in ways that unfairly constrain her agency. The epistemic wrong lies in the preemption of the one who is requested to understand from bringing this constraint on her agency to light. In such cases it is worth noticing that there is something peculiarly epistemically violent about situations where someone is forced or even asked to understand the world in ways that asymmetrically limit her agency.[11]

In spite of this violence, the possibility of simultaneously animating multiple worlds provides resistant opportunities to play on and within those understandings in strategically useful ways. Using Lugones's language, by occupying multiple worlds simultaneously, one can use dominant meanings in oppressive worlds to accomplish vital tasks in worlds of resistance, as, for example, when slave songs were used to navigate the Underground Railroad during the time of slavery in the United States.[12] Alternately, one can in some cases refuse to understand another without refusing to engage with her in ways that hold open possibilities for ethical epistemic relations. Using Lugones's language, refusing to follow the sense of worlds that limit my possibilities is not the same thing as simply disengaging, since multiply present worlds are always 'intertwined materially and semantically.' A refusal to engage from within a world of domination can be an invitation to others to experience that world from within an alternate world that resists domination. In other words, a refusal to engage from within a particular background by way of not understanding can call attention to that background; it can also provide a starting point for re-coordinating our ways of thinking and acting together. Clearly, this kind of refusal might not prove productive when engaging with persons who have no expressed interest or commitment to dismantling worlds of oppression. However, it can be a way of snapping those who do hold such interest out of bad habits of thinking and toward creatively animating new ones. Such refusals are one way to begin the work of dismantling worlds that oppressively construct some inhabitants as partial agents and welcoming knowers to worlds that do not.

[11]This kind of knowledge produces what W.E.B. DuBois referred to as 'double consciousness': 'It is a peculiar sensation, this double-consciousness, this sense of always looking at oneself through the eyes of others, of measuring one's soul by the tape of the world that looks on in amused contempt and pity. One ever feels his twoness, – an American and a Negro; two souls, two thoughts, two unreconciled strivings; two warring ideals in one dark body, whose dogged strength alone keeps it from being torn asunder' (1989, 3).

[12]Allsion Bailey's 'Strategic Ignorance' (2007) details these types of cases. My own suggestion below differs insofar as it provides a way of thinking about the destruction of oppressive worlds, whereas Bailey's suggestion is for considering how resistant subjects can get things done in spite of the continued existence of oppressive worlds.

11.4 Strategic Refusals

Consider Brison's confusion (feigned or unfeigned) when she says, 'I would have thought that a murder attempt plus a sexual assault would require more, not less, of an explanation than a murder attempt by itself' (2001, 3). Brison's refusal to understand why the addition of a sexual assault to a murder attempt would give sense to the latter brings into relief the ground within which others (including ourselves) are able to make sense here. In a world in which a random murder attempt makes no sense at all, but a random sexual assault followed by a murder attempt makes perfect sense, women's agency is constrained in a way that would be unthinkable for men. By calling our attention to this fact, Brison's refusal calls on us to consider how the world we live in allows this situation to continue. It also asks us to think about what it would mean to live in a world in which rape and sexual assault made no sense at all. To think in this way is to consider the conditions for the possibility of female agency and to expose the ways in which common habits of sense making intertwine with material conditions to constrict women's possibilities.

Brison's refusal to understand is not a denial of the actual and real oppressive world in which women's agency is under both material and discursive constraint. Instead, her refusal positions her in relation to that world in a way that refuses that world's terms; in other words, at the moment of 'not understanding' Brison refuses to rely on sexist attitudes or habits of thought as the ground within which she makes sense of things. For that reason her refusal does not deny the reality of sexual assault, but brings into focus the gestures and attitudes that maintain sexual assault as nothing peculiar. This type of refusal foregrounds the violence of those material forms of life that sustain worlds in which sexual assault does make sense while simultaneously reaching toward the possibility of worlds in which it does not.

A strategic refusal to understand does not completely dismantle an oppressive world, since worlds are maintained not only by our comportments and habits of mind, but also by the practices and institutions within which those comportments and habits have a place. For this reason, so long as oppressive institutions and practices exist, the maintenance of oppressive worlds continues. Nonetheless, a strategic refusal to understand can help us to illuminate how those institutions and practices work by bringing them out of the background and to the fore. Moreover, such refusals affirm that oppression is not necessary, but actively maintained by our interactions with one another, even on the most basic level of how we approach the world. Lastly, a refusal to understand can be not only a way of animating resistance, but also of lovingly inviting others to interact in ways that make the conditions for more equitable agency possible. In other words, Brison's refusal to understand invites the reader to consider why and how a person might find sexual assault incomprehensible and to further reflect on how our institutions and practices might be arranged so as to make that incomprehensibility the norm.

Contrasting Brison's refusal to understand with a refusal that fails to open women's agency helps to further define the strategic use of refusing to understand. For example, Kimberlé Crenshaw cites evidence that women jurors are often the *last* to recognize that rape survivors are not responsible for being raped, precisely because doing so reveals female jurors' own vulnerability (1996, 371). Similarly, we could imagine a refusal to understand a murder attempt plus a sexual assault in a way that hints at the possibility that the victim of such an attack did something to provoke it, so as to assure oneself that one is not vulnerable to attack in the way the victim was. In effect this kind of refusal would posit an all too familiar, 'That makes no sense. What did she do to end up in that situation?' Notice, however, that this refusal to understand remains firmly implanted in a world where *sexual assault itself* makes sense. In contrast, Brison's strategic refusal calls into question why sexual assault would make any sense at all, no matter what women do or do not do. When used strategically, refusals to understand highlight the ways in which oppressive worlds constrain agency. Such refusals are *political* insofar as they forward the aim of the feminist project to call into question the conditions of women's oppression. At the same time, such refusals can be *ethically and epistemically productive* insofar as they ask us to move toward new ways of making sense within which we might discover new possibilities for acting in and thinking about the world together.

Meaningful action is something that does not begin and end with the person acting. How others understand the world and make sense of their experiences has an impact on what we are able to do and how we make sense of our own experiences. Moreover, how we all understand and make sense of our experiences is held in place not only by each other, but by the institutions and practices within which we find ourselves living and acting. These institutions and practices are perpetuated by our use of them to understand ourselves and each other. Consequently, the contexts within which we make sense of, or refuse to make sense of, our own and others' reasoning have significant implications for our ethical and epistemic lives.

Acknowledgments Thank you to the women in the Midwest Society for Women in Philosophy for their suggestions and encouragement with regard to this essay. Thanks also to Madelyn Detloff and Heidi Grasswick whose input has greatly improved the clarity of my writing and ideas.

References

Ahmed, Sara. 2006. *Queer phenomenology: Orientations, objects, others*. Durham: Duke University.

Bailey, Alison. 2007. Strategic ignorance. In *Race and epistemologies of ignorance*, ed. Shannon Sullivan and Nancy Tuana, 77–95. New York: State University of New York Press.

Brison, Susan. 2001. *Aftermath: Violence and the remaking of a self*. Princeton: Princeton University Press.

Code, L. 1995. *Rhetorical spaces: Essays on gendered locations*. New York: Routledge.

Crenshaw, Kimberlé. 1996. Mapping the margins: Intersectionality, identity politics, and violence against women of color. In *Critical race theory: The key writings that formed the movement*, ed. Kimberlé Crenshaw et al., 357–383. New York: The New Press.

DuBois, W.E.B. 1989. *The souls of black folk*. New York: Bantam.

Lugones, María. 2003. *Pilgrimages/peregrinajes: Theorizing coalition against multiple oppression*. Lanham: Rowman & Littlefield.

Scheman, Naomi. 1996. Forms of life: Mapping the rough ground. In *The Cambridge companion to Wittgenstein*, ed. Hans Sluga and David Stern, 383–410. Cambridge: Cambridge University Press.

Turner, Victor. 1974. *Dramas, fields, and metaphors*. Ithaca: Cornell University Press.

Wendell, Susan. 1996. *The rejected body: Feminist philosophical reflections on disability*. New York: Routledge.

Williams, Patricia. 1992. *The alchemy of race and rights*. Cambridge: Harvard University Press.

Wittgenstein, Ludwig. 1965. *The blue and brown books*. New York: Harper.

Wittgenstein, Ludwig. 1969. *On certainty*. New York: Basil Blackwell.

Chapter 12
Liberatory Epistemology and the Sharing of Knowledge: Querying the Norms

Heidi E. Grasswick

Abstract Feminist epistemology is a form of liberatory epistemology, and as such is focused on the role of knowledge production in the generation and maintenance of oppression and the effects of oppression on the production of and possibilities for knowledge. I argue that as part of this project, feminists need to take up the question: with whom do we share knowledge, and with whom should we share knowledge? To answer this, we must examine how knowledge-sharing norms function, particularly in contexts of oppression. Knowledge-sharing norms capture the expectations within a community or relationship concerning what knowledge ought to be voiced and thus shared across particular parties, and what knowledge either ought not, or need not be shared. I argue that, surprisingly, from the perspective of a liberatory epistemology, we cannot assume that increased knowledge sharing is always a good thing, but rather must assess the function and value of knowledge sharing and particular knowledge-sharing norms within localized contexts. Nevertheless, criteria for such assessments can be outlined, in accordance with the goals of a liberatory epistemology.

Keywords Feminist epistemology • Knowledge sharing • Liberatory epistemology • Secrecy

12.1 The Circulation of Knowledge: A Feminist Issue

In many poor villages in Bangladesh, lack of telecommunications has left remote villagers at the mercy of corrupt middlemen who come to their villages and set exceptionally low prices for crops and products, contributing to a cycle of poverty for the locals. The Village Phone program, one of the very successful microcredit programs (programs that offer small loans to supply capital for a business opportunity) makes

H.E. Grasswick (✉)
Department of Philosophy, Middlebury College, Middlebury, VT 05753, USA
e-mail: grasswick@middlebury.edu

H.E. Grasswick (ed.), *Feminist Epistemology and Philosophy of Science: Power in Knowledge*, DOI 10.1007/978-1-4020-6835-5_12,

cell phones available to poor village women, allowing them to operate small village businesses of cell phone use where there was no phone at all before.[1] With cell phone access, villagers can now protect themselves from exploitive buyers by calling to check current market prices for their products prior to a sale. The ability to access this type of information has brought significant economic benefits to the villagers.

This example offers a case where access to a certain kind of information or knowledge can make a significant difference to one's livelihood. It illustrates one of the key reasons feminists have been interested in the circulation and distribution of knowledge: they have recognized the importance of access to knowledge for one's quality of life. If access to important forms of knowledge is difficult for members of marginalized groups, life will in turn be more difficult for these groups. Patterns of differential access to knowledge that form along the lines of gender and race (to give just two examples) can disadvantage such groups, and they constitute a symptom of and contributor to oppression. This is also why literacy is such an important benchmark for the well-being of marginalized groups. Literacy is a crucial tool in knowledge accessibility.[2]

But the interest of many feminist epistemologists in the circulation of knowledge also runs deeper than the material and socio-political implications of access to knowledge. Quite obviously, questions of knowledge circulation, distribution and use are all *epistemic* in nature, simply in the sense that they are questions directly concerned with knowledge. But for feminists who have developed social approaches to epistemology, questions about the circulation and distribution of knowledge are also deeply *epistemological*, in a traditional philosophical sense, because these questions direct our attention to how we know and in some cases reveal whether or not we know. For example, Helen Longino's critical contextual empiricism sets out critical exchange as necessary for justification and consequently the production of knowledge, while theorists such as Lorraine Code (1991, 1995) and Miranda Fricker (2007) have focused on the diminishment of women's epistemic agency that results when women lack access to certain forms of knowledge about themselves in situations of oppression and marginalization. Within a socially-framed epistemology that recognizes our epistemic dependence on others, including the importance of testimony as a source of knowledge, questions concerning the circulation, distribution and use of knowledge are all central, bearing on analyses of how and what we know, both communally and individually.

Feminists have repeatedly argued for the centrality of the question 'knowledge for whom?' in epistemology (Code 1991; Harding 1991). In this paper I argue that feminists also need to attend to a different, but related question: 'with whom do we share knowledge, and with whom *should* we share knowledge?' Norms of knowledge sharing form an important feature of epistemic communities and their practices,

[1]One of the major sponsors of such microcredit programs has been the Grameen bank. http://www.grameen-info.org/grameen/gtelecom/.

[2]By noting the importance of literacy for knowledge accessibility, I do not mean to deny that there are also important forms of knowing which do not require literacy.

yet they have thus far received scant attention. Whether or not particular forms of knowledge are easy or difficult to access from others, and whether such knowledge is shared freely, makes a difference to what other kinds of knowledge can be generated in a community as well as what social goals can be achieved. For example, union building is difficult in a workplace where the cultural norm is for no one to share salary information.[3] Under such circumstances, salary discrepancies cannot be identified, knowledge of the injustices captured by such discrepancies cannot be generated, and such knowledge cannot then be used to motivate workers to work together for a common cause of improving the situation. Identifying such cultural norms of knowledge sharing and withholding is crucial to understanding some of the many obstacles to overcoming oppression, including the difficulties of generating specific kinds of knowledge necessary to overcome oppression. If norms of knowledge sharing make a difference to what other kinds of knowledge can be generated in a community, then feminists' concerns about access to knowledge, the mechanisms through which it circulates, and the social forces that construct knowers' credibility when viewed within a socially-framed epistemology, will not be simply ethical questions. Rather, to understand how knowledge is produced, and how we should engage to know well, we will need to understand these circulatory functions of knowledge. Examining how knowledge sharing norms function and change will be an important part of a liberatory epistemology, that is an epistemology interested in generating forms of knowledge that explain the how oppression operates, and envision possibilities for social change. However, I also argue that surprisingly, from the perspective of a liberatory epistemology, we cannot assume that increased knowledge sharing is always a good thing, but rather must assess the function and value of knowledge sharing and particular knowledge-sharing norms within localized contexts. Nevertheless, criteria for such assessments can be outlined, in accordance with the goals of a liberatory epistemology.

12.2 Feminist Epistemology as Liberatory Epistemology

As noted by Alcoff and Potter as early as 1993, feminist epistemology can no longer be conceived as primarily about women. Rather, developments in feminist epistemology have emphasized the need to understand gender as 'a component of complex interrelationships with other systems of identification and hierarchy' such as class, race, sexuality, culture and age (Alcoff and Potter 1993, 3). As they note,

---

[3] Section 7 of the National Labor Relations Act grants employees the right 'to self-organization, to form, join, or assist labor organizations, to bargain collectively through representatives of their own choosing and to engage in other concerted activities for the purpose of collective bargaining or other mutual aid or protection'(*Basic Guide to the National Labor Relations Act* 1997). This right prevents an employer from banning the sharing of salary information amongst workers. I thank Dave Saldana for drawing my attention to these connections with and details of the Act.

'because gender as an abstract universal is not a useful analytical category and because research has revealed a plethora of oppressions at work in productions of knowledge, feminist epistemology is emerging as a research program with multiple dimensions. And feminist epistemology should not be taken as involving a commitment to gender as the primary axis of oppression, in any sense of "primary," or positing that gender is a theoretical variable separable from other axes of oppression and susceptible to a unique analysis' (Alcoff and Potter 1993, 3–4).

As these excerpts from Alcoff and Potter reveal, a constant concern with oppression and its epistemological dimensions is visible throughout the many historical shifts in feminist epistemology. Feminist epistemology is perhaps best described as a *liberatory epistemology* (Scheman 2001, 26; Tuana 2001, 18). I define a liberatory epistemology as an epistemology that seeks to understand the connections between oppression and knowledge, including the connections between liberatory social change and knowledge. More specifically, a liberatory epistemology focuses on the role of knowledge production in the generation and maintenance of oppression as well as the effects of oppression on the production of and possibilities for knowledge. A liberatory epistemology seeks to develop epistemological tools – concepts, theories, and understandings of knowledge – that will help generate the kind of knowledge required to bring about positive (liberatory) social change.[4]

As such, a liberatory epistemology will focus on contexts of oppression and have a particular interest in analyzing the kinds of knowledge and knowledge production practices directly implicated in both the maintenance of and resistance to oppression. A liberatory epistemology will direct its energies toward certain forms of knowledge and particular circumstances of knowledge production pertinent to conditions of oppression. At the same time, a liberatory epistemology will likely carry ramifications for how we understand various areas of knowing beyond contexts of oppression as well. It is unlikely to serve as simply a limited add-on to standard epistemological accounts. This is because a liberatory epistemology will demand that any adequate epistemological theorizing, whatever its focus, must at least be conceptually capable of making visible the epistemological dimensions of oppression. Liberatory epistemology need not claim that every realm of knowledge has connections with oppression (perhaps the knowledge I have that I am currently sitting on a chair does not), but it will challenge the adequacy of any epistemology that is incapable of making such relations visible when and where they exist.[5]

[4] My use of the term liberatory epistemology should not be understood as having any connection to the liberation theology movement. Nor should it be understood in any sense of one group working to secure the liberation of another. Rather, many feminist epistemologists, including myself, have adopted the term 'liberatory epistemology' simply to make clear first that feminist epistemologies are concerned with oppression in *all* its guises (not only those pertaining to 'women') and its links to knowledge, and second that these epistemologies aim towards positive social change.

[5] In a different context, Helen Longino makes a similar point regarding how feminists ought to select from amongst theoretical virtues guiding inquiry. She claims that a bottom line requirement for feminist theoretical virtues is that they be capable of revealing gender. The choice of feminist theoretical virtues means that 'inquiry guided by these virtues is more likely to reveal it [gender] or less likely to preserve its invisibility than the traditional virtues' (1997, 50).

Numerous feminist epistemologists have critiqued the adequacy of mainstream approaches to epistemology that set out *general* conditions of knowing without specifically addressing contexts of oppression on exactly such grounds of their conceptual inability to make visible the epistemological dimensions of oppression (Code 1991; Harding 1991; Potter 1993).

In this paper I focus on just one element of knowledge production and distribution that I take to be important within a liberatory epistemology: knowledge-sharing norms. Not only is it important to identify and understand the functioning of knowledge-sharing norms if we are to fully understand the relations between oppression, knowledge and social change, but within particular contexts we can assess the knowledge-sharing norms in operation according to the goals of a liberatory epistemology. I argue that surprisingly, feminists should not assume that more knowledge sharing always serves liberatory goals. Rather a more nuanced and contextual assessment of knowledge sharing and its norms is required, especially for a liberatory epistemology. Before such an argument can be made, however, a clearer explication of what knowledge-sharing norms are and how they function is needed.

12.3 Knowledge-Sharing Norms (KSNs): Definition and Function

Knowledge-sharing norms (KSNs) are those expectations within a community or relationship concerning what knowledge ought to be voiced and thus shared across particular parties, and what knowledge either ought not, or need not be shared. They describe how and to what extent we hold each other *accountable* for both sharing and withholding knowledge.

For any piece of knowledge I have, relative to a given audience and context, it can be placed on a continuum with the one pole representing knowledge that definitely ought not be shared with this particular audience, and the other pole representing knowledge that definitely ought be shared with this audience. Police need to read one's Miranda rights to a person when arresting them, and doctors must (in general) not reveal the content of conversations with their patients to outside parties. Somewhere in the middle, we place knowledge that can be shared or withheld, with no sense of obligation either way. I am free to tell my employer what I did on my day off, but I am under no obligation to do so. Understanding the particular norms of knowledge sharing of a community helps us identify where on this continuum the knowledge in question lies in any particular set of circumstances. Though I have termed them knowledge-*sharing* norms, it is important to note that KSNs include the norms of appropriate withholding of knowledge. They capture the scope and degree of expectations of knowledge sharing and withholding.

A caveat is also required. Throughout my discussions I refer to knowledge sharing, but more accurately, I mean to capture the norms of sharing and withholding that which *we think we know*. The appropriateness of the norms of knowledge sharing cannot turn on whether or not we are mistaken in what we take to be our knowledge.

There will, of course, be times when we are mistaken: we take something to be the case and later realize that it isn't so – we had held what turns out to have been a false belief. Once we recognize our error, another set of norms will come into play: norms of spreading or not spreading (known) falsehoods, and norms of actively correcting falsehoods in the public realm.[6] But as long as we have reason to think we have knowledge, the issues at hand will be whether or not that 'knowledge' ought or ought not to be shared.[7]

KSNs are part of the everyday functioning of our social epistemic lives. They are sometimes formalized, but more are often not. Formalized knowledge-sharing norms include laws of disclosure, such as laws requiring that a house seller reveal to the buyer known problems with the property, or controversial HIV/AIDS disclosure laws. Of course, because KSNs include norms concerning the withholding of knowledge, they also include formalized laws related to privacy issues for individuals as well as the rights of employers to maintain corporate or government secrecy. Many formalized KSNs lack legal status but are formalized at the policy level, such as corporate gag orders against speaking with the press, or the demands of health insurance companies that medical information be revealed to them prior to offering coverage. Such policies and the KSNs underwriting them are enforced by actions such as employment dismissals and insurance coverage denials.

But in spite of the ease with which one can generate examples of formalized KSNs, far more common are the wealth of unformalized KSNs with which we engage in our day to day interactions. Good social skills require an implicit understanding of the norms in operation in localized contexts, so that one doesn't develop a reputation as a gossip by detailing the goings-on of one's neighbors on one's daily stop at the local post-office, or be considered ungenerous for being unwilling to share a recipe with one's guests. As social norms, KSNs form part of highly localized cultures.

For the significant portion of our KSNs that are unformalized, it is much more difficult to identify them and we often recognize these norms only when we notice a breach, or when it becomes clear that a particular norm is failing to serve our purposes well (epistemic or otherwise). Where there is no identifiable problem, we simply go on practicing with these norms, mostly unaware of them.

When one recognizes a breach in these norms, one senses an affront, a certain kind of offense. Consider Anne, who is going through relationship difficulties and both confides in her friend Karen and turns to Karen to try to work out what needs to be done to resolve the issues in Anne's primary relationship. Karen takes their

---

[6]Recognizing the falsehood of a claim isn't the only way we might be led to reject a claim of 'knowledge.' We might take a belief to be well supported, at least well enough to call it knowledge, and then later be faced with countervailing evidence that also needs to be accounted for. Faced with such countervailing evidence, our confidence in the original belief and its overall support may be shaken, and we may no longer be willing to call it knowledge or claim 'I know this', even if we still maintain some doxastic commitment to the claim and haven't quite given it up.

[7]There will also be additional norms of responsible knowing that come into play: how sure do we expect people to be when they claim knowledge?

friendship seriously and is committed to putting in the time and effort to help Anne sort through her difficulties. After a significant period of time, however, Karen discovers that Anne has been engaged in an affair leading up to and throughout these difficult times, but in spite of turning to her friend Karen for help in sorting through her long-term relationship with her partner, has not revealed the affair to Karen. Karen feels an affront, as though Anne has mistreated her. One interpretation of Karen's sense of affront is simply her disappointment over the fact that her friendship with Anne is not what Karen thought it was: she expected Anne would share such knowledge with her. But I take it that there is more to Karen's sense of an affront than this. At least part of the affront is grounded in Karen's sense that she has committed time and energy to what she took to be a joint epistemic project – trying to figure out the challenges and solutions to Anne's difficulties in her relationship with her partner. Karen takes Anne's ongoing affair to be a significant piece of information, crucially relevant to the conversations they are engaged in, and Anne's secret (from Karen) has stymied those efforts.

Anne's breach may well affect their ongoing friendship and its epistemic dimensions. Karen may continue to engage with Anne as a friend, but may not trust her to the same degree. She may recognize that at least on certain personal matters, Anne may be an untrustworthy knower – that is, untrustworthy in her ability to recognize what knowledge is significant enough to the joint epistemic project that it ought to be shared with her friend. Or, if we suppose that Anne did recognize the significance of the fact of her affair to her discussions with Karen, we might interpret her as untrustworthy in her character – lacking the strength of will to share this potential embarrassing information with her friend.[8] In either case, she has proved herself to be an untrustworthy partner in this particular context of knowledge-seeking, and this affects Karen's potential for succeeding in joint knowing projects with her. Anne may be able to offer her friend explanations for her behavior and reasons which dissipate the affront; perhaps she had an agreement with her partner that should one of them ever have an affair, they would tell each other first, or perhaps she felt the need to protect the person with whom she was having the affair.[9] But such explanations would be required in order to alleviate the affront, and a certain amount of healing and repair to the friendship may still be required, even in the presence of such explanations.

The case of Karen and Anne draws attention to the implications of such breaches of knowledge-sharing norms within close relationships. But breaches of knowledge-sharing norms also occur in more public settings, and within larger communities. Again, such breaches enable us to see these norms clearly when perhaps we had been oblivious to them before. For example, in cases of whistleblowing, someone from within an organization or institution calls public attention to some wrongdoing occurring within that organization, believing the revelation of the wrongdoing to be in the public interest. In many cases, the wrongdoing revealed involves the

[8] See Daukas (this volume) for further discussion of trustworthy knowers in joint epistemic projects.

[9] I thank Ann Garry and Carla Fehr for offering these plausible explanations.

organization (or key members of the organization) failing to share knowledge that the whistleblower believes ought to have been shared or communicated to those outside the organization (either a larger community with an interest in the organization's knowledge, or the public at large). The knowledge-sharing whistleblower makes public knowledge that someone has failed to share – knowledge that has been hidden from either the public or a relevant stakeholder (Grasswick 2010). So for example, in the early 1990s, Jeffray Wigand, a former chief executive in the tobacco industry made headlines when on the television show *60 Minutes* he claimed that big tobacco knew (that is, had scientific results to show) that cigarettes were addictive (Johnson 2003).[10] In another case that Lorraine Code (2006) discusses extensively, Dr. Nancy Olivieri, a medical researcher at Toronto's Hospital for Sick Children became concerned about the risks of a therapeutic drug she was researching and, in violation of her confidentiality agreement with the drug company sponsors of the research, published her findings and insisted on revising the consent forms for the patients involved in the drug trials.

Moments of attention granted to whistleblowers constitute moments for public assessment with respect to knowledge-sharing norms: does the public accept the whistleblower's claim that such a piece of knowledge ought to have been shared? Does the public accept that the whistleblower was correct in their judgment that the knowledge needed to be shared? Whistleblowers are rarely successful, and they are not always received (by the public) in a positive light. But when knowledge-sharing whistleblowers are successful, it is because it is recognized that a violation of the norms of knowledge sharing has occurred.[11] Their actions offer moments of recognition of some of these important norms of knowledge sharing.

12.4 Variation in Knowledge-Sharing Norms

Knowledge-sharing norms vary immensely across communities and contexts. This is to be expected, since a particular set of epistemic goals embraced by a particular community will in part define the appropriateness of particular knowledge-sharing norms. For example, in the context of take-home exams, where norms of knowledge sharing will be formalized, students are expected not to share answers with each other or discuss the contents of the exam. Yet in another context, students may be given explicit instructions to work together on practice problems or other assignments, freely discussing and sharing their understandings of the material. Quite different pedagogical purposes drive these two exercises, and though they potentially occur within the same educational community, they are localized according to the goals of the particular epistemic project at hand.

---

[10]Wigand's story was popularized in the 1999 movie *The Insider*.

[11]I discuss cases of whistleblowers and their role in revealing breaches of trust between scientific and lay communities extensively in 'Scientific and Lay Communities: Earning Epistemic Trust through Knowledge Sharing' (Grasswick 2010).

Additionally, many of our knowledge-seeking activities are deeply intertwined in our other social practices and activities, and non-epistemic goals are mixed with epistemic goals in the development and maintenance of particular KSNs. For example, in a professional setting, it may be considered more acceptable to keep knowledge of one's personal life private, whereas in a social setting one might be expected to be more forthcoming in light of certain social goals of the activities. Cultural features of particular communities may help shape such norms. In some cases the non-epistemic reasons for the norms may bear more weight than the epistemic. When I first started teaching at a small residential liberal arts college in a rural area, it was made clear to me that I was *expected* to share my home phone number with my students so that they would have a means of contacting me anytime. Colleagues at larger institutions were shocked when I described this culture of faculty accessibility. At their large urban institutions, the norm was not to share such information. As many at large institutions viewed the situation, it was not just an issue of privacy, but also an issue of security (particularly for female faculty), where easy access to faculty might lead to harassment. The culture of my institution suggested that security was a lower risk in the context of our small community, and that it was reasonable to trade off that small risk for the positive effects of faculty accessibility – of fostering a close parenting-type relationship with our students, *regardless* of whether they ever did call us at home. The student-teacher relationship was taken to be in some part shaped by their understanding that they *could* reach us at home. (In fact, I rarely received phone calls at home except in emergencies). Here, although the result was a form of greater knowledge sharing than at institutions with more restrictions on the sharing of personal information such as phone numbers, it was motivated less by an epistemic goal (that of increasing the students' knowledge of their professors' lives) and more by a social goal of fostering a particular kind of institutional culture and faculty-student relationship.[12]

12.5 Assessing Knowledge-Sharing Norms: Knowledge Sharing as an Intrinsic Good

Anita Allen reflects

> The web of accountability relationships is both flexible and sticky. ...The web is sticky in the sense that socially determined and reinforced expectations impel us. Expectations impel us, for example, to tell our mothers certain things, to explain certain things to our friends, and to justify much to our employers. The web is flexible in the sense that we have a good deal of freedom to stretch and mold these connections to suit individual taste (Allen 2003, 197–198).

[12]Of course, because it is an educational setting, there are further epistemic implications of fostering this particular institutional culture. Presumably one argument for developing an institutional culture of close faculty-student relations is that the students will learn better within such a community.

Allen notes the flexibility within given accountability relationships at an individual level. The social norms of knowledge sharing are also flexible and can change. For example, positive public reaction to knowledge-sharing whistleblowers can sometimes lead to increased expectations of knowledge sharing from corporations. Though I have described knowledge-sharing norms as features of the social practices of knowing that have pull on us, ranging widely across different contexts and communities, we can reflect on these norms, and evaluate them according to how well they serve our epistemic and our related practical needs. Some will work well, and others will not. Such assessment will be the first step in our ability to envision knowledge-sharing norms embedded in social practices that are both epistemically fruitful and liberatory.

Thus far, I have described KSNs as *sociological* norms, core components of our social practices. We identify, understand and describe them by observing what people do and what people expect of each other. But a critical assessment of such sociological norms engages us in normative evaluation (not just description), considering how well they function in serving our goals. As part of a liberatory epistemology, I suggest that such an evaluation of KSNs should attend to both the epistemic and ethico-political outcomes of their functioning, noting that these won't always or even commonly be separable. In what follows, I focus on the epistemic ramifications of KSNs – that is, the knowledge produced or hindered as a result of KSNs – considering them in relation to the epistemological goals of liberatory epistemology. Do particular knowledge-sharing norms operating in specific contexts help or hinder liberatory goals through the knowledge they produce? Can the fostering of certain KSNs aid some of our anti-oppressive strategies by producing important knowledge?

As a first attempt at evaluating KSNs, it might seem reasonable to adopt a very simple principle that increased knowledge sharing has value in itself, and in an ideal world we would maximize our knowledge sharing. From this principle, we could surmise that KSNs that are more demanding with respect to the sharing of knowledge would be *prima facie* preferable. At least two different arguments can be given for this principle of increased knowledge sharing. The first argument is not specific to a liberatory epistemology, though the second is. First, one might be tempted to think that because knowledge is a good in itself (a position common amongst epistemologists), sharing knowledge is also a good in itself. After all, the point of seeking knowledge is to *have* it (and perhaps then use it) and assuming knowledge is a good, sharing knowledge increases the number of people who have the good. According to this general principle, KSNs that capture high expectations of sharing knowledge amongst people and thus foster a high level of knowledge sharing would be judged better than more restrictive KSNs. On such a view, perhaps time pressures and non-epistemic or practical goals limit our abilities to share all of our knowledge with each other in particular cases, but in an ideal world, we would share everything, giving us the most knowledge to work with in our respective epistemic endeavors. In cases where time is not a limiting factor, and specific practical goals do not intervene, we should work toward evolving cultures that increase knowledge sharing amongst us.

But a second argument for the principle of increased knowledge sharing makes the case even stronger by taking up the question explicitly from the perspective of a liberatory epistemology. For those interested in overturning forces of oppression, it is important to note that hiding knowledge from others is often a sign of power as well as an effective use of power. For example, the privacy that comes with the privilege of wealth contributes to the ability of an abusive husband to hide the domestic violence in his household. The power of the Ted Turners and the Conrad Blacks of the world to wield influence over the media permits a very effective measure of public and political control, by shaping and limiting what knowledge reaches public forums. Knowledge sharing and the free exchange of ideas are understood to be both core features of democracy and important elements in the building of democracy. It is a small move from recognizing the important political implications of controlling access to knowledge to claiming that increasing the norms and expectations of sharing knowledge is hands-down a positive and liberatory move across the board. If knowledge brings power, then sharing knowledge is likely to result in a sharing of or wider dispersal of power. More specifically, if the oppressed and the marginalized need access to knowledge to undertake social change, and if it is harder to maintain oppression within open and transparent cultures of knowledge production, norms of knowledge sharing that demand a high level of transparency will score high according to a liberatory epistemology.

Both these arguments suggest that increased knowledge seeking is *prima facie* good and preferable, though of course this claim and the arguments for it are not incompatible with finding exceptions. Just as many epistemologists claim knowledge as a general good while acknowledging that in certain circumstances we may have an interest in not knowing something (to protect someone from psychological hurt for example by shielding them from painful knowledge), the claim that knowledge sharing is a *prima facie* good allows for extenuating circumstances to override that good. Similarly we could accept the argument that increased knowledge sharing is generally liberatory, while acknowledging that there may be particular sets of circumstances where this principle fails to hold.

12.6 Assessing Knowledge-Sharing Norms: Rejecting Knowledge Sharing as an Intrinsic Good

Clearly these arguments for the *prima facie* value of increased knowledge sharing, and overall knowledge sharing as an ideal, at first appear quite attractive for liberatory epistemologists. Taking them seriously would place the burden of proof on those who argue for the benefits of restricting knowledge sharing in particular cases. This would offer some degree of protection against justifications of oppressive practices that involve the withholding of knowledge. For example, given the principle of the *prima facie* value of knowledge sharing, we need to be convinced that confidentiality agreements in employment situations that keep knowledge from the public are warranted, rather than accepting them without discussion. The burden of proof would lie on the side of restricting knowledge sharing.

Unfortunately, tempting as this view may be for liberatory epistemologists, as I show below, it burdens us with an unrealistic and confused view of the importance and role of knowledge sharing, carrying with it problematic ramifications for oppressive situations. I argue that liberatory epistemologists should not take knowledge sharing across the board to be the ideal, nor adopt a principle of the *prima facie* value of increased knowledge sharing, for several interrelated reasons.

First, although the principle of knowledge sharing as a *prima facie* good is not incompatible with finding exceptions, in order for the principle to stand up, the exceptions found must be infrequent enough to constitute genuine exceptions rather than the norm. If the exceptions are frequent, the principle will not be very useful as a guide to assessing knowledge-sharing norms. Further, the more exceptions there are, the more likely it is that the principle of the *prima facie* value of increased knowledge sharing is just plain wrong.[13] As I show below, it is in fact very easy to generate abundant examples of contexts where increased knowledge sharing is problematic, either given general epistemic goals, or the more specific goals of liberatory epistemology.

Key to recognizing the problem with asserting the *prima facie* value of knowledge sharing is that increased knowledge sharing can frequently inhibit the production of certain forms of knowledge. So there can be *epistemological* reasons for restricting knowledge sharing. It is not just practical and non-epistemic goals that intervene to override the value of knowledge sharing (though these will play an important role too). If this is the case, the original argument of moving from the good of knowledge to the (obvious) good of knowledge sharing, does not hold. Sometimes, knowledge sharing and the production of further knowledge will operate as *conflicting* values.

Double-blind studies are perhaps the quintessential example of a context where the withholding of knowledge (in this case, not revealing to subjects or researchers who belongs to the control group and who the test group) is fundamental to the production of a certain kind of scientific knowledge – determining the effectiveness of whatever is being tested. Knowledge of the effectiveness of the therapy cannot be ascertained, or cannot be ascertained as well, in the absence of double-blind studies. Thus, limiting access to some kinds of knowledge can actually *foster* the production of other kinds of knowledge. Similarly, the academic practice of anonymous review is designed to improve objective assessment of the quality of academic work, generating a kind of objective knowledge about that work, by withholding information about authors and reviewers. These practices are striking because they are employed in science, a pursuit which has long taken knowledge sharing and

[13] Although I do not discuss it here, my reasoning has similarities with naturalized approaches to epistemology, suggesting that epistemological principles need to be derived from our actual practices rather than put forth as abstract and ideal principles without concern for whether or not they can be instantiated in our practices.

the free exchange of ideas to be central to its success.[14] For example, the public accessibility of scientific results is an important requirement for the replicability of results, and replicability is well-recognized as a core feature of science (Harding 2000, 125). Yet even in science, there are common restrictions on the sharing of certain kinds of knowledge, specifically for the purpose of improving the production of certain other kinds of knowledge.

From the perspective of a liberatory epistemology, withholding knowledge can also often serve liberatory social and political goals directly, and norms of increased knowledge-seeking can easily work against liberatory goals. In the realm of science, the practices of anonymous review mentioned above that involve withholding information about the authors can be especially important for women and other underrepresented groups. For example, Virginia Valian argues that 'the gender schemas that we all share result in our overrating men and underrating women in professional settings' (Valian 2005, 198). Supported by empirical data, Valian's claims about the widespread and unintentional nature of these gender schemas suggest that practices such as anonymous review are absolutely necessary in order to ensure an unbiased reading of the quality of women's academic work and their ability to advance their careers by being given appropriate credit (through publication) in their fields.[15] Additionally, invoking norms of increased knowledge sharing generally decreases spheres of privacy, and this can work directly against liberatory social and political goals. In a country such as the United States with privatized health care, expectations and requirements of turning over information about pre-existing conditions to insurance providers can lead to further discrimination and poorer health care for those who are already challenged by medical conditions. In the post 9–11 age, laws introduced such as the Patriot Act giving the United States government greater access to wire-tapping and knowledge about the private lives of individuals threaten the well-being of many marginalized groups of people such as recent immigrants.[16] Taking increased knowledge sharing to be a *prima facie* good is an abstract principle that misses entirely any understanding of how sharing knowledge can make those most vulnerable in society even more vulnerable. The same premise that drove the liberatory argument for accepting the principle of increased knowledge sharing – that withholding knowledge from others can give one power – suggests that in the case of the oppressed, withholding knowledge from their oppressors might well either give them more power, or at least make them less vulnerable to the will and interests of those in power. Feminists and race theorists have frequently appealed to such reasoning in the justification of oppositional secrets, where marginalized or oppressed groups keep secrets from

[14]In the case of anonymous review, of course, the practice is employed across many academic disciplines, not just science.

[15]I thank Phyllis Rooney for making this connection to Valian's work.

[16]Regardless of the security interests of the United States' population which such laws may or may not serve, these conditions are certainly not conducive to many other liberatory goals of segments of the population who are affected by the laws.

their oppressors as acts of resistance and survival. Catherine Hundleby cites examples such as the Underground Railway, the location of women's shelters and lesbians passing as straight women as cases where survival can be directly tied to the keeping of oppositional secrets (Hundleby 2005).[17] From the perspective of the oppressed then, withholding knowledge is frequently an important strategy.

So far I have offered first, examples of epistemic reasons for restricting knowledge sharing in science that do not appeal specifically to a liberatory epistemology, and second, examples concerning the value of privacy that directly invoke liberatory social and political goals, without reference to epistemic goals. But there is also an argument to be made for the restriction of knowledge sharing specifically on epistemic grounds within the framework of a liberatory epistemology. Earlier, I stated that a liberatory epistemology is particularly interested in specific forms of knowledge: those necessary to understand and overcome oppression. According to a liberatory epistemology, more restrictive norms of knowledge sharing are likely preferable in cases where increases in knowledge sharing would work against the production of knowledge needed to understand or overcome oppression. In the context of oppression, such instances might be widespread. Alison Bailey discusses how the oppressed can use what she calls 'strategic ignorance' as 'a way of expediently working with a dominant group's tendency to see wrongly' (Bailey 2007, 88). For example, the oppressed might 'play dumb' (not sharing their knowledge) conforming to the dominant group's expectations, in order to gain more information which could be helpful to both survival and resistance. Consciousness raising, a key epistemological tool for feminists, offers another example. By coming together in consciousness-raising groups, and sharing with each other experiences of discrimination, sexism and marginalization, women have been able to come to understand those experiences as symptomatic of oppression, rather than as individual or personal problems. They come to realize both the systematic nature of the oppression, and its injustice. But importantly, it is a mistake to interpret the value of consciousness raising as wholeheartedly endorsing the sharing of knowledge. While the sharing of experiences is the form of knowledge sharing indicative of the consciousness-raising process, it is also crucial that this sharing take place in a safe environment, by restricting the group to those who have had similar experiences. Consciousness-raising groups for women would hardly be very successful if men were present in the room, particularly those viewed as perpetrators of women's oppression. While women often come to proclaim publicly their understandings of oppression later, to actually produce the understanding in the first place requires a trusting environment and thus a limited audience or community.[18] The same reasoning applies to caucus groups,

[17] Hundleby acknowledges the direct political value of such oppositional secrets, though what she is interested in exploring is the epistemological justification for holding them.

[18] As a result of new insights derived from consciousness-raising groups, 'speak-outs' were often organized by feminists to publicly break the silence surrounding the oppression of women. Such stories are recounted in Susan Brownmiller's *In Our Time*: *Memoir of a Revolution* (1990), as cited in Miranda Fricker's *Epistemic Injustice* (2007, 150).

such as women of color caucuses within feminist groups, and junior faculty meetings as distinct from all-faculty meetings.

Consciousness-raising and caucus groups reveal a second reason for a rejection of the principle of increased knowledge sharing as a *prima facie* good. Not only are the exceptions to the principle quite frequent when we start looking, but cases of consciousness-raising and caucus groups remind us that knowledge sharing always occurs within a particular context, and is directed at a *particular* audience. Part of the value of the knowledge sharing comes in the choices we make concerning with whom and when we share. This explains in part why the exceptions to the principle of increased knowledge sharing are so frequent: we never share knowledge of everything with everyone, and it would be an odd idea to think of doing so. When there is a good to be had through knowledge sharing, it is through knowledge sharing with a particular individual or group of people. We build social and epistemic relations by sharing with some people and forming and maintaining epistemic communities through that knowledge sharing. As Cynthia Townley notes, 'We count an epistemic colleague as one who will generally be trustworthy and discreet, who roughly shares our norms of disclosure and revelation, or whose deviations can be challenged and reviewed' (Townley 2003, 109).[19] If the value of knowledge sharing comes from the development of specific epistemic relations in which we are counting on our partners in epistemic pursuits to be discreet with others, then a principle of the *prima facie* good of increased knowledge sharing cannot be right.

A third reason to reject a general principle of the *prima facie* good of increased knowledge sharing comes from the recognition that limiting access to knowledge is frequently necessary in order to achieve our specific epistemic goals within a particular context. There is simply too much knowledge, and sharing it all can get in the way of not only our practical goals, but also our epistemic goals. This is in part a result of the limits and design of our cognitive capacities. A politician expects her staff to provide her with briefing notes, not the entire body of research they have uncovered on the topic on which she will be interviewed. In some cases, strategies of swamping people with information are used to make sound reflective understanding more difficult to achieve. For example, lawyers for a corporation engaged in a battle over an environmentally questionable development project may employ a strategy of swamping the courts with hundreds of pages of environmental impact assessments, not all of them high quality, with the intent of making it impossible to sift through and come to a reasonable judgment on the matter. When it comes to being on the receiving end of knowledge sharing, we constantly depend on others to operate as knowledge filtering systems in order to allow us to do a better job of the knowing tasks at hand. Many epistemologists have rejected the idea that the accumulation of knowledge (or truths) *per se* is appropriately thought of as the goal of knowledge seeking. Rather, they have recognized that it is really *significant* knowledge that we are after when we engage in

[19] I thank Phyllis Rooney for directing me to Townley's work and seeing the connections with the themes of this paper.

epistemic pursuits (Anderson 1995; Kitcher 2001). As Elizabeth Anderson explains 'Theoretical inquiry does not just seek any random truth. It seeks answers to questions. What counts as a significant truth is any truth that bears on the answer to the question being posed' (Anderson 1995, 39). Similarly in the case of knowledge sharing, its good derives not from random knowledge sharing, but the sharing of knowledge *significant* to the project at hand. The proposed principle of increased knowledge sharing as a *prima facie* good ignores our epistemic need to limit our knowledge intake to that which is significant for the project at hand. Returning to the case I discussed earlier of Anne and Karen, the breach of trust in this relationship occurred not because there existed a piece of knowledge that was not shared, but rather because the knowledge of the affair withheld was considered by one of the parties to be *significant* to the joint epistemic pursuit they were engaged in.

Recognizing that the value of knowledge sharing really applies only to the *significant* knowledge in question for the project at hand, and is relative to the particular recipient or audience (in many times a participant in a joint epistemic project) helps explain why exceptions to any *prima facie* principle of increased knowledge sharing are so frequent. Surprisingly, in spite of its initial appeal, a principle of the *prima facie* good of increased knowledge sharing must be rejected. A liberatory epistemology must look to a more contextually-based assessment of the value of knowledge sharing, and relatedly, the appropriateness of particular knowledge-sharing norms.

12.7 Assessing Knowledge-Sharing Norms: Returning to the Goals of Liberatory Epistemology

With respect to the assessment of KSNs from within the framework of a liberatory epistemology, there are several lessons to be drawn from my discussions above. First, KSNs will in part need to be assessed according to traditionally conceived non-epistemic ethico-political goals such as the well-being of the oppressed and their opportunities to resist oppression. I have described several ways in which KSNs may directly aid or inhibit the goals of positive social change that are an important part of a liberatory epistemology. Insofar as liberatory epistemology has an interest in revealing the connections between knowledge seeking and the maintenance and resistance to oppression, it will be critical of those KSNs that contribute to the maintenance of oppression (such as a culture of acceptable far-reaching government and corporate secrecy), and it will be suggestive of KSNs that help foster resistance. In some contexts the ethico-political goals of liberatory epistemology will imply a need to widen the expectations of knowledge sharing, while in others it will imply narrowing the expectations of knowledge sharing.

The second lesson is that it is also appropriate to assess KSNs according to how well they contribute to the process of knowledge production. Future knowledge production is a key epistemic goal. Having argued against the view that the value of knowledge sharing lies only in the value of that knowledge itself, I have shown how in particular contexts, a certain degree of knowledge withholding can foster the production of particular kinds of knowledge better than increased knowledge sharing. It is a question of determining the appropriate expectations of knowledge sharing and withholding that will best serve the epistemic interests of further knowledge production for a liberatory epistemology. We need not deny that there is epistemic merit to be found in sharing the (significant) knowledge previously generated, yet it is clear that this is not where our epistemic goals end. The extent to which KSNs can foster future knowledge production must also be taken into account.

The third lesson to be drawn is particularly significant in terms of reconceptualizing the value of knowledge sharing. My examples of the ramifications of KSNs for the production of knowledge focus on specific forms of knowledge. To assess the appropriateness of particular KSNs then, we will need to consider how well they contribute not just to the production of knowledge generally, but to the production of those forms of knowledge we are interested in generating (recall the case of double-blind studies). Given how I outlined a liberatory epistemology above, as being especially interested in particular forms of knowledge necessary for understanding and overcoming oppression, we can now see that a liberatory epistemology will be interested in fostering KSNs that specifically aid the development of our knowledge and understanding of oppression. As I noted early on in this paper, feminist epistemologists have focused on the question '*knowledge for whom*?' The answer to this question, and correspondingly the answer to the question '*what kind of knowledge do we want to produce?*' will be crucial to determining what kind of KSNs we deem appropriate according to a liberatory epistemology.

Miranda Fricker's discussion of the hermeneutical injustice characteristic of the experiences of the oppressed offers one of the clearest examples of how certain contexts of knowledge sharing are necessary to develop forms of knowledge important for the oppressed. According to Fricker, hermeneutical injustice describes the cognitive disadvantage that arises when one is unable to understand one's experience because of a 'gap in the collective hermeneutical resource' (Fricker 2007, 151).[20] Fricker draws on Susan Brownmiller's book *In Our Time*: *Memoir of a Revolution* (1990) to describe such cases of hermeneutical injustice and the way in which consciousness raising amongst women was used to generate the necessary hermeneutical tools. In particular, Fricker describes the story of Carmita Wood and the unwanted sexual advances she and others experienced in the workplace during a time when there was no concept of 'sexual harassment.' By telling their stories in a safe environment, the women came to see the commonality of their experiences, and came to identify it as 'sexual harassment,' a previously unknown phenomenon. As Fricker describes the process:

> If we look at the history of the women's movement, we see that the method of consciousness raising through 'speak-outs' and the sharing of scantly understood, barely articulate experiences was a direct response to the fact that so much of women's experience was obscure,

---

[20] Less germane to my discussion here but nevertheless important are the details of what makes it an *injustice*. Fricker specifies that although in a sense all suffer the cognitive disadvantage of the absence of these hermeneutical resources, the one whose experience is misunderstood is unduly burdened when the misunderstanding represents a significant area of their social experience (2007, 154).

> even unspeakable, for the isolated individuals, whereas the process of sharing these half-formed understandings awakened hitherto dormant resources for social meaning that brought clarity, cognitive confidence, and increased communicative facility.... Women were collectively able to overcome extant *routine* social interpretive habits and arrive at *exceptional* interpretations of some of their formerly occluded experiences; together they were able to realize resources for meaning that were as yet only implicit in the social interpretive practices of the time (Fricker 2007, 148).

By creating communities and safe spaces where the women could share such personal (and often painful) experiences with each other, they were able to generate knowledge of specific forms of their oppression. Importantly, as I noted in the last section, sharing such personal knowledge needed first to be done in women-only spaces (essentially forming a norm of withholding their experiential knowledge from men, yet being open with the women in the group) though once the knowledge of the systematic nature of the phenomenon was generated and 'sexual harassment' was named, speak-outs could be organized to break the public silence on the issue. But for the generation of the knowledge itself, communities needed to be formed with KSNs that fostered trust and knowledge sharing within. Of course, there need not be a direct correlation between stringent norms of knowledge sharing and the actual knowledge sharing that occurs. As I discuss below, high burdens of knowledge sharing that pressure individuals to share may fail to generate the atmosphere of trust necessary for productive knowledge sharing. In the case of consciousness raising, there is an expectation of sharing within the confines of the group, but if those expectations are set too high, the sense of safety and trust within the group that is necessary to encourage people to speak about their experiences may be jeopardized.

These three lessons all set out goals against which we can evaluate how well our knowledge-sharing norms are serving us. However, there is a fourth lesson of a different nature that is also crucial to understanding the value of knowledge sharing.

12.8 Positionality and Trust

A fourth and final lesson to be gleaned from my discussions is that *positionality* is an important variable in assessing KSNs. People do not come to the knowledge-sharing table from equal positions of power, and the impact of specific KSNs on a person will vary, depending on one's social situation. A liberatory epistemology needs to consider who is expected to share knowledge with whom. The relevance of positionality formed part of my argument against adopting a principle of the *prima facie* value of increased knowledge sharing: since knowledge sharing often increases vulnerability, the demands of KSNs can be particularly harmful to those already marginalized, either directly (through political actions changing the material conditions of the marginalized) or indirectly (through new knowledge generated as a result of the sharing that then has political implications). We cannot assume anything like a principle of the *prima facie* value of increased knowledge sharing when the effects of increased knowledge sharing vary so widely across social position. Unlike the first three lessons, positionality does not offer us a specific goal according to which KSNs can be assessed. But it does provide a framework through which to understand those goals: rather than assuming that a particular set of KSNs will serve everyone equally well (whether by ethico-political standards or epistemic standards) a liberatory epistemology will consider again the feminist question of *knowledge for whom*, assessing how well a set of KSNs functions for those in particular social locations.

Taking the relevance of positionality seriously leads us to a clearer understanding of the difficulties of harnessing epistemic efforts across power differentials, something with which the feminist movement is familiar. Discussing the relationship between women of color and white women, María Lugones writes 'I keep secrets. Even though I am told over and over by white feminists that we must reveal ourselves, open ourselves, I keep secrets. Disclosing our secrets threatens our survival' (Lugones 2003, 11). Lugones here draws our attention to the vulnerability of feminists of color in relation to white feminists, even while committed to common feminist goals. Her description captures how the KSNs within feminist communities place pressure on women of color to share their understandings with white women. It is arguably too much to expect such disclosure on behalf of women of color, since disclosing knowledge about themselves can then be used against them. This could happen even without malicious intent if white women do not fully understand the position and vulnerabilities of women of color.[21] Indeed, Lugones's comments suggest that the very fact that white women are operating with these KSNs, expecting such knowledge sharing on behalf of women of color, reveals that they do not fully appreciate the situation of women of color. This is a case where the KSNs in place are not working well in producing an epistemically productive community, not simply because secrets are being kept which might be relevant to the epistemic project at hand, but because the presence of the KSNs themselves, taken to be inappropriate by women of color, is further damaging the relations of trust with potentially far-reaching implications.

As this case illustrates, KSNs do more than produce the phenomena of sharing and withholding knowledge. They also create social pressures for knowers to share or withhold, and as such they help shape the relations between knowers and levels of trust within a community and across social positions. Understanding how positionality plays into the creation of relations of trust within epistemic communities

---

[21] As Sarah Hoagland points out, good intentions on behalf of the relatively privileged aren't enough. Speaking of the position of the relatively privileged within the feminist community, she writes: 'even when we seek in friendship the openings and unexpected connections that situated knowledges make possible, we can be dangerous. To whom are we addressing ourselves, to whom are we offering information, and why?' (2001, 138).

reveals that there is no reason to assume that the best KSNs will be reciprocal across parties. Contexts can be envisioned where the goals of liberatory epistemology will best be served by adopting KSNs which apply to some positions but not all within the community. For example, given the situation Lugones describes, the best way to develop the trusting environment within the feminist community might well be to shift the relations of accountability, adopting KSNs where those who are more vulnerable due to social position are not held to the same expectations of knowledge sharing as those who are more privileged. There is of course much more to developing relations of trust than just establishing appropriate KSNs. However, recognizing the role of KSNs in developing the cultures of trust necessary for knowledge production helps us understand that the assessment of such norms is very complex, particularly in cases of power differentials where cultures of trust are more difficult to establish. We must take into account both the epistemic effects of the knowledge circulation itself as well as the social effects of the norms in developing epistemically productive cultures.

12.9 Conclusions

KSNs, particularly informal ones, are complex and difficult to identify. Yet they perform an important role in epistemic practices, guiding our knowledge sharing and knowledge withholding in ways that can foster or hinder epistemically productive and politically sound epistemic communities. As I have argued, the knowledge circulation enabled by KSNs has ethico-political effects as well as epistemic effects concerning future knowledge production, yet these effects differ according to social location. The complexity of these effects offer evidence against the plausibility of any general principle asserting the *prima facie* value of increased knowledge sharing, in spite of the potential appeal of such a principle for liberatory epistemologists. Instead, a liberatory epistemology must examine the specific contexts in which our KSNs are operating, and assess them according to both ethico-political and epistemic goals, with a sensitivity to the power differentials that predominate in contexts of oppression.

Though I have rejected any general principle asserting the *prima facie* value of increased knowledge sharing, it remains reasonable to expect that in a great many cases, norms that foster transparency within (and across) epistemic communities will be epistemically fruitful and liberatory. But when they are so, it will not be because there is any direct link between increased knowledge sharing and increased knowledge, or increased knowledge sharing and liberation. Rather, it will be because in the particular context assessed, the conditions are such that norms encouraging broad transparency increase rather than decrease the culture of trust necessary to generate liberatory knowledge. Finding the right level of knowledge-sharing norms can be difficult, particularly in contexts of oppression, and my discussion simply raises the bar for the kind of complex assessment required to understand how particular knowledge-sharing norms can help us know well, particularly in our feminist liberatory pursuits.

Acknowledgments This paper has benefited from the comments of audiences who heard early versions at the CSWIP (Canadian Society for Women in Philosophy) annual conference in Edmonton, Alberta (October 2007), the Society for Analytical Feminism Conference in Lexington Kentucky (April 2008), and a Middlebury College Life of the Mind lecture (April 2008). I am especially indebted to Carla Fehr and Phyllis Rooney who read drafts of this paper in detail. Others who have offered helpful comments include Lorraine Code, Nancy Daukas, Ann Garry, Sandra Harding, Victor Nuovo, Dave Saldana, Susan Sherwin, Alexis Shotwell and Ilya Storm.

References

Alcoff, Linda, and Elizabeth Potter. 1993. Introduction: When feminisms intersect epistemology. In *Feminist epistemologies*, ed. Linda Alcoff and Elizabeth Potter, 1–14. New York: Routledge.

Allen, Anita. 2003. *Why privacy isn't everything: Feminist reflections on personal accountability*. Lanham: Rowman and Littlefield.

Anderson, Elizabeth. 1995. Knowledge, human interests, and objectivity in feminist epistemology. *Philosophical Topics* 23(2): 27–58.

Bailey, Alison. 2007. Strategic ignorance. In *Race and epistemologies of ignorance*, ed. Shannon Sullivan and Nancy Tuana, 77–94. Albany: State University of New York Press.

Basic Guide to the National Labor Relations Act. 1997. Retrieved September 2009. From www.nlrb.gov/nlrb/shared_files/brochures/basicguide.pdf.

Brownmiller, Susan. 1990. *In our time: Memoir of a revolution*. New York: Dial Press.

Code, Lorraine. 1991. *What can she know? Feminist theory and construction of knowledge*. Ithaca: Cornell University Press.

Code, Lorraine. 1995. *Rhetorical spaces: Essays on gendered locations*. New York: Routledge.

Code, Lorraine. 2006. *Ecological thinking: The politics of epistemic location*. Oxford: Oxford University Press.

Fricker, Miranda. 2007. *Epistemic injustice: Power and ethics in knowing*. Oxford: Oxford University Press.

Grasswick, Heidi. 2010. Scientific and lay communities: Earning epistemic trust through knowledge sharing. *Synthese* 177: 387–409.

Harding, Sandra. 1991. *Whose science? Whose knowledge? Thinking from women's lives*. Ithaca: Cornell University Press.

Harding, Sandra. 2000. Should philosophies of science encode democratic ideals? In *Science, technology, and democracy*, ed. Daniel Lee Kleinman, 121–138. Albany: State University of New York Press.

Hoagland, Sarah L. 2001. Resisting rationality. In *Engendering rationalities*, ed. Nancy Tuana and Shannon Sullivan, 125–150. Albany: State University of New York Press.

Hundleby, Catherine. 2005. The epistemological evaluation of oppositional secrets. *Hypatia* 20(4): 44–58.

Johnson, Roberta Ann. 2003. *Whistleblowing: When it works – and why*. Boulder: Lynne Rienner Publishers.

Kitcher, Philip. 2001. *Science, truth, and democracy*. Oxford: Oxford University Press.

Longino, Helen E. 1997. Cognitive and non-cognitive values in science: Rethinking the dichotomy. In *Feminism, science, and the philosophy of science*, ed. Lynn Hankinson Nelson and Jack Nelson, 39–58. Dordrecht: Kluwer.

Lugones, Maria. 2003. *Pilgrimages/peregrinajes: Theorizing coalition against multiple oppressions*. Lanham: Rowman and Littlefield.

Potter, Elizabeth. 1993. Gender and epistemic negotiation. In *Feminist epistemologies*, ed. Linda Alcoff and Elizabeth Potter, 161–186. New York: Routledge.
Scheman, Naomi. 2001. Epistemology resuscitated. In *Engendering rationalities*, ed. Nancy Tuana and Sandra Morgen, 23–52. Albany: State University of New York Press.
Townley, Cynthia. 2003. Trust and the curse of Cassandra (an exploration of the value of trust). *Philosophy in the Contemporary World* 10(2): 105–111.
Tuana, Nancy. 2001. Introduction. In *Engendering rationalities*, ed. Nancy Tuana and Sandra Morgen, 1–20. Albany: State University of New York.
Valian, Virginia. 2005. Beyond gender schemas: Improving the advancement of women in acadamia. *Hypatia* 20(3): 198–213.

Index

H.E. Grasswick (ed.), *Feminist Epistemology and Philosophy of Science: Power in Knowledge*, DOI 10.1007/978-1-4020-6835-5,